Smart Factory Navigator

Lukas Budde · Roman Hänggi · Thomas Friedli ·
Adrian Rüedy

Smart Factory Navigator

Identifying and Implementing the Most Beneficial Use Cases for Your Company—44 Use Cases That Will Drive Your Operational Performance and Digital Service Business

Springer

Lukas Budde
Institute of Technology Management
University of St. Gallen
St. Gallen, Switzerland

Thomas Friedli
Institute of Technology Management
University of St. Gallen
St. Gallen, Switzerland

Roman Hänggi
IPEK
OST Eastern Switzerland University
of Applied Sciences
Rapperswil, Switzerland

Adrian Rüedy
IPEK
OST Eastern Switzerland University
of Applied Sciences
Rapperswil, Switzerland

ISBN 978-3-031-17253-3 ISBN 978-3-031-17254-0 (eBook)
https://doi.org/10.1007/978-3-031-17254-0

This Springer imprint is published by the registered company Springer Nature Switzerland AG
The registered company address is: Gewerbestrasse 11, 6330 Cham, Switzerland

Preface

Digitalization, Industry 4.0, or smart factory have become important terms when it comes to implementing digital technologies in industrial companies. The implementation is extremely complex and is often associated with a lot of new topics, challenges, and a need for additional competences for a company. Different companies have turned to us for advice on how to implement these new approaches to optimize their businesses or factories. In assisting these companies, we have often been confronted with significant management confusion and questions regarding the right process for implementation. We have often experienced that the same questions are asked: What is Industry 4.0 and how is it different from the smart factory? Which technologies are relevant to us? What are the benefits of certain technologies, such as IoT, for our plants? Which competences do we need in the future to cope with these new approaches? How does the smart factory fit with our lean approach? How can we digest all these necessary changes and strengthen our competencies in this area?

These and related questions have challenged us in many research and consulting projects and formed the basis for our "Smart Factory Framework." While we were working on our framework we came up with the following definition of a smart factory:

The smart factory is a fully connected manufacturing system, combining human competencies and technological capabilities to conduct all required tasks and decisions for producing goods efficiently and enabling innovative digital services.

It can be said that a smart factory has a lot to do with technology. The key, however, is that it is only the combination of technology and humans that makes it successful. In addition, goal orientation is part of the smart factory. We implement the smart factory approach in order to increase the operational performance of the factory. Technology is only an enabler. So how did we go about developing an approach for the implementation of the smart factory? Firstly, we extensively researched published and implemented smart factory use cases. The focus was not only on the use cases themselves but also on their predicted outcomes. We categorized these use cases into 44 distinct use cases that make up the Smart Factory Framework. In our book, we describe these use cases in detail, including implemented real-world examples, and the implications in terms of quality, cost, delivery, and flexibility. In addition, we have mapped the use cases to the required technologies that can potentially be applied. Our research has identified 25 technologies relevant to smart factory implementation, the use of which depends on the different use cases.

The Smart Factory Implementation Process is a usable approach for the concrete realization of the smart factory in your company. This seven-step process starts with the selection of promising use cases, followed by a prioritization step, adaptation to the company-specific context and a detailing of the information needs for the different management levels. A stringent project setup and execution process helps the users to achieve positive results that can then be rolled out on a global scale. This process has been developed interactively with dozens of companies and applied in many research and consulting projects. It has proven to be a realistic and results-oriented approach for the smart factory journey. This book is the summary of a four-year process involving intensive research and project work with many different companies in different industries. The manifold discussions, challenges, and presentations helped to build this into a robust approach. This journey would not have been possible without the support of various individuals who challenged us along the way. We would like to thank our research assistants Kay Beeler and Leonardo Laglia for their support. We also appreciate the illustration skills of Roland Siegenthaler. We have also applied this approach in various lectures and project work at the University of St. Gallen (HSG) and the University of Applied Sciences St. Gallen (OST), thereby receiving valuable input from our students which has been incorporated into our approach. In addition, our research partner Cognizant, a global information technology services and consulting company, supported us shape our approach through its project work and its research practice.

We are confident that our book will give you the needed guidance on your own digitalization journey increasing your operational excellence and competitiveness.

St. Gallen, Switzerland — Lukas Budde
Rapperswil, Switzerland — Roman Hänggi
St. Gallen, Switzerland — Thomas Friedli
Rapperswil, Switzerland — Adrian Rüedy

Contents

About the Authors

Lukas Budde, Post-Doc for Production Management
Lukas Budde studied mechanical engineering at the Technical University of Darmstadt and received his Ph.D. from the University of St. Gallen. Since 2015, he has been a post-doc and division deputy in the field of production management at the Institute for Technology Management at the University of St. Gallen and manager of the Complexity Management Academy in Aachen. His research focuses on smart manufacturing and digital platform ecosystems in manufacturing companies. Lukas has supported companies in many digital transformation projects. He is a lecturer for Operations Management courses and for custom programs of the Executive School of the University of St. Gallen. Since 2018, he has also been the managing director of INOS, the largest regional innovation network for SMEs in Switzerland.

Roman Hänggi, Professor of Production Management
After studying engineering and earning a doctorate in economics, Roman Hänggi started his career in industry in the early 1990s by optimizing an optics manufacturing facility at Leica with Lean Production practices. His enthusiasm for lean processes and operational excellence has accompanied him throughout his professional life. He implemented further successful lean and digital projects in the production processes at Bosch, Hilti, and Arbonia, including new digital service business models. Curiosity and the desire to pass on his extensive knowledge from industry led him to become professor of

production management at OST Eastern Switzerland University of Applied Sciences. In this position, he motivates students in Rapperswil and St. Gallen in lectures on lean practices and digitalization in industry. Additionally, practice is also important to him. This is why he supports industrial companies on their way to becoming lean champions and Industry 4.0 winners, as well as serves on the board of industrial companies. He also teaches as a lecturer in executive programs at the University of St. Gallen. In his free time, Roman can be found on the ski slopes in Appenzell (Switzerland).

Thomas Friedli, Professor of Production Management
Thomas Friedli is a professor of production management at the University of St. Gallen (Switzerland). His main research interests are in the field of managing operational excellence, global production management, and management of industrial services. He is a lecturer in the (E)MBA programs in St. Gallen, Fribourg, and Salzburg. He spent several weeks as adjunct associate professor at Purdue University in West Lafayette, USA. Prof. Friedli leads a team of 15 researchers who develop new management solutions for manufacturing companies in today's business landscape. He is also the editor, author, or co-author of 14 books and various articles.

Adrian Rüedy, Lecturer Industry 4.0
After completing his bachelor's degree in industrial engineering, Adrian started at OST Eastern Switzerland University of Applied Sciences industrial engineering as a project manager for digitization in industry with a focus on integrated data flow in manufacturing. With his practical experience of five years in special machinery and plant engineering as a design engineer, he has a deep understanding of the processes and challenges in manufacturing. In 2020, he completed a master's degree in business information technology while working on various industrial and government-funded research projects. At DigitalLab@OST, he uses his experience in digital transformation to advise companies on the path to digitization. By combining applied research and consulting experience with knowledge in dealing with complexity and change management, he loves to empower students on their future path in industry.

Abbreviations and Definitions

4Vs of Big Data	Volume, Variety, Velocity, Veracity
5G	5th generation in wireless cellular network technology
Actuators	Receive an electrical signal and create a prescribed physical change by motion, force, etc.
AGVs	Automated guided vehicles are driverless, automatically steered, and fixed area means of conveyance with their own drive
AI	Artificial intelligence grants machines and systems the ability to perform tasks traditionally accomplished exclusively by the human mind
AM	Additive manufacturing technology makes it possible to print 3D objects
AMRs	Autonomous mobile robots are similar to AGVs but additionally use AI
ANNs	Artificial neural networks are biologically inspired simulation programs
APS	Advanced planning and scheduling, software used to manage production planning and shop floor scheduling
AR	Augmented reality serves to enhance the user's perception of reality by combining visual cues from the physical and digital worlds
B2B	Business-to-businesses is a business model where a company sells products/services to other companies
B2C	Business-to-consumer is a business model where a company sells products/services directly to end-users
BI	Business intelligence
Big Data Analytics	Includes the process of collecting, storing, processing, and analyzing large amounts of data

Blockchain	Blockchain is a shared, immutable ledger that facilitates the process of recording transactions and tracking assets in a business network
CAD	Computer-aided design software supports the digital drawing of 2D and 3D models
CAM	Computer-aided manufacturing is the use of software and computer-controlled machinery to automate a manufacturing process
Cloud Computing	Enables the delivery of computing services over the Internet (e.g., databases, software, and intelligence)
CNC	Computer numerical control is one in which the functions and motions of a machine tool are controlled by a prepared program
Collaborative Robotics	Work in collaboration with people as opposed to in place of them
CPS	Cyber-physical systems can be seen as an autonomous factory floor where the machines needed for given processes communicate and cooperate thanks to data generated by numerous sensors throughout the factory
CSR	Corporate social responsibility refers to practices and policies that help a company be socially responsible to itself, its stakeholders and the public
Data Matrix Code	A Data Matrix is a 2D barcode that encodes data in black and white or contrasting dark and light cells arranged in a grid and contains less information than a QR code but has an error correction of 30%
Digital Twin	A digital twin is a digital representation of the physical reality; it is often used to simulate options and improvements based on the digital world and relate the learnings to the physical reality
DMS	Decision-making system
Drones	An unmanned aerial vehicle (UAV) is commonly known as a drone
ECO	Engineering change order
Edge Computing	Brings computation and data storage closer to the sources of data
ENT	Medical abbreviation for ears, nose, and throat
ERP	Enterprise resource planning, comprehensive IT system for the flow of goods and value in the company
GHG	Greenhouse gas absorbs and emits radiant energy within the thermal infrared range, causing the greenhouse effect

HCPS	Human-cyber-physical system is the term for semi-automated systems that incorporate human insight into the control loop, resulting in intelligent manufacturing systems
HCS	Human-cyber-systems describe the development of intelligent systems to increase the efficiency of interaction between humans and cyber-systems (e.g., computers)
HDDs	Hard disk drives are electro-mechanical long-term storage devices for data
HMDs	Head-mounted displays project information directly into the wearer's eyes
HPS	The human-physical system describes traditional manufacturing systems in which operations were entirely manual and information gathering, analysis, decision making, and operational control were primarily the responsibility of humans
IaaS	Infrastructure as a service is a cloud computing service model that refers to the on-demand availability of nearly indefinitely scalable computing resources as Internet-based services
IIoT	Industrial Internet of things describes the use of Internet of things technologies for the factory or for industrial processes and is often used synonymously with the German term Industrie 4.0 in the USA
ILO	International labor organization
Industry 4.0	The term goes back to a research project in Germany in 2011 the term "Industrie 4.0" was established, in which industrial production is to be interlinked with modern information and communications technology
IoT	Internet of things describes the network of physical objects that are embedded with sensors, software, and other technologies for the purpose of connecting and exchanging data with other devices and systems over the Internet
IT	Information technology includes computers, ancillary equipment, software, firmware and similar procedures, services, and related resources
Kanban	Just-in-time pull production system, where production or logistics orders are placed through simple card logic
KPI	Key performance indicator refers to a set of quantifiable measurements used to gauge a company's overall long-term performance

LAN	Local area network is a collection of devices connected in one physical location
LoRaWAN	Long range wide area network, technology that enables data to be transmitted wirelessly over long distances
M2M Communication	Machine-to-machine communication is the automated exchange of information between technical systems such as machines through wired or wireless communication channels
MES	Manufacturing execution system are integrated, modular, and real-time information processing systems that cover the entire manufacturing process
ML	Machine learning is the learning component of AI, which processes large amounts of information to look for patterns to learn previously unknown inputs
Mobile Computing	Technology used for transmitting voice and data through small, portable devices using wireless enabled networks
MRO	Maintenance, repair, and overhaul
MRP	Material requirements planning is the core functionality in an ERP system for planning plant and production orders, taking into account optional constraints
NFC	Near field communication enables communication between two electronic devices over a distance of 4 cm or less
OEE	Overall equipment efficiency measures the utilization, performance, and quality of a process in one number
OEM	Original equipment manufacturer is a company that manufactures and sells products or parts of a product to another company
OTIF	On-time in-full
PaaS	Platforms as a service provides a complete and flexible cloud platform for developing, running, and managing applications
PLM	Product lifecycle management system provide the context, security, traceability, and processes needed across enterprise teams and systems to ensure product data is accessible and trustworthy
QR-Code	Quick response code is a machine-readable two-dimensional type of barcode
RAM	Random access memory is a temporary storage space for data and programs

RFID	Radio-frequency identification uses electromagnetic fields to automatically identify and track tags attached to objects
ROI	Return on investment, profitability metric used to evaluate how well an investment has performed
RUL	Remaining useful life estimates the number of remaining years for items, components, or systems
SaaS	Software as a service, software licensed on a subscription basis and is centrally hosted
SC	Supply chain is the entire system involved in producing and delivering a product or service
SCADA	Supervisory control and data acquisition enables field data to be collected and transmitted back to a master station via communication technology
SCM	Supply chain management can be defined as the coordination and integration of business processes within an organization and across the key members of the supply chain
SCRUM	Scrum is a process model of project and product management, especially for agile software development
Sensors	Sensors are objects to detect and measure physical variables and convert them into electrical, magnetic, or mechanical output signals
Simulation	Imitation or enactment, as of something anticipated or in testing
SKU	Stock-keeping unit is a unique number assigned by a company to its products to keep track of stock levels internally
Smart Assets	Assets such as machines, transportation systems and tools, reflecting the physical layer in a factory that becomes smart through the equipment with sensors and communication technology
Smart Factory	Is a fully connected manufacturing system, combining human competences and technological capabilities to conduct all required tasks and decisions for producing goods efficiently and enabling innovative digital services
Smart Manufacturing	An adaptable system where flexible lines adjust production processes automatically for multiple types of products and changing conditions
Smart Products	Communication technologies embedded in products which can provide feedback from the user cycle back to the company's production and product development cycles

Smart Supply Chain	Encompasses the exchange of information and integration of suppliers and customers in the supply chain
Smart Working	Technology and humans work together to support better decision making and production steps
SMEs	Small and mid-size enterprises
Solution Ecosystems	The design and the participation of companies in ecosystems, making them more meaningful for their customers and for the society as a whole
STEM	Science, technology, engineering, and maths
Track & Trace Technologies	Allows for the identification of current and past location of objects
TRU	Traceable resource unit, a traceable entity in which information is stored in a traceability system
UAS	Unmanned aircraft system
UAV	Unmanned aerial vehicle
USP	Unique selling proposition is a singular characteristic of a company that sets it apart from others
UWB	Ultra-wideband is a wireless technology that uses pulse patterns
Vision Technologies	A set of technologies which rely on image capturing and analysis
VR	Virtual reality is a realistic three-dimensional world that takes into account the visual, auditory, and tactile senses
Wearables	Wearable computing devices, e.g., smartwatch
WLAN	Wireless local area network allows devices to connect and communicate within a limited area over a wireless network
WMS	Warehouse management systems provide visibility of inventory and manages supply chain fulfillment operations

List of Figures

Chapter 1
Introduction

When taking a look at economic data, it becomes more than evident that production in high-cost countries still matters. Despite contrary opinions, production is not leaving the "rich countries" (Ferdows, 2021) and never has been. Several high-cost developed countries even have the highest per capita manufacturing value added in the world, e.g., Switzerland's per capita manufacturing value add is 5.5 times higher than China's, and Germany's is 3.2 times higher. With regard to the current discussions about the resilience, robustness and responsiveness of global supply chains, there is no indication that this could change in the near future. These discussions will further highlight the importance of these countries (see Fig. 1.1).

Covid19 led to several issues in global supply chains, such as broken delivery streams, and revealed downsides of cost-dominant operational strategies leading to always more fragmented and globally dispersed supply chains. Global political disruptions will increase the focus on risk management of your supply chain. As a consequence, robustness and resilience made a comeback as top priorities by the re-evaluation of global production footprints (Luban & Hänggi, 2022).

Additionally, ecological and social sustainability will drive and change the organization of production systems. The ecological footprint will foster urban and local manufacturing, the sharing of resources (e.g., machines) to maximize utilization, and remanufacturing and re-assembly infrastructures to increase product life cycles and resource reuse. The increasing relevance of the circular economy will drive new production set-ups, thereby changing existing manufacturing networks.

Customization pressure is another factor further influencing the design of products and manufacturing processes. To stay in control of the always more increasing complexity and related costs, late-stage customizations will be one specific trend impacting footprints and site specializations. Industrial companies and particularly production locations in high-cost countries have been under significant cost pressure for years due to exchange rate fluctuations, global competition, and technological changes.

These constantly applied external forces have made these companies look continuously for productivity improvements. Companies have been undergoing a constant

L. Budde et al., *Smart Factory Navigator*,
https://doi.org/10.1007/978-3-031-17254-0_1

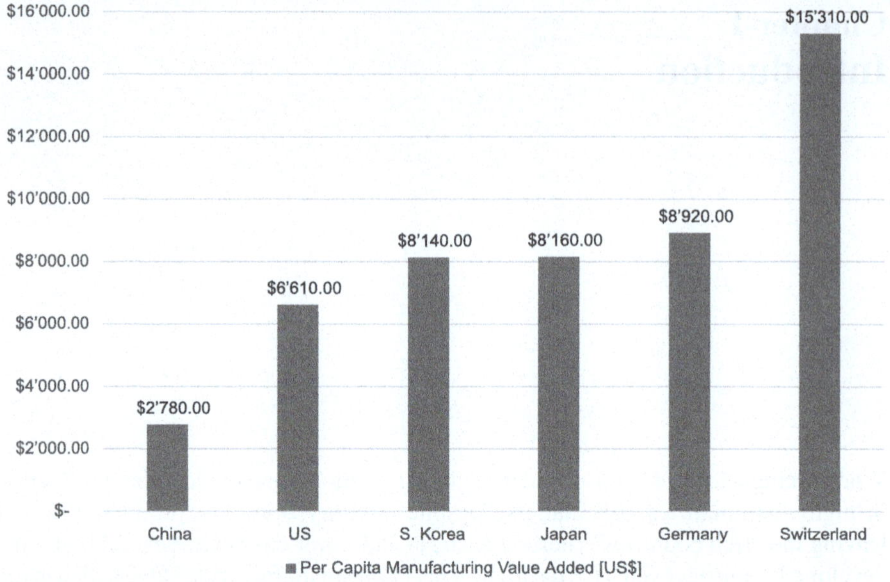

Fig. 1.1 Per Capita Manufacturing Value Added (US$, 2019). Data from World Bank (Ferdows, 2021)

fitness program. Productivity measures are, therefore, already implemented widely, e.g., lean production is used intensively and is still continuing to evolve (Hänggi et al., 2021). Automation has played an important role for years. These constant challenges to productivity and the ensuing productivity improvements have played a key role in sustaining manufacturing in high-cost countries over the last decades. Access to low-cost production has also helped to keep the overall network performance balanced.

Nevertheless, the constant drive to optimize will continue in high-cost countries as well as on the global production network level (Friedli et al., 2021). We believe that, in particular, new digital technologies will create the next level of opportunities for further productivity increase. The potential of these available technologies is significant. Understanding and implementing these production innovations will be a future success factor for high-wage locations. We go even further and believe that the implementation of new digital technologies will be relevant for any industrial company to become a leader. However, this has to be tailored to the respective needs: "more" is not always "better", despite typical Industry 4.0 maturity models.

The approach described in this book guides you through the entire process from ideation to road mapping and implementation and demonstrates, how you can systematically take advantage of digitalization to further strengthen your operations, and what you need to manage the required transformation. The implementation of new technologies can be challenging and complex. New competencies might be needed, a deep understanding of the as-is and the to-be is vital, and new, so far unknown,

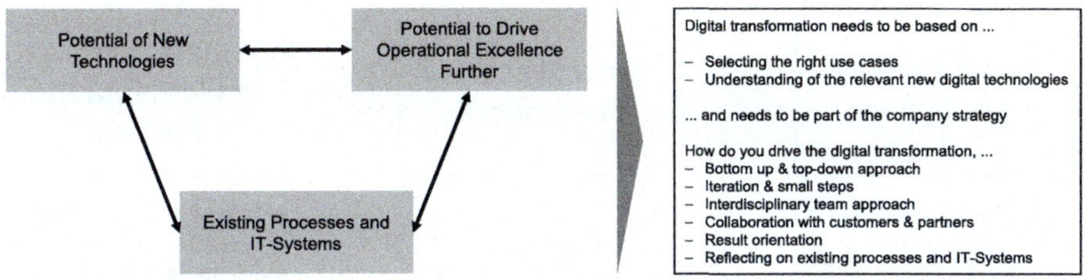

Fig. 1.2 How to drive the digital transformation?

technology partners may be required. The term "digital transformation" is used to highlight this shift and to delineate it from one-time IT projects.

Often companies focus on implementing these new technologies without a clear understanding of the aims and benefits of the technological change. This often-missing goal orientation leads to frustration and ultimately stops initiatives. Our approach is to focus on the "what" without neglecting the strength and change potential of new technologies. We are focusing in our book on documented use cases in factories that have already been piloted and show a positive impact. Some new technologies have the potential to fundamentally change manufacturing and its processes, e.g., utilizing augmented reality for assembly process or using of drones collecting data or supporting inspection processes (Maghazei & Netland, 2019). However, if the application is designed purely technology driven without considering the business case a lot of uncertainties are attached to this approach. Our focus therefore lies on proven benefits to provide support in the short und midterm.

There is a need to understand the potential of the new technologies. What can they bring to the table? What do they offer as added value to the current operations? Which processes and structures will benefit the most by implementing new digital technologies?

Usually, companies do not start at ground zero. They have a legacy of systems and approaches already implemented. Therefore, it is often necessary to rework existing processes and systems before new technologies can be applied. These digital changes need to be an integral part of a company's strategy otherwise financial investments, resource allocation and focus will not persist. Digital transformation is more than just a project; it is a constant focus and reconsideration of all aspects of operations, regarding digital potentials (see Fig. 1.2).

We identified in our research several success factors necessary for digital transformation, such as bottom-up & top-down approach, iteration & small steps, interdisciplinary team approach, collaboration with customers & partners, result orientation, and reflecting on existing processes and IT-Systems. These elements are essential to drive change successfully. [1]

[1] Budde, L., Benninghaus, C., Hänggi, R. & Friedli, T. (2022). Managerial Practices for the Digital Transformation of Manufacturers. Digital 2 (4), 463–483. https://www.mdpi.com/2673-6470/2/4/25

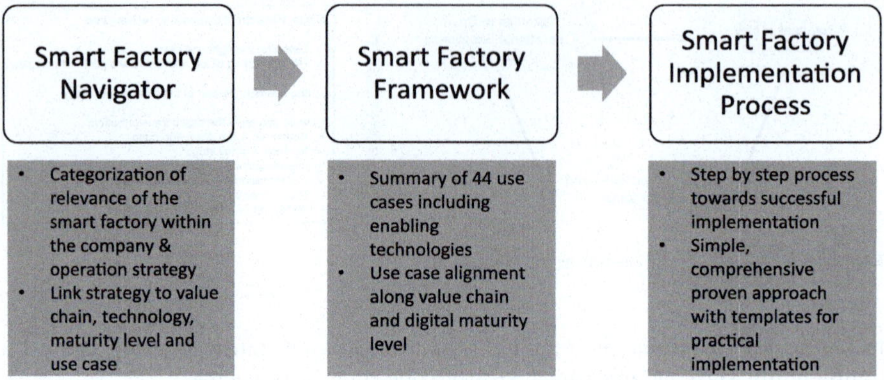

Fig. 1.3 Book structure

Figure 1.2 highlights the fact that everything starts by selecting the right use cases and understanding the relevant new technologies. According to our definition, use cases describe concrete applications of the smart factory with technologies as enablers that achieve added value in connection with the available database and under the company-specific application context. This characteristic is more comprehensive than the definition used in the software industry because it includes the business perspective (Jacobson, 1992). It is critical to understand what to focus on and to identify appropriate use cases that will have a real impact. There will be no success without a clear benefit. Companies find this process rather difficult. The identification of potential use cases has to be done strategically, and it is fundamental to align the limited resources.

Our book helps to achieve this. It starts with an introduction to the relevant technologies for a smart factory. For us, a smart factory is defined by the application of different combinations of digital technologies to further advance operational excellence. Often smart factory initiatives are also referred to as "digital factories" or "intelligent factories" initiatives. We, however, prefer to use the term "smart factory".

We selected 25 technologies that are relevant for a smart factory. The potential of these technologies will be demonstrated in 44 use cases. These 44 generic use cases summarize our research over the last years. We collected and clustered over 500 published smart factory use cases and built 44 categories out of them. This comprehensive overview provides the opportunity to navigate through the complex Industry 4.0 environment and ultimately identify the applications with the highest potential for a respective company. A structured implementation process derived from direct interactions with leading companies complemented the approach. The book is structured into three main parts (see Fig. 1.3).

To support the implementation process in practice, we have summarized the 44 use cases in Sects. 3.2–3.9. Our described implementation process will give you a procedure on how you can select and prioritize your use cases and a structured approach to manage the implementation successfully.

References

Ferdows, K. (2021). Perspectives on global manufacturing. *Die Unternehmung, 75*(4), 469–475. https://doi.org/10.5771/0042-059X-2021-4-469

Friedli, T., Lanza, G., & Remling, D. (Eds.). (2021). *Global manufacturing management: From excellent plants toward network optimization.* Springer International Publishing. https://doi.org/10.1007/978-3-030-72740-6

Hänggi, R., Fimpel, A., & Siegenthaler, R. (2021). LEAN Production – einfach und umfassend: Ein praxisorientierter Leitfaden zu schlanken Prozessen mit Bildern erklärt. *Springer*. https://doi.org/10.1007/978-3-662-62702-0

Jacobson, I. (1992). *Object-oriented software engineering: A use case driven approach.* ACM Press.

Luban, K., & Hänggi, R. (Eds.). (2022). *Erfolgreiche Unternehmensführung durch Resilienzmanagement: Branchenübergreifende Praxisstudie am Beispiel der Corona-Krise.* Springer. https://doi.org/10.1007/978-3-662-64023-4

Maghazei, O., & Netland, T. (2019). Drones in manufacturing: Exploring opportunities for research and practice. *Journal of Manufacturing Technology Management, 31*(6), 1237–1259. https://doi.org/10.1108/JMTM-03-2019-0099

Chapter 2
Smart Factory Navigator

2.1 Smart Factory—The Future of Manufacturing

In recent years we have experienced the acceleration of the introduction and application of emerging technologies—Internet of Things (IoT), cloud computing, augmented reality and artificial intelligence (AI), to name a few—to the manufacturing context. To maintain competitiveness in a globalized world, manufacturing operations have looked to these digital technologies for a way to increase efficiency and flexibility in their production. This emerging trend of digitalization, or digital transformation, of both products and production systems has been coined "Industry 4.0" in Germany and "smart manufacturing" or "Industrial Internet of things (IIoT)" in the United States (Porter & Heppelmann, 2015; Thoben et al., 2017). All concepts of the phenomenon are based on a set of core technologies, methodologies, and trends, outlined in recent years by industry and academic research.

The scale and implication of this transformation has led to a great deal of academic attention being paid to the subject matter. Numerous authors have already discussed the origin (Almada-Lobo, 2016) examples (Lasi et al., 2014), impact (Schmidt et al., 2015) and future paths of Industry 4.0 in many diverse publications (Qin et al., 2016). Industrial companies have created their own paths toward implementing digital technologies for productivity gains. However, the various perspectives and the many different directions in practice with which the subject matter can be approached—the research has ranged from technical particularities to business process perspectives—have rendered the current state of research disjointed and heterogenous (Lee & Lee, 2015).

Broadly, the smart factory vision ought to be understood as a holistic approach to increase productivity and efficiency—through the integration of the abovementioned emerging technologies—at every step of the value chain. While the manufacturing dimension is surely a central one, the smart factory transformation also extends the evolution to other, adjacent dimensions. The role of human operators within smart manufacturing, for instance, is a topic of great importance, leading to the development of a plurality of smart working approaches (Longo et al., 2017). At the

L. Budde et al., *Smart Factory Navigator*,
https://doi.org/10.1007/978-3-031-17254-0_2

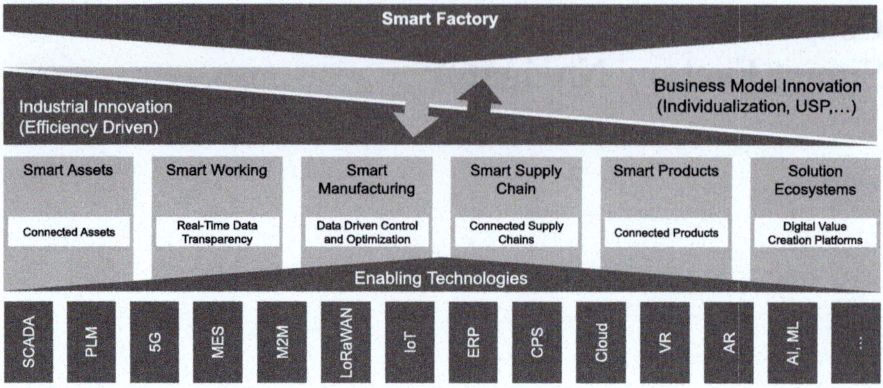

Fig. 2.1 Smart factory dimensions (based on Frank et al., 2019)

same time, traditional supply-chain processes are also incurring transformation as digital technologies allow for the "exchange of information and integration of the (smart) supply-chain, synchronizing production with suppliers to reduce delivery times and information distortions" (Frank et al., 2019, p. 16). Finally, technology is also being embedded more and more in products to provide customers with additional services (Dalenogare et al., 2018). Thus, the smart factory vision extends across the value chain of a company towards more service-oriented manufacturing.

The smart factory dimension is driven, or is naturally influenced, by six elements: smart assets, smart working, smart manufacturing, smart supply chain, smart products and solution ecosystems. These elements strongly influence the smart factory vision (see Fig. 2.1).

Smart Assets

These are assets such as machines, transportation systems and tools that become smart through the use of sensors and communication technology. This allows equipment to generate, communicate and process data. The communication can be between machines (machine to machine) or between other control systems in a factory (e.g., MES, ERP) that enable humans to make better decisions and actions.

Smart Working

Real-time data transparency about assets, processes, as well as key performance indicators (KPIs) (e.g., quality of produced products) and other supporting new technologies has changed the way humans and machines team up and is known as "smart working". Technology supports humans in making better decisions (by analyzing tons of data and intelligent algorithms) concerning such issues as when to change the settings of the machines or in which sequence production orders should be executed. Moreover, technology and humans can work together to execute certain production or assembly steps (e.g., collaborative robotics, augmented reality). Although this makes work for humans in a factory more pleasant, it also makes it more complex due to the many connected interdependencies.

Smart Manufacturing

Smart manufacturing has its origin in the advanced manufacturing concept and is understood as an adaptable system where flexible lines adjust production processes automatically for multiple types of products and changing conditions (Schuh et al., 2017; Wang et al., 2016). This enables companies to increase quality, productivity, and flexibility, producing more customized products in a sustainable way with better resource consumption (Dalenogare et al., 2018).

Smart Supply Chain

The smart supply chain encompasses the exchange of information and integration of suppliers and customers in the supply chain (upstream to suppliers and downstream to customers). It provides better synchronization and alignment of real demand and production volumes along all steps of the value chain. It reduces waste, delivery times and information distortions that produce bullwhip effects between companies.

Smart Products

Smart products are also part of the bigger smart factory or Industry 4.0 concept. Communication technologies embedded in products can provide feedback (e.g., about customer usage, errors, etc.) from the user cycle back to a company's production and product development cycles. Additionally, and of even more importance, is the ability to provide new services and more customized solutions. New business models such as product-service systems create new opportunities for value creation.

Solution Ecosystems

The design of and the participation in ecosystems open up new opportunities for companies to become even more meaningful for their customers and for society as a whole. To solve the real issues of customer and social concerns (e.g., resource consumption, low utilization, etc.) along the entire value chain requires the collaboration of very different stakeholders in one system. Therefore, companies must rethink traditional roles and business models in collaboration with new stakeholders.

From a life cycle point of view, three main cycles are connected in the smart factory logic and enhance it (see Fig. 2.2): the development cycle, the user cycle and the production cycle. Data and information flow between these different cycles to generate value in three areas over the whole life cycle of a product (Nyffenegger et al., 2018):

- Value for the product. Products are developed (e.g., the company will know how products or machines are used in reality)
- Value for the company. Production and design of processes are optimized (e.g., the company will know what led to failure in the field)
- Value for the user (e.g., the user receives more customized services).

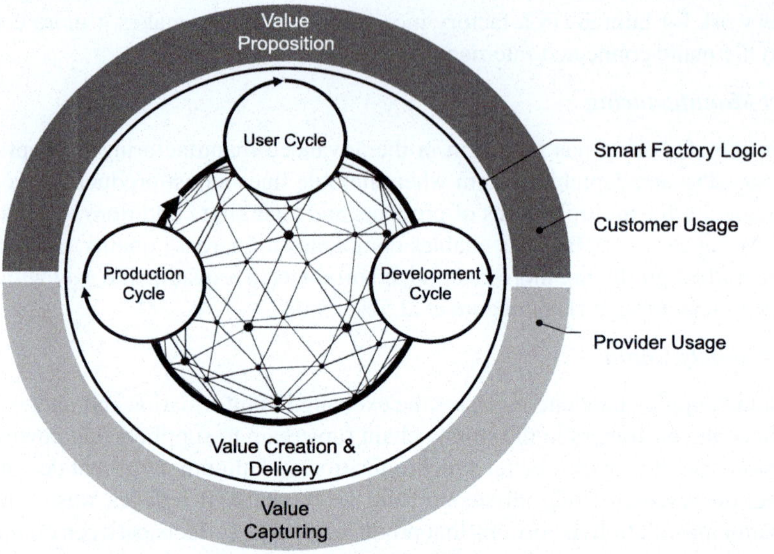

Fig. 2.2 Smart factory cycles (Schuh et al., 2017)

Consequently, the smart factory is more than just addressing efficiency gains and increasing sustainability. It lays the foundation for developing new service-oriented business models based on available data over the whole life cycle of a product. Two pathways can be distinguished for manufacturing companies in making this transformation into a more service and data oriented industrial approach (see Fig. 2.3). The first pathway is based on the operational strategy of companies and focuses on manufacturing operations exploiting the potential of effectivity and efficiency within all the relevant manufacturing-related activities and decision-making processes in the factories, such as production planning, sourcing and supply, by making use of new technologies. The second pathway focuses on the customer's perspective and enables the design of new digital services to exploit new business opportunities. Both the smart factory as well as the digital services are strongly connected due to the fact that the generate knowledge through data enables companies to transform these new capabilities into innovative service-oriented manufacturing business models.

The manufacturing aspect, however, remains vital to the overall vision, and the abovementioned dimensions are closely interconnected to it (Frank et al., 2019). The smart manufacturing approach is built around the idea of the smart factory, also coined the intelligent or digital factory (Lucke et al., 2008; Stock & Seliger, 2016). Not surprisingly, the first conceptualizations of an intelligent factory preceded the early Industry 4.0 initiatives. Industry has always been under pressure to further improve productivity. Therefore, the introduction of new technologies to the manufacturing environment is as old as organized work in plants. Industry 4.0, therefore, was rather an umbrella term coined, for marketing reasons, to better communicate this fact and bring industry back to the agenda of politicians. Lucke et al. (2008) first defined the

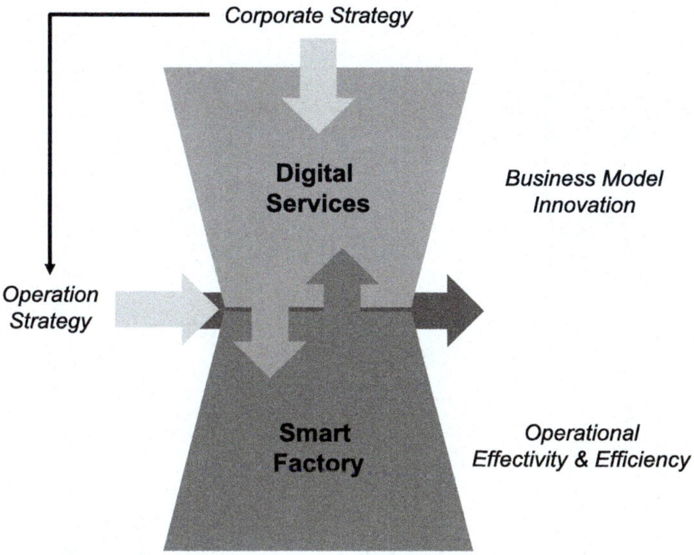

Fig. 2.3 Smart factory and digital services

smart factory vision as a manufacturing environment in which humans and production activities are complemented by intelligent, computer-based systems guaranteeing an uninterrupted flow of production with higher performance and quality. In today's context, however, such a conceptualization does not go far enough in terms of the degree of automation and the absence of human labour in smart production processes. More recent publications, such as Stock and Seliger (2016) or Suginouchi et al. (2017), offer a vision of an autonomous factory floor in which the machines needed for given processes communicate and cooperate thanks to data generated by numerous sensors throughout the factory. It is also known as the cyber physical system. This primary focus on productivity and efficiency as the primary goal while implementing digital technologies is now in practice the shared and agreed concept. New digital products and services will build stable and efficient processes with positive results regarding productivity.

The common ground for all the different concepts is data. Data from products (e.g., machines in the field), data from suppliers (e.g., production status) and data from production (e.g., order status) have to be gathered, integrated and analyzed in a way to create added value. Figure 2.4 summarizes the smart factory vision. New technologies in automation such as cyber-physical systems or mobile computing are implemented into a factory and make it smart. These technologies are the driving force for the availability of a massive amount of important data. These data then allow to learn through AI, machine learning or simulation technologies. This is true for a single manufacturing site as well as a global production network.

To intelligently control production processes, data from physical assets in the factory need to be captured and transferred to upper automation levels. The smart

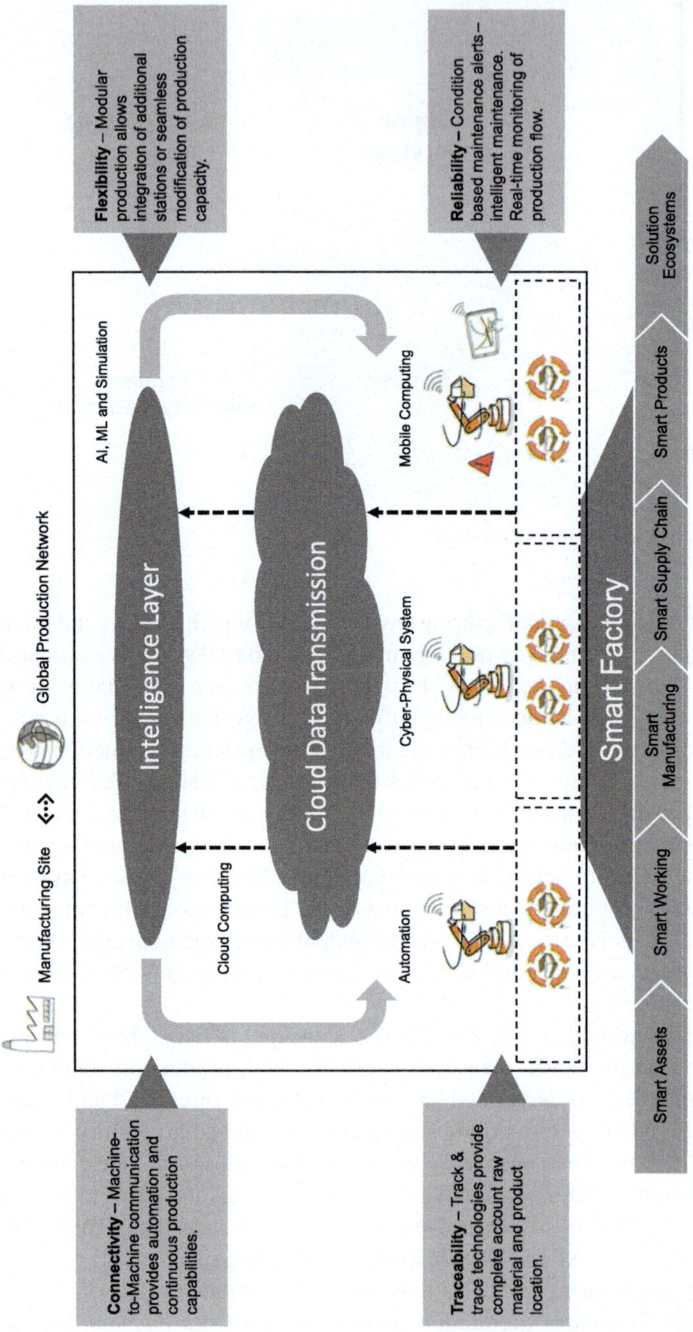

Fig. 2.4 Smart factory vision

factory is a factory characterized by the vertical integration of technologies into four distinct layers. Figure 2.5 (Osterrieder et al., 2019) illustrates these four layers: physical, data, cloud, intelligence and the control layer. The physical layer can be understood as the traditional conception of the shop floor, in which machines and operators physically operate. The data layer is where the data on the production process is collected. Sensors and actuators in the physical layer send and receive data to and from the cloud, while intelligent software controls the three Vs of "big data": variety, velocity and volume. Consequently, these data are stored in the third layer of the smart factory, the cloud and intelligence layer, where they can be sorted and processed by intelligent analytics. The final and highest layer of the smart factory is the control layer, where the master program directing the factory is supervised and tweaked by human decision-makers.

Other smart factory concepts share this vision of autonomy and connectivity. In many versions, smart factories operate with machines equipped with sensors and actuators, which allow the collection, communication, and processing of data that act accordingly (Stock & Seliger, 2016; Wang et al., 2016). These machines are configured with machine to machine communication (M2M) to set up a system of machines capable of automatically carrying out predetermined tasks (Suginouchi

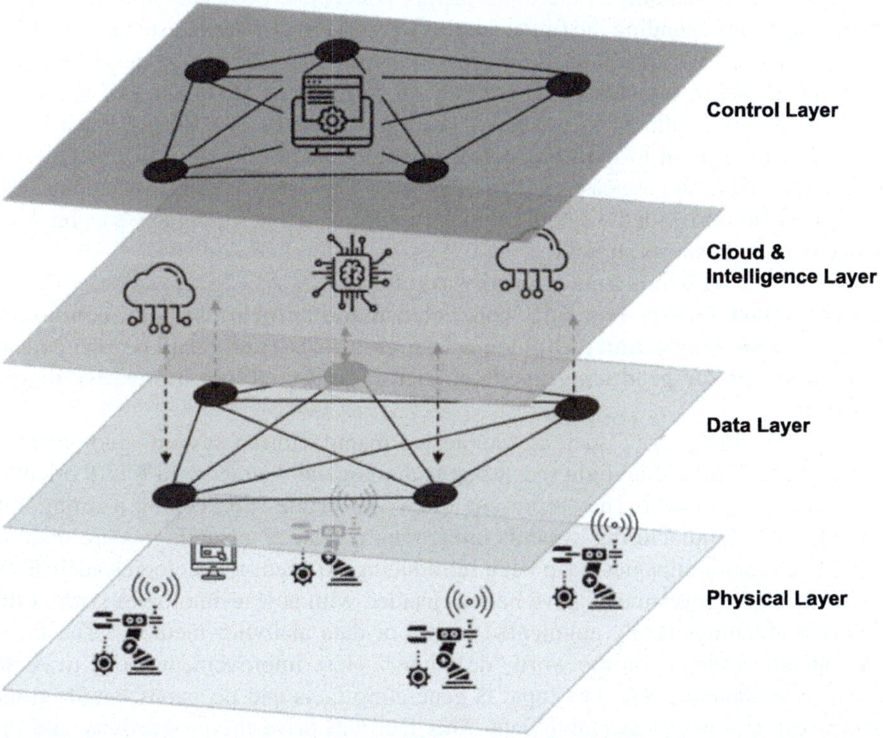

Fig. 2.5 Vertical integration of a smart factory (Osterrieder et al., 2019)

et al., 2017; Tang et al., 2016). Such a system is directed by some sort of higher controlling entity, which can be human or intelligent software (Lucke et al., 2008; Zuehlke, 2008). Therefore, the latest conceptions of the smart factory have transformed the notion of the factory shop floor from a one-dimensional environment to a multi-layered one (Qin et al., 2016; Wang et al., 2016).

A smart factory has to address all these different dimensions to stay competitive in the future. The collaboration of companies along the value chain (up- and downstream) requires the real-time exchange of data. Today we have observed the emergence of many digital platforms in the manufacturing and supply chain context. These platforms have different goals. There are platforms for optimizing sales and order processes by bringing together buyers and suppliers. These platforms digitalize resource availabilities and prices. As such, transactional costs (e.g., the search) can be reduced. There are also digital platforms that offer connectivity, integrating a factory into an open ecosystem where other companies (complementary companies) can offer applications (apps). It is like an open app store for manufacturing, where companies can choose from or offer and sell their own applications (e.g., a production planning app). Additionally, there are also platforms that are focused on connecting suppliers along the value chain in a specific branch with a certain goal (e.g., decarburization), and platforms that are built to improve customer connections.

A clear understanding of the smart factory concept, as presented above, is fundamental to comprehending the full technology perspective of the smart manufacturing transformation. While technical and specific research is the key to developing the concept further, a broader and more general strategic understanding of the smart factory is also of utmost importance. The aim of this book is to present a set of use cases in relation to their respective strategic business value. Various types of smart manufacturing use cases will be rigorously analyzed in terms of what business goals they support with the aim of ultimately helping to link a company's technology strategy to its business strategy.

Therefore, we define a smart factory as follows:

The smart factory is a fully connected manufacturing system, combining human competencies and technological capabilities to conduct all required tasks and decisions for producing goods efficiently and enabling innovative digital services.

In this context, terms such as "connected manufacturing system" and "technological capabilities" highlight the strong technological demand of a future oriented manufacturing system. They represent a new way while still keeping a strong link to the past. Manufacturing systems integrating advances technologies are nothing new. For years companies have been introducing different technologies such as IT systems. These investments have been expanded with new technologies such as the Internet of Things (IoT), augmented reality or data analytics methods. The focus in a smart factory is on the word "decisions". Here improvements need to occur in many companies. The key topic is generating facts and decisions based on the increasing amount of available data. This fact will drive the productivity and the overall manufacturing vision further. Stability is the key in a smart factory. Only stable processes allow a factory to digitize effectively and efficiently. Companies need to know where digitalization helps. Thus, waste reduction in processes is a priority and targets investments to the right area.

2.2 The Smart Factory Navigator Model

The smart manufacturing transformation is intrinsically rooted in efficiency and effectivity gains. The emerging technologies listed are being applied to various manufacturing process stages with the ultimate aim to optimize and generate additional value. Different technologies, however, can be used in various applications and will have a series of different impacts on business objectives according to their application and implementation. In previous research into the concrete application of smart manufacturing technologies, we observed that companies struggle to identify use cases relevant to their business, processes, and context. This impedes their efforts to bring their business and operations to a higher level (Classen et al., 2018). Thus, a more comprehensive understanding of how individual technologies can contribute to operational and strategic business objectives can help companies identify which smart manufacturing solutions might be best suited to their respective goals.

In the last five years, we screened more than 500 smart manufacturing and service use cases in practice and theory to create insights and explore patterns of implementation. The generic use cases—defined as concrete applications of one or more technologies towards a specific function or process step—were linked directly to the potential business benefits they brought. In this way, we aimed to provide a way for companies to combine their respective corporate and operational strategy.

The use cases are the relevant means to drive the operational strategy and the smart factory further. Given that digital technologies are the cornerstone of any operational strategy, these use cases are fundamental. The operational strategy can be split into two distinct elements or phases. First, the basics need to be addressed—operational processes and how these can be improved. These improvements are mainly directed toward cost reduction and efficiency increase. These operational improvements can only be achieved by interlinking with relevant data. The availability of data is becoming an important part of a company's operational strategy. The availability of accurate and timely data can be used to generate additional services (see Fig. 2.3). These smart services, e.g., preventive maintenance plans for customers, create a link between "operations" and "product". This shows the increasing relevance of the operation strategy due to the importance of these operational key data. If these data are not available, smart services cannot be developed.

The implementation of smart manufacturing in industrial companies involves a complex transformation. Different technologies, related interdependencies, unclear application areas as well as unknown effects on operational performances make the strategic decision-process concerning which technologies to be implemented a difficult task. Although there exist many maturity models and analyses about potential impacts, these approaches are mainly focused on technologies and neglect context-related business perspectives. A valid question is always whether the high maturity suggested by most models is really the best alternative. Depending on the business context, less digitalization could, in some cases, be the better choice.

Guidance for managers concerning the selection of use cases and related technologies that support the achievement of superior performance levels is missing. This hinders implementation in factories. The Smart Factory Navigator fills this gap. The navigator guides managers to select use cases based on strategic priorities and application areas within their context. The navigator works like a dynamic steering wheel (see Fig. 2.6). Based on the selected strategic targets (e.g., quality, flexibility), a turn of the inner circles allows access to use cases as well as related technologies that enable higher performance levels in the respective application areas (such as, e.g., production scheduling or quality management). The Smart Factory Navigator also points out that the use cases develop over time. Base use cases form the foundation. Building on this foundation descriptive, diagnostic, predictive and prescriptive use cases can be considered. The human is always in the centre of the smart factory. The smart factory is represented by an intelligent team of humans and machines that are able to adapt to new circumstances immediately.

At the heart of the Smart Factory Navigator are the users and stakeholders of a particular use case. The human centricity with the technologies that act as enablers for the use cases form the core of the Smart Factory Navigator. The following two chapters focus on this core of the model and thus form the basis for Chap. 3 and

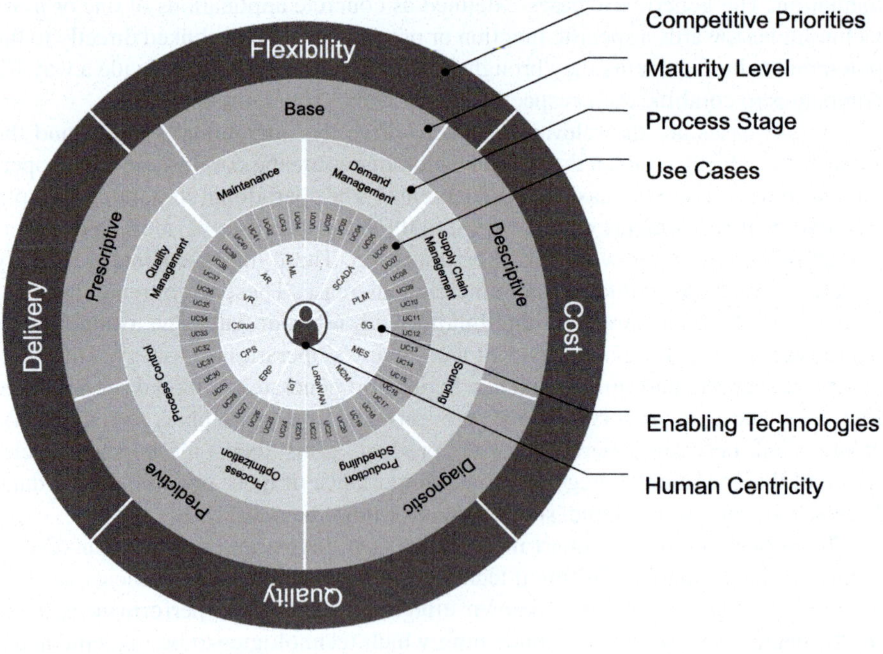

Fig. 2.6 Smart factory navigator

the Smart Factory Framework, in which the dimensions of process stages, maturity levels and competitive priorities are discussed on the basis of the 44 use cases.

2.3 Human Centricity

In the last few years Industry 4.0 was primarily driven from a technology perspective. However, as every introduction of new technology potentially impacts the interactions between humans and machines, the understanding and modelling of the role of humans is decisive for the sustainable development of intelligent manufacturing systems (Saabye et al., 2022). Take as an example the vision of ultra-flexible smart factories which are able to self-adapt production processes due to demand or product changes (Lu et al., 2020). The wide acceptance of this digital vision by the workforce as well as the integration of the relevant implicit shop-floor knowledge are vital elements for success. If these elements are integrated into a smart factory implementation process a strong collaboration between machines, algorithms and people can be achieved. "Future machines can establish trustworthy relationships with humans via proactive communication, empathic understanding, and need-driven collaborations, which can lead to high performance human beings and productive yet flexible manufacturing processes" (Lu et al., 2021).

Humans as sensors, actuators, communicators, and decision-makers interact with physical objects (e.g., machines) thereby becoming intelligent building similar competencies. These physical objects are connected through information and data streams in the cyber systems in order to exploit the potentials of both worlds (Cimini et al., 2020; Zhou et al., 2018). Figure 2.7 illustrates how a well-orchestrated architecture and thought-through approach for a smart factory needs to be built in order to support the decision making of the human operator who activates the physical devices such as machines or logistics systems.

Only if the combination of human, technological and machine abilities and strengths are holistically considered, can a productivity explosion for manufacturing companies be possible. This is why "humans" are at the centre of our Smart Factory Navigator and our definition of smart factory (see also Sect. 2.1). Five key criteria should be considered when starting to transform to the smart factory.

Firstly, for the identification and implementation of the smart factory use cases human implicit knowledge is required. This is knowledge that is normally not captured by any system or tool. Only workers that are familiar with processes, machines and interdependencies have this knowledge and thus can enable the systematic identification process of potential hypothesis. Thus, already in the first step of the identification of potential use cases relevant knowledge about processes and machines needs to be integrated. This implicit knowledge is also helpful for later in the process of a digital transformation where it can be used to improve data-driven models for the smart factory. In order to make this implicit knowledge transparent cooperation with the workers and their buy-in are key requirements.

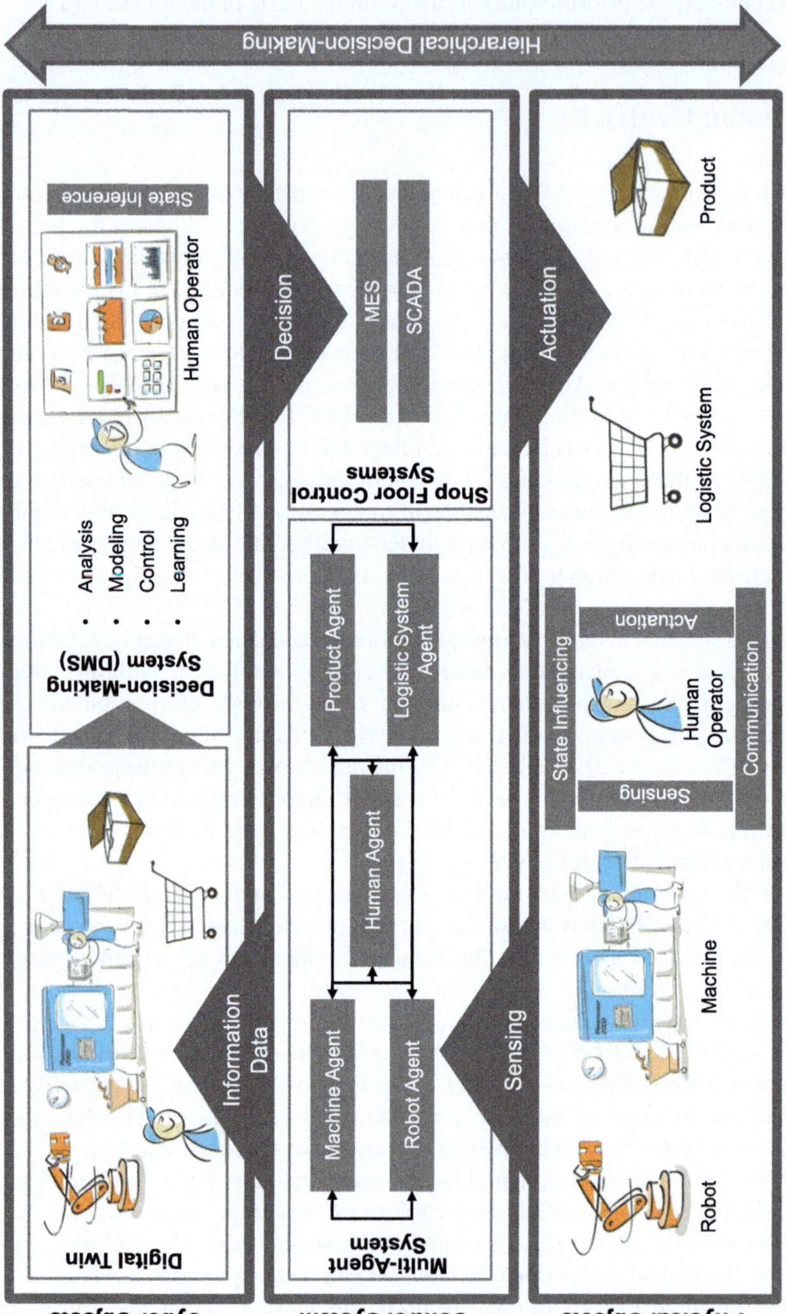

Fig. 2.7 Social cyber-physical manufacturing system architecture (based on Cimini et al., 2020)

Secondly, only if it is clear, which concrete problems in current processes, human actions or decisions exist and are experienced and highlighted by workers, should these problems be followed up. An agreed upon and discussed problem is key for the buy-in of the workforce.

Thirdly, workers need to be involved in the solution development and become part of it, like an internal customer. Only if the company ensures the active involvement and considers the views of the worker, will solutions be accepted and integrated into the daily business. The adaption to new technical solutions will be carried out in the organization.

Fourthly, this factor is associated with a dual perspective on problem solving. The enabling of smart factory technologies and human capabilities need to be complementary and designed in such a way, so as to create an optimal team composition. Most significant performance improvements can be achieved when humans and machines join forces to form collaborative intelligence (Wilson & Daugherty, 2018).

Fifthly, smart factory technologies can support human work in the factory in different ways. Well-being and the satisfaction of humans can be significantly improved. The lifting of heavy weights or simple repetitive routine work should be a concept from the past. In addition, routine visual controls of parts by employees should be replaced by camera-based solutions. Consequently, quality as well as employee satisfaction will rise.

The involvement of the five criteria previously stated will lead to a new generation of intelligent manufacturing system, where lots of brainwork is required to create a human cyber physical system. In Fig. 2.8 the evolution is described from Human Physical Systems (HPS), within traditional manufacturing systems, to a new-generation of intelligent manufacturing systems—Human Cyber Physical Systems (HCPS). It is also shown in the figure, how the cyber system mediates the interaction and combination of physical systems with humans to achieve a higher level of intelligence in manufacturing operations.

Only a proper integration of the different subsystems enables the achievement of superior performance of the whole system. Therefore, it is important to understand

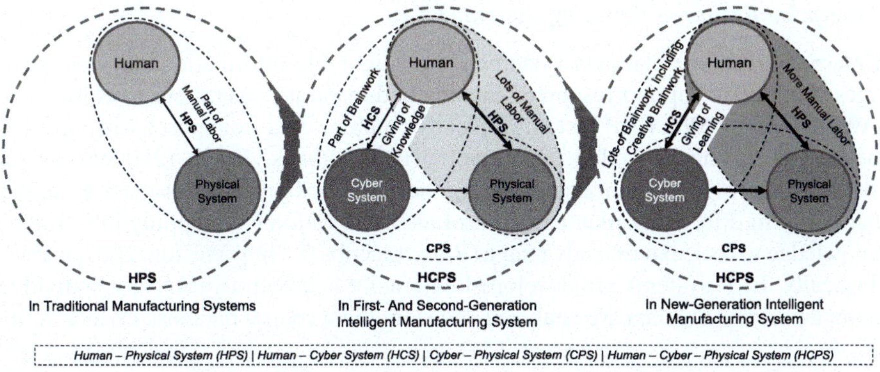

Fig. 2.8 Evolution from the HPS to the new-generation HCPS (based on Zhou et al., 2018)

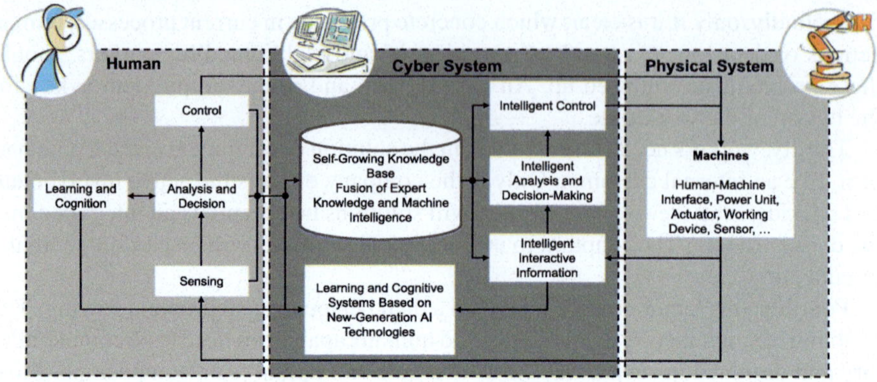

Fig. 2.9 Mechanism of new-generation intelligent manufacturing systems (based on digital technologies—the technology perspective and toolbox Zhou et al., 2018)

the interactions between humans and intelligent manufacturing equipment, as well as the involvement of humans in the intelligent orchestration of the manufacturing environment (Cimini et al., 2020). In Fig. 2.9 a summary is presented of this mechanism between humans, cyber systems, and physical systems.

2.4 Enabling Technologies

As alluded to in the previous section, smart manufacturing is based on a distinct set of emerging, 'disruptive' technologies. These technologies are the building blocks of the generic use-cases and collectively make up the technological toolbox required for the smart factory. For a detailed and comprehensive understanding of the various technology applications presented in the following Smart Factory Framework (see Chap. 3), each technology is briefly defined and explained below.

Enterprise Resource Planning System (ERP)

Enterprise resource planning systems are configurable informational system packages intended to support resource planning and integrate several business functions (Wu & Wang, 2006). ERP systems typically combine data from purchasing, production, sales, finance, or other departments (Benninghaus, 2018). ERP systems act as "frameworks for organizing, defining, and standardizing the business processes necessary to effectively plan and control an organization (…) tapping into internal knowledge to seek external advantage (Robert Jacobs & 'Ted' Weston, 2007, p. 357). Typically, ERP systems are developed externally and then tweaked for individual companies, and the data are analyzed and stored in a central database (Wu & Wang, 2006).

Manufacturing Execution System (MES)

Manufacturing execution systems are the link between the shop floor and resource planning (ERP) (Gerberich, 2011). They are integrated, modular and real-time information processing systems that cover the entire manufacturing process (Gerberich, 2011; Kletti et al., 2015). Data are integrated and stored in a central database (Gerberich, 2011). The latest guideline of the Association of German Engineers (VDI) 5600 lists the following operational tasks for MES software: order, equipment, materials, human resources, quality, information and energy management, detailed scheduling and process control, data acquisition and performance analysis (Verein Deutscher Ingenieure, 2016).

Product Lifecycle Management System (PLM)

Product lifecycle management systems allow for the effective management of a product along its lifecycle (Saaksvuori & Immonen, 2008). It can be understood as an information management system that combines processes, data, humans and other resources (Pagoropoulos et al., 2017). Relevant product information is provided such as bills of material, change orders, production procedures, or test specifications to stakeholders (Benninghaus, 2018). PLM platforms have enabled the "creation, organization and dissemination of product-related knowledge across the enterprise" (Jupp & Nepal, 2014, p. 31).

Supervisory Control and Data Acquisition System (SCADA)

A supervisory control and data acquisition system in the factory allows for the collection of field data and its communication back "to a master station via communications technology" (Bailey & Wright, 2003). The SCADA software is placed on top of a hardware architecture e.g. of all production machines with which it is interfaced via programmable logic controllers and remote terminal units (Daneels & Salter, 1999). Once the field data is communicated to the central site, data analysis is carried out to supervise the performance of the processes and relay back any necessary control actions to the machines via operator displays (Bailey & Wright, 2003). Typical functions of SCADA software include trending, alarm handling, logging/archiving, report generation and automation (Daneels & Salter, 1999).

Cloud Computing

Cloud computing offers the possibility of storing data and information in real-time in a virtual cloud. This kind of data management solution offers great flexibility and scalability as centralized information is accessible from any location with access authority (Bruque Cámara et al., 2015). Cloud computing also allows for greater customization of IT services, replacing local data storage, service or software with services offered by external cloud providers (Benninghaus, 2018). Aside from data and information storage, cloud services offer infrastructure as a service (IaaS), "platforms as a service" (PaaS) and "software as a service" (SaaS) (Bruque Cámara et al., 2015). Recent developments include the application of the on-demand principles of cloud computing to the production context (cloud manufacturing) (Ren et al., 2015).

Edge Computing

Edge computing refers to the shift of certain computing tasks—primarily data processing and storage—from the cloud to the edge of a network (Shi & Dustdar, 2016). The network edge can be defined as "any computing and network resource along the path between data sources and cloud data centers" (Shi et al., 2016, p. 638). Often, the data sources are themselves the edge of the network, generating the data to be used for several applications. With edge computing, computing resources are transferred closer to the source of the data, and some workload is shifted from the cloud data repositories. The introduction of edge computing can address the growing concern that cloud computing is not able to cope with the rapidly increasing volume of data being generated (Shi & Dustdar, 2016; Shi et al., 2016). With this transfer of tasks, edge computing can contribute to alleviating the cloud's latency, bandwidth, security and privacy issues (Shi & Dustdar, 2016).

Long Range Wide Area Network (LoRaWAN)

Long Range Wide Area Network technology allows for the wireless transmission of data across long distances through LoRa modulation (Haxhibeqiri et al., 2018). The networks are usually organized in a stars-of-stars topology where end-devices communicate raw data packets to gateways that, in turn, forward the data to a central server—which can be either a local or a cloud server (de Carvalho Silva et al., 2017; Haxhibeqiri et al., 2018). The solution is an attractive possibility for IoT networks due to its "long range, low power and low bit-rate" and is already applied in a variety of industrial use-cases (de Carvalho Silva et al., 2017).

5G

5G is the latest generation in wireless cellular network technology designed to meet increasing capacity demand generated by "massive data growth over the last ten years" (Shafi et al., 2017). 5G has been applied to a variety of areas, namely, enhanced mobile broadband, ultra-reliable and low latency communications (URLLC), and massive machine-type communications (Shafi et al., 2017). In particular, URLLC can be very useful for the implementation of smart factory solutions, facilitating the wireless control of industrial equipment. 5G is becoming increasingly important for IoT applications (Li et al., 2018).

M2M Communication

M2M communication consists of the automated exchange of information between technical systems such as machines. Machines communicate through technologies such as fieldbus, wireless local area network (WLAN), hypertext transfer protocols, transmission control protocols, or serial interfaces (Benninghaus, 2018). Thus, a key prerequisite for M2M is communication standards (Siepmann & Graef, 2016).

Mobile Computing

Commercial mobile devices (i.e., tablets, smartphones) have many applications for the manufacturing environment (Morkos et al., 2012). Applications include communication (audio & visual), readers for barcodes, visualization of manuals and instructions, and mobile notification centers (Benninghaus, 2018).

Internet of Things (IoT)

The Internet of Things is best defined as a term for objects that are connected and share information with the internet (Benninghaus, 2018). This type of technology allows for solutions which are increasingly integrated, from standalone devices in a network to an intelligent object network in which physical worlds and virtual worlds interact (Kache & Seuring, 2017).

Sensors and Actuators

Sensors are objects to detect and measure physical variables and convert them into electrical, magnetic or mechanical output signals. Parameters are measured such as pressure, acoustic, light, force, voltage, or temperature (López et al., 2011). Actuators receive an electrical signal and create a prescribed physical change by motion, force, etc. Both are key technologies for other digital technologies such as robotics or wearables (Benninghaus, 2018).

Track & Trace Technologies

Track and trace technology allows for the identification of the current and past locations of objects (Mittal et al., 2019). Such information can be traced continuously in real-time (Benninghaus, 2018). For this purpose, Near field communication (NFC) and radio frequency identification (RFID) are two widely used wireless communication technologies in manufacturing contexts. On the other hand, QR-codes, data matrix code, and 2D barcodes are additional, more simplistic solutions for tracking and tracing activities.

Vision Technologies

Vision technologies—commonly referred to as computer or machine vision—can be understood as a set of technologies that rely on "image capturing and analysis" to perform a variety of functions across the manufacturing process (Nandini et al., 2016). The vision technology systems can be classified into three different categories: PC-based vision systems, embedded systems, and smart cameras (Liu et al., 2015). The performance and complexity of these systems will vary according to the associated hardware and software setup. The primary aim of vision technologies is to act as a supporting element of automated processes. For this reason, such technologies are usually coupled with artificial intelligence and machine learning algorithms (Golnabi & Asadpour, 2007; Tai-shan & Du-wu, 2006). The most common applications of these technologies are automated visual inspection, process control, parts identification and robotic guidance and control (Golnabi & Asadpour, 2007).

Cyber-Physical Systems (CPS)

Cyber-physical systems are often referred to simply as digital technologies (Hozdić, 2015; Kang et al., 2016; Lee et al., 2015; Monostori, 2014; Obermaier, 2016). Such systems can be understood as "collaborating computational entities which are in intensive connection with the surrounding physical world and its on-going processes, providing and using, at the same time, data-accessing and data-processing services" (Monostori, 2014, p. 9). Thus, CPS represent an "integration of computation with physical processes" able to communicate with other CPS or the internet (Lee, 2008, p. 363). Importantly, CPS do not necessarily need to exchange data with the internet (Wang & Wang, 2018).

Big Data Analytics

Big data can be interpreted as a term for "large, diverse, complex, longitudinal, and/or distributed data sets" (Floridi, 2012). This approach to data was developed from decision support systems and business intelligence (Chen et al., 2012). With big data, organizations can analyze data with respect to the 4Vs (volume, variety, velocity, veracity). Volume represents the large quantity of data collected and in need of storage or processing. Variety reflects the diversity in format and source of data, and velocity addresses the frequency of data generation (Gandomi & Haider, 2015). Further accounts also include value (obtaining benefits from data) and veracity (reliability of data sources) as key factors (Wamba et al., 2015). Therefore, big data analytics is "the ability to acquire, store, process and analyze large amount of (…) data in various forms and deliver meaningful information to users to discover business values and insights in a timely fashion" (Wang et al., 2018 p. 6). Hence, big data analytics aim to support decision-making, create transparency, discover variability, or improve performance (Wamba et al., 2015; Gandomi & Haider, 2015, p. 140). The methods utilized in the analysis of big data include data mining, database management systems, data warehousing, clustering, regression, anomaly detection, neural networks, genetic algorithms, multivariate statistical analysis, or heuristic search (Chen et al., 2012).

AI, ML & Simulation

There is no widely accepted standard definition of artificial intelligence, but the following conceptualization by Rich is widely cited in literature: "Artificial Intelligence is the study of how to make computers do things at which, at the moment people do better" (Ertel, 2017, p. 2). Today, the term serves as a broader label for different tools such as neural networks, fuzzy logic, inductive learning, genetic algorithms, discrete event simulation or knowledge-based systems (Sanders & Gegov, 2013). Ultimately, AI grants machines and systems the ability to perform tasks traditionally accomplished exclusively by the human mind. In this context, the learning component (machine learning) is a pivotal component of AI. Machine learning processes large amounts of information and looks for patterns to learn previously unknown inputs (Schuld et al., 2015).

Collaborative Robotics

Collaborative Robotics—a subset of smart robotics—can be understood as robots working in collaboration with humans as opposed to in place of them. Collaborative robots rely on artificial intelligence and machine learning in order to perceive their environments, take decisions and act accordingly to avoid collisions with humans

(Benninghaus, 2018; Weng et al., 2009). The interaction and collaboration with humans is strictly controlled and monitored—through collision detection and predefined levels of interaction—to avoid safety issues.

Automated Guided Vehicles (AGV)

Automated guided vehicles are driverless vehicles whose movements can be controlled by wires, strategically positioned radars, or special tapes (Benninghaus, 2018). They can provide automated loading, transportation, and unloading capabilities (Ventura et al., 2015). Automated guided vehicles are able to plan their route in advance and analyze the environment dynamically through the use of laser scanners and sensors (Benninghaus, 2018). This process is facilitated by sensor techniques, network models, artificial intelligence, and track & trace technology (Benninghaus, 2018). Applications in the manufacturing context include tuggers, unit loaders and forklifts (Benninghaus, 2018).

Additive Manufacturing (AM)

AM can be defined as the "process of joining materials to make objects from 3D model data, usually layer upon layer, as opposed to subtractive manufacturing technologies" (F42 Committee, 2012, p. 2). The process is also often referred to as "3D printing". AM techniques can be differentiated into solid-based, powder-based and liquid-based processes (Hopkinson & Dickens, 2006). Hence, different materials such as polymers, metals, ceramics or synthetic resin can be utilized (Strange & Zucchella, 2017). Depending on the application, AM advantages include the reduction of waste, the ability to create complex and functionally integrated parts, rapid design changes, accelerated process development, and no required tooling and customized production (Benninghaus, 2018). The further development of this technology has allowed for a shift from rapid prototyping to the production of final products in small quantities or spare parts (Kang et al., 2016).

Blockchain

Blockchain is a shared, immutable ledger that facilitates the recording of transactions and tracking of assets across an enterprise network. By definition, an asset can be tangible (a house, a car, cash, land) or intangible (intellectual property, patents, copyrights, trademarks). Practically anything that has value can be tracked and traded on a blockchain network, reducing risk and lowering costs for all parties involved (Gupta, 2020). Blockchain solutions have so far been known and successfully used primarily in the financial sector. Applications for blockchain technologies are seen in the increasing transparency of supply chains, the tracking of goods in logistics or the exchange of relevant information (e.g., smart contracts) between parties (Ditsche & Streichfuss, 2021).

Wearables

Wearables are defined as "computing devices small enough and light enough to be worn on one's body without causing discomfort" (Barfield, 2015, p. 3). They are intended to assist the operator with context-specific information. These technologies are classified as on or in-body wearables (Barfield, 2015). Further classification for the on-body technologies are hand-worn (i.e., smartwatches), foot-worn (i.e., shoes with a tracker), head-mounted (i.e., AR goggles) or body dressed (i.e., sensor shirts) (Benninghaus, 2018).

Augmented Reality (AR)

Augmented reality can be understood as a visualization technology which serves to enhance the user's perception of reality by combining visual cues from both the physical and digital world (Mekni & Lemieux, 2014; Syberfeldt et al., 2016). As formulated by Kipper and Rampolla, AR takes digital or computer-generated information—such as images, audio, video and touch or haptic sensations—and lays it over a real-time environment (Kipper & Rampolla, 2013). In this way, AR provides additional information to the worker and assists tasks in real-time. Possible devices include smart glasses, head-mounted displays, traditional monitors, windowpanes, or handheld devices (Benninghaus, 2018). Newer developments include spatial solutions that directly project information on an object without any display (Benninghaus, 2018). In contrast to virtual reality, augmented reality merely extends the real world—as opposed to replacing it (Mittal et al., 2019). For manufacturing contexts, AR can assist assembly or maintenance operations.

Virtual Reality (VR)

Virtual reality allows for the creation of a realistic three-dimensional world in which the optical, auditory and tactile senses of the users are accounted for. In VR, "users can look at, move around, and interact with these worlds as if they were real" (Mujber et al., 2004). Therefore, the environment and characteristics of the world are entirely modelled. For instance, a realistic representation of the production process could be recreated and simulated in an interactive manner (Siepmann & Graef, 2016). VR systems can be classified in terms of levels of immersion: non-immersive, semi-immersive and fully immersive (Benninghaus, 2018).

Drone

An unmanned aerial vehicle (UAV), commonly known as a drone, is an aircraft without any human pilot, crew, or passengers on board. UAVs are a component of an unmanned aircraft system (UAS), which includes adding a ground-based controller and a system of communications with the UAV (Hu, J.). There is a massive market for drones today. Often being used privately to take pictures and films, but special market segments such as surveying are growing fast. The key market is still military application. In recent years drones pilot applications in logistics are expanding. The expectations are high, but the breakthrough "killer applications" has not been reached up to now. Regulation restrictions are not helping the growth (Hu et al., 2022).

References

Almada-Lobo, F. (2016). The industry 4.0 revolution and the future of manufacturing execution systems (MES). *Journal of Innovation Management, 3*(4), 16–21. https://doi.org/10.24840/2183-0606_003.004_0003

Bailey, D., & Wright, E. (2003). *Practical SCADA for industry*. Elsevier.

Barfield, W. (2015). Wearable computers and augmented reality: Musings and future directions. In W. Barfield (Ed.), *Fundamentals of wearable computers and augmented reality* (2nd ed., pp. 3–12). CRC Press. https://doi.org/10.1201/b18703-3

Benninghaus, C. (2018). *Impact of digitalization on the strategic management of international manufacturing networks: A configurations perspective*. University of St.

Bruque Cámara, S., Moyano Fuentes, J., & Maqueira Marín, J. M. (2015). Cloud computing, Web 2.0, and operational performance: The mediating role of supply chain integration. *The International Journal of Logistics Management, 26*(3), 426–458. https://doi.org/10.1108/IJLM-07-2013-0085

Chen, Chiang, & Storey. (2012). Business intelligence and analytics: From big data to big impact. *MIS Quarterly, 36*(4), 1165. https://doi.org/10.2307/41703503

Cimini, C., Pirola, F., Pinto, R., & Cavalieri, S. (2020). A human-in-the-loop manufacturing control architecture for the next generation of production systems. *Journal of Manufacturing Systems, 54*, 258–271. https://doi.org/10.1016/j.jmsy.2020.01.002

Classen, M., Blum, C., Budde, L., Hänggi, R., & Friedli, T. (2018). Digitalisierung à la carte: Individuelle Auswahl von Use-Cases nach strategischen Prioritäten. *ZWF Zeitschrift Für Wirtschaftlichen Fabrikbetrieb, 113*(12), 850–854. https://doi.org/10.3139/104.112016

Dalenogare, L. S., Benitez, G. B., Ayala, N. F., & Frank, A. G. (2018). The expected contribution of Industry 4.0 technologies for industrial performance. *International Journal of Production Economics, 204*, 383–394. https://doi.org/10.1016/j.ijpe.2018.08.019

Daneels, A., & Salter, W. (1999). *What is SCADA?*

de Carvalho Silva, J., Rodrigues, J. J., Alberti, A. M., Solic, P., & Aquino, A. L. (2017). *LoRaWAN—A low power WAN protocol for Internet of Things: A review and opportunities*, 1–6.

Ditsche, J., & Streichfuss, M. (2021). *Massive potential for industry: Blockchain beyond the cryptocurrencies*. Roland Berger. https://www.rolandberger.com/en/Insights/Publications/Massive-potential-for-industry-Blockchain-beyond-the-cryptocurrencies.html

Ertel, W. (2017). *Introduction to artificial intelligence* (2nd ed.). Springer International Publishing: Imprint: Springer. https://doi.org/10.1007/978-3-319-58487-4

F42 Committee. (2012). Terminology for additive manufacturing technologies. *ASTM International*. https://doi.org/10.1520/F2792-12

Floridi, L. (2012). Big data and their epistemological challenge. *Philosophy & Technology, 25*(4), 435–437. https://doi.org/10.1007/s13347-012-0093-4

Frank, A. G., Dalenogare, L. S., & Ayala, N. F. (2019). Industry 4.0 technologies: Implementation patterns in manufacturing companies. *International Journal of Production Economics, 210*, 15–26. https://doi.org/10.1016/j.ijpe.2019.01.004

Gandomi, A., & Haider, M. (2015). Beyond the hype: Big data concepts, methods, and analytics. *International Journal of Information Management, 35*(2), 137–144. https://doi.org/10.1016/j.ijinfomgt.2014.10.007

Gerberich, T. (2011). Lean oder MES in der Automobilzulieferindustrie. *Gabler*. https://doi.org/10.1007/978-3-8349-6754-1

Golnabi, H., & Asadpour, A. (2007). Design and application of industrial machine vision systems. *Robotics and Computer-Integrated Manufacturing, 23*(6), 630–637. https://doi.org/10.1016/j.rcim.2007.02.005

Gupta, M. (2020). *Blockchain for dummies* (3rd ed.). Wiley.

Haxhibeqiri, J., De Poorter, E., Moerman, I., & Hoebeke, J. (2018). A Survey of LoRaWAN for IoT: From technology to application. *Sensors, 18*(11), 3995. https://doi.org/10.3390/s18113995

Hopkinson, N., & Dickens, P. (2006). Emerging rapid manufacturing processes. In N. Hopkinson, R.J.M. Hague, & P.M. Dickens (Eds.), *Rapid manufacturing* (pp. 55–80). Wiley. https://doi.org/10.1002/0470033991.ch5

Hozdić, E. (2015). Smart factory for industry 4.0: A review. *Journal of Modern Manufacturing Systems and Technology, 7*, 28–35.

Hu, J., Niu, H., Carrasco, J., Lennox, B., & Arvin, F. (2022). Fault-tolerant cooperative navigation of networked UAV swarms for forest fire monitoring. *Aerospace Science and Technology, 123*, 107494. https://doi.org/10.1016/j.ast.2022.107494

Jupp, J. R., & Nepal, M. (2014). BIM and PLM: Comparing and learning from changes to professional practice across sectors. In S. Fukuda, A. Bernard, B. Gurumoorthy, & A. Bouras (Eds.), *Product lifecycle management for a global market* (Vol. 442, pp. 41–50). Springer. https://doi.org/10.1007/978-3-662-45937-9_5

Kache, F., & Seuring, S. (2017). Challenges and opportunities of digital information at the intersection of Big Data Analytics and supply chain management. *International Journal of Operations & Production Management, 37*(1), 10–36. https://doi.org/10.1108/IJOPM-02-2015-0078

Kang, H. S., Lee, J. Y., Choi, S., Kim, H., Park, J. H., Son, J. Y., Kim, B. H., & Noh, S. D. (2016). Smart manufacturing: Past research, present findings, and future directions. *International Journal of Precision Engineering and Manufacturing-Green Technology, 3*(1), 111–128. https://doi.org/10.1007/s40684-016-0015-5

Kipper, G., & Rampolla, J. (2013). *Augmented reality: An emerging technologies guide to AR* (1st ed). Syngress/Elsevier.

Kletti, J., Deisenroth, R., Diesner, M., Kletti, W., Lübbert, J.-P., Schumacher, J., & Strebel, T. (2015). MES als Werkzeug für die perfekte Produktion. In J. Kletti (Ed.), *MES—Manufacturing execution system* (pp. 19–29). Springer. https://doi.org/10.1007/978-3-662-46902-6_2

Lasi, H., Fettke, P., Kemper, H.-G., Feld, T., & Hoffmann, M. (2014). Industry 4.0. *Business & Information Systems Engineering, 6*(4), 239–242. https://doi.org/10.1007/s12599-014-0334-4

Lee, E. A. (2008). *Cyber physical systems: Design challenges*. 2008 11th IEEE International Symposium on Object and Component-Oriented Real-Time Distributed Computing (ISORC), 363–369. https://doi.org/10.1109/ISORC.2008.25

Lee, I., & Lee, K. (2015). The Internet of Things (IoT): Applications, investments, and challenges for enterprises. *Business Horizons, 58*(4), 431–440. https://doi.org/10.1016/j.bushor.2015.03.008

Lee, J., Bagheri, B., & Kao, H.-A. (2015). A cyber-physical systems architecture for Industry 4.0-based manufacturing systems. *Manufacturing Letters, 3*, 18–23. https://doi.org/10.1016/j.mfglet.2014.12.001

Li, S., Xu, L. D., & Zhao, S. (2018). 5G Internet of Things: A survey. *Journal of Industrial Information Integration, 10*, 1–9. https://doi.org/10.1016/j.jii.2018.01.005

Liu, Z., Ukida, H., Niel, K., & Ramuhalli, P. (2015). Industrial inspection with open eyes: Advance with machine vision technology. In Z. Liu, H. Ukida, P. Ramuhalli, & K. Niel (Eds.), *Integrated imaging and vision techniques for industrial inspection* (pp. 1–37). Springer. https://doi.org/10.1007/978-1-4471-6741-9_1

Longo, F., Nicoletti, L., & Padovano, A. (2017). Smart operators in industry 4.0: A human-centered approach to enhance operators' capabilities and competencies within the new smart factory context. *Computers & Industrial Engineering, 113*, 144–159. https://doi.org/10.1016/j.cie.2017.09.016

López, T. S., Ranasinghe, D. C., Patkai, B., & McFarlane, D. (2011). Taxonomy, technology and applications of smart objects. *Information Systems Frontiers, 13*(2), 281–300. https://doi.org/10.1007/s10796-009-9218-4

Lu, Y., Adrados, J. S., Chand, S. S., & Wang, L. (2021). Humans are not machines—Anthropocentric human-machine symbiosis for ultra-flexible smart manufacturing. *Engineering, 7*(6), 734–737. https://doi.org/10.1016/j.eng.2020.09.018

Lu, Y., Xu, X., & Wang, L. (2020). Smart manufacturing process and system automation—A critical review of the standards and envisioned scenarios. *Journal of Manufacturing Systems, 56*, 312–325. https://doi.org/10.1016/j.jmsy.2020.06.010

Lucke, D., Constantinescu, C., & Westkämper, E. (2008). Smart factory—A step towards the next generation of manufacturing. In M. Mitsuishi, K. Ueda, & F. Kimura (Eds.), *Manufacturing systems and technologies for the new frontier* (pp. 115–118). Springer. https://doi.org/10.1007/978-1-84800-267-8_23

Mekni, M., & Lemieux, A. (2014). *Augmented reality: Applications, challenges and future* trends.

Mittal, S., Khan, M. A., Romero, D., & Wuest, T. (2019). Smart manufacturing: Characteristics, technologies and enabling factors. *Proceedings of the Institution of Mechanical Engineers, Part b: Journal of Engineering Manufacture, 233*(5), 1342–1361. https://doi.org/10.1177/0954405417736547

Monostori, L. (2014). Cyber-physical production systems: Roots, expectations and R&D challenges. *Variety Management in Manufacturing, 17*, 9–13. https://doi.org/10.1016/j.procir.2014.03.115

Morkos, B., Taiber, J., Summers, J., Mears, L., Fadel, G., & Rilka, T. (2012). Mobile devices within manufacturing environments: A BMW applicability study. *International Journal on Interactive Design and Manufacturing (IJIDeM), 6*(2), 101–111. https://doi.org/10.1007/s12008-012-0148-x

Mujber, T. S., Szecsi, T., & Hashmi, M. S. J. (2004). Virtual reality applications in manufacturing process simulation. *Proceedings of the International Conference on Advances in Materials and Processing Technologies: Part, 155–156*, 1834–1838. https://doi.org/10.1016/j.jmatprotec.2004.04.401

Nandini, V., Deepak Vishal, R., Arun Prakash, C., & Aishwarya, S. (2016). A review on applications of machine vision systems in industries. *Indian Journal of Science and Technology, 9*(48). https://doi.org/10.17485/ijst/2016/v9i48/108433

Nyffenegger, F., Hänggi, R., & Reisch, A. (2018). A reference model for PLM in the area of digitization. In P. Chiabert, A. Bouras, F. Noël, & J. Ríos (Eds.), *Product lifecycle management to support industry 4.0* (Vol. 540, pp. 358–366). Springer International Publishing. https://doi.org/10.1007/978-3-030-01614-2_33

Obermaier, R. (2016). Industrie 4.0 als unternehmerische Gestaltungsaufgabe: Strategische und operative Handlungsfelder für Industriebetriebe. In R. Obermaier (Ed.), *Industrie 4.0 als unternehmerische Gestaltungsaufgabe: Betriebswirtschaftliche, technische und rechtliche Herausforderungen* (pp. 3–34). Springer Fachmedien Wiesbaden. https://doi.org/10.1007/978-3-658-08165-2_1

Osterrieder, P., Budde, L., & Friedli, T. (2019). The smart factory as a key construct of industry 4.0: A systematic literature review. *International Journal of Production Economics*. S0925527319302865. https://doi.org/10.1016/j.ijpe.2019.08.011

Pagoropoulos, A., Pigosso, D. C. A., & McAloone, T. C. (2017). The emergent role of digital technologies in the circular economy: A review. *9th CIRP IPSS Conference: Circular Perspectives on PSS, 64*, 19–24. https://doi.org/10.1016/j.procir.2017.02.047

Porter, M. E., & Heppelmann, J. E. (2015). How smart, connected products are transforming companies. *Harvard Business Review, 93*(10), 96–114. Scopus.

Qin, J., Liu, Y., & Grosvenor, R. (2016). A categorical framework of manufacturing for industry 4.0 and beyond. *Procedia CIRP, 52*, 173–178. https://doi.org/10.1016/j.procir.2016.08.005

Ren, L., Zhang, L., Tao, F., Zhao, C., Chai, X., & Zhao, X. (2015). Cloud manufacturing: From concept to practice. *Enterprise Information Systems, 9*(2), 186–209. https://doi.org/10.1080/17517575.2013.839055

Robert Jacobs, F., & 'Ted' Weston, F. C. (2007). Enterprise resource planning (ERP)—A brief history. *Special Issue Evolution of the Field of Operations Management SI/ Special Issue Organisation Theory and Supply Chain Management, 25*(2), 357–363. https://doi.org/10.1016/j.jom.2006.11.005

Saabye, H., Kristensen, T. B., & Wæhrens, B. V. (2022). Developing a learning-to-learn capability: Insights on conditions for Industry 4.0 adoption. *International Journal of Operations & Production Management, 42*(13), 25–53. https://doi.org/10.1108/IJOPM-07-2021-0428

Saaksvuori, A., & Immonen, A. (2008). Product lifecycle management. *Springer*. https://doi.org/10.1007/978-3-540-78172-1

Sanders, D., & Gegov, A. (2013). AI tools for use in assembly automation and some examples of recent applications. *Assembly Automation, 33*(2), 184–194. https://doi.org/10.1108/01445151311306717

Schmidt, R., Möhring, M., Härting, R.-C., Reichstein, C., Neumaier, P., & Jozinović, P. (2015). Industry 4.0—Potentials for creating smart products: Empirical research results. In W. Abramowicz (Ed.), *Business information systems* (pp. 16–27). Springer International Publishing.

Schuh, G., Prote, J. P., & Dany, S. (2017). Reference process for the continuous design of production networks. *IEEE International Conference on Industrial Engineering and Engineering Management (IEEM), 2017*, 446–449. https://doi.org/10.1109/IEEM.2017.8289930

Schuld, M., Sinayskiy, I., & Petruccione, F. (2015). An introduction to quantum machine learning. *Contemporary Physics, 56*(2), 172–185. https://doi.org/10.1080/00107514.2014.964942

Shafi, M., Molisch, A. F., Smith, P. J., Haustein, T., Zhu, P., De Silva, P., Tufvesson, F., Benjebbour, A., & Wunder, G. (2017). 5G: A tutorial overview of standards, trials, challenges, deployment, and practice. *IEEE Journal on Selected Areas in Communications, 35*(6), 1201–1221. https://doi.org/10.1109/JSAC.2017.2692307

Shi, W., Cao, J., Zhang, Q., Li, Y., & Xu, L. (2016). Edge computing: Vision and challenges. *IEEE Internet of Things Journal, 3*(5), 637–646. https://doi.org/10.1109/JIOT.2016.2579198

Shi, W., & Dustdar, S. (2016). The promise of edge computing. *Computer, 49*(5), 78–81. https://doi.org/10.1109/MC.2016.145

Siepmann, D., & Graef, N. (2016). Industrie 4.0—Grundlagen und Gesamtzusammenhang. In A. Roth (Ed.), *Einführung und Umsetzung von Industrie 4.0: Grundlagen, Vorgehensmodell und Use Cases aus der Praxis* (pp. 17–82). Springer. https://doi.org/10.1007/978-3-662-48505-7_2

Stock, T., & Seliger, G. (2016). Opportunities of sustainable manufacturing in industry 4.0. *Procedia CIRP, 40*, 536–541. https://doi.org/10.1016/j.procir.2016.01.129

Strange, R., & Zucchella, A. (2017). Industry 4.0, global value chains and international business. *Multinational Business Review, 25*(3), 174–184. https://doi.org/10.1108/MBR-05-2017-0028

Suginouchi, S., Kokuryo, D., & Kaihara, T. (2017). Value co-creative manufacturing system for mass customization: Concept of smart factory and operation method using autonomous negotiation mechanism. *Procedia CIRP, 63*, 727–732. https://doi.org/10.1016/j.procir.2017.03.313

Syberfeldt, A., Holm, M., Danielsson, O., Wang, L., & Brewster, R. L. (2016). Support systems on the industrial shop-floors of the future—Operators' perspective on augmented reality. *Procedia CIRP, 44*, 108–113. https://doi.org/10.1016/j.procir.2016.02.017

Tai-shan, Y., & Du-wu, C. (2006). *The method of intelligent inspection of product quality based on computer vision.* 2006 7th International Conference on Computer-Aided Industrial Design and Conceptual Design, 1–6. https://doi.org/10.1109/CAIDCD.2006.329469

Tang, D., Zheng, K., Zhang, H., Sang, Z., Zhang, Z., Xu, C., Espinosa-Oviedo, J. A., Vargas-Solar, G., & Zechinelli-Martini, J.-L. (2016). Using autonomous intelligence to build a smart shop floor. *The 9th International Conference on Digital Enterprise Technology—Intelligent Manufacturing in the Knowledge Economy Era, 56*, 354–359. https://doi.org/10.1016/j.procir.2016.10.039

Thoben, K.-D., Wiesner, S., & Wuest, T. (2017). "Industrie 4.0" and smart manufacturing—A review of research issues and application examples. *International Journal of Automation Technology, 11*(1), 4–16. https://doi.org/10.20965/ijat.2017.p0004

Ventura, J. A., Pazhani, S., & Mendoza, A. (2015). Finding optimal dwell points for automated guided vehicles in general guide-path layouts. *Decision Models for the Design, Optimization and Management of Warehousing and Material Handling Systems, 170*, 850–861. https://doi.org/10.1016/j.ijpe.2015.03.007

Verein Deutscher Ingenieure. (2016). *Fertigungsmanagementsysteme (Manufacturing Execution Systems—MES)* (TECHNICAL RULE No. 5600). Verein Deutscher Ingenieure. https://www.beuth.de/en/technical-rule/vdi-5600-blatt-1/251310997

Wamba, S. F., Akter, S., Edwards, A., Chopin, G., & Gnanzou, D. (2015). How 'big data' can make big impact: Findings from a systematic review and a longitudinal case study. *International Journal of Production Economics, 165*, 234–246. https://doi.org/10.1016/j.ijpe.2014.12.031

Wang, L., & Wang, X. V. (2018). *Cloud-based cyber-physical systems in manufacturing* (1st ed.). Springer International Publishing:Imprint: Springer. https://doi.org/10.1007/978-3-319-67693-7

Wang, S., Wan, J., Li, D., & Zhang, C. (2016). Implementing smart factory of industry 4.0: An outlook. *International Journal of Distributed Sensor Networks, 12*(1), 3159805. https://doi.org/10.1155/2016/3159805

Wang, Y., Kung, L., & Byrd, T. A. (2018). Big data analytics: Understanding its capabilities and potential benefits for healthcare organizations. *Technological Forecasting and Social Change, 126*, 3–13. https://doi.org/10.1016/j.techfore.2015.12.019

Weng, Y.-H., Chen, C.-H., & Sun, C.-T. (2009). Toward the human-robot co-existence society: On safety intelligence for next generation robots. *International Journal of Social Robotics, 1*(4), 267. https://doi.org/10.1007/s12369-009-0019-1

Wilson, H. J., & Daugherty, P. R. (2018, July 1). Collaborative intelligence: Humans and AI are joining forces. *Harvard Business Review*. https://hbr.org/2018/07/collaborative-intelligence-humans-and-ai-are-joining-forces

Wu, J., & Wang, Y. (2006). Measuring ERP success: The ultimate users' view. *International Journal of Operations & Production Management, 26*(8), 882–903. https://doi.org/10.1108/01443570610678657

Zhou, J., Li, P., Zhou, Y., Wang, B., Zang, J., & Meng, L. (2018). Toward new-generation intelligent manufacturing engineering, *4*(1), 11–20. https://doi.org/10.1016/j.eng.2018.01.002

Zuehlke, D. (2008). Smart factory—From vision to reality in factory technologies. *17th IFAC World Congress, 41*(2), 14101–14108. https://doi.org/10.3182/20080706-5-KR-1001.02391

Chapter 3
Smart Factory Framework

3.1 Structure of the Use Cases

This Smart Factory Framework consists of a detailed presentation and analysis of the 44 generic use cases. These use cases are relevant for a company's operational strategy. A detailed understanding of the use cases is vital for selection and prioritization. Each individual use case is described and summarized in a dedicated entry containing the following information:

1. A description along the Smart Factory Framework categories
2. The relevant technologies required
3. The impact on competitive priorities, cost, quality, delivery, flexibility
4. The innovators/exemplary use cases in practice
5. When and how to apply this use case?
6. Questions to ask

The entries are designed to provide information on the nature and application of the individual use cases as well as their possible business impacts. They are organized first by process step and subsequently by degree of analytical foresight. The generic use case entries are preceded by the description of four base use cases—data capturing, data transmission and data storage—deemed to be essential for the development and application of the other technological applications. The use case "information visualization" is also presented alongside the base use cases. While it is not considered essential to the subsequent use cases, it can serve as a tool to complement and augment the potential of each individual application and is therefore presented alongside the base use cases. Below is an overview of the 44 use cases presented (Fig. 3.1).

The description of the use cases is oriented along the process stages relevant for production. The numbering is intended to simplify the search for a specific use case. If a use case spans several process stages, it is described in the first process stage in which it occurs. The ascending numbering within a process stage shows the digital maturity level of the use case in a certain process stage. Use cases assigned to the

L. Budde et al., *Smart Factory Navigator*,
https://doi.org/10.1007/978-3-031-17254-0_3

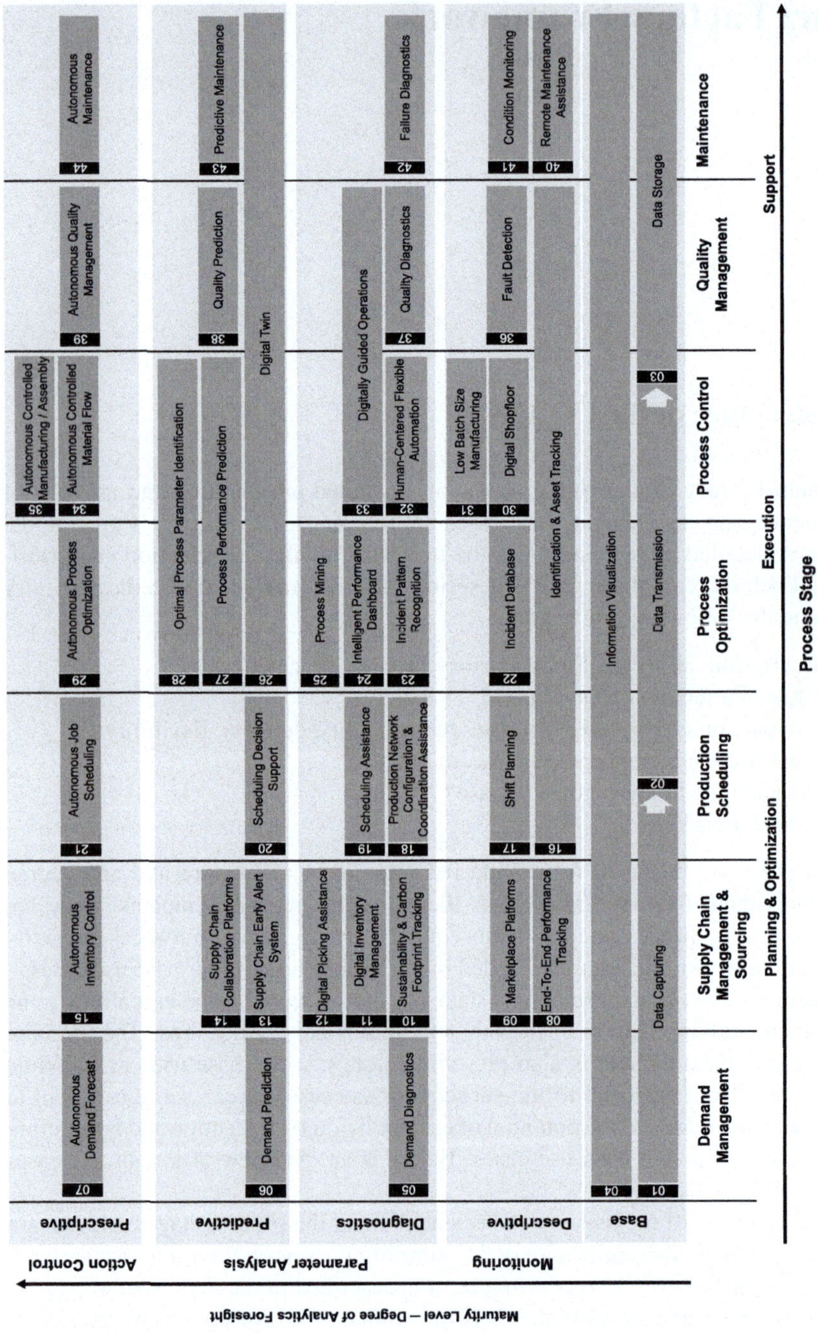

Fig. 3.1 Smart factory framework with 44 generic use cases

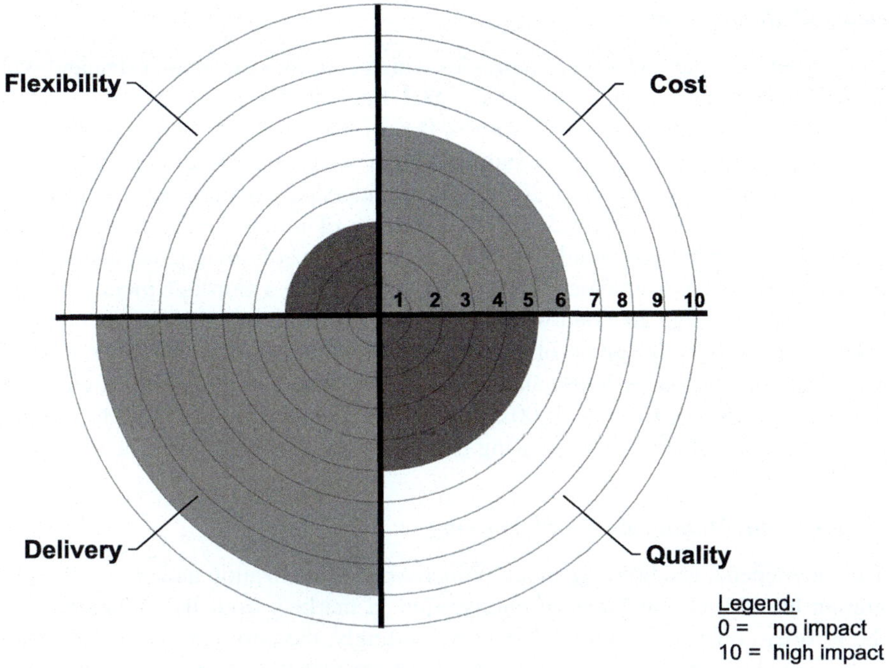

Fig. 3.2 Impact of the use case on the four competitive priorities

base are considered separately from the others. For this reason, the numbering of the "base" does not correspond to the explanations above.

When discussing and selecting potential use cases, an early strategic check to your strategic aims of your company is vital. Therefore, we are providing a view on how each use case impacts the four competitive priorities cost, quality, delivery, and flexibility (see Fig. 3.2).

In this example, the use case has a strong impact on the delivery situation, which means that if you want to optimize the delivery situation in your company, you need to address this specific use case. The impact categorization is based on our understanding of how this use case generally impacts the four competitive priorities. Of course, this may differ from your company-specific assessment. It should be seen as a starting point for discussion.

3.1.1 Process Stage

In Fig. 3.1 the description of the use cases is oriented along the process stages relevant for production. We have split the industrial value chain into the following process stages.

Demand Management

Demand management is a set of activities which coordinate customer demand with supply chain capabilities (Croxton et al., 2002; Lambert & Cooper, 2000; Rexhausen et al., 2012). The importance of agility—"the ability of firms to align with suppliers and customers to streamline operations"—is a well-established way of building competitive advantage (Gligor, 2014). Thus, understanding customer demand and aligning it with the supply-side of a business is of utmost importance. Demand management can help companies be more proactive towards anticipated demand and more reactive towards unanticipated demand by building a thorough picture of what current demand looks like and what future demand could look like (Croxton et al., 2002). With this better grasp of customer demand, companies can work towards optimizing the inventory levels and mitigating the inherent variability of customer demand (Lambert & Cooper, 2000). In this light, demand management techniques can prove very effective in increasing the agility and performance of a company's supply chain (Gligor, 2014).

Supply Chain Management and Sourcing

The interdependence brought about by increasing globalization and economic integration has shifted the locus of competition in markets from individual firms to the broader concept of supply chains. Accordingly, the performance of individual companies largely depends on the effective harmonization of the various tiers of their supplier network. Supply chain management & sourcing can be defined as the coordination and integration of business processes within an organization and across the key members of the supply chain (Lambert & Cooper, 2000; Mentzer et al., 2001). This integration of supply chains works towards reducing frictions and waste of valuable resources. Ultimately, SCM practices aim to "provide products, services and information that add value for customers and other stakeholders" (Lambert & Cooper, 2000).

Production Scheduling

Production scheduling concerns the optimal allocation of available production resources over a given period of time with the goal of optimizing a certain goal (Graves, 1981; Pinedo, 2012). Typical objectives include "minimization of makespan, lateness and earliness, as well as the minimization of total cost" (Maravelias & Sung, 2009). Given certain constants such as production facility data, production recipes, production costs and production targets (orders), scheduling problems are primarily concerned with how to best allocate resources and in what order to execute the planned tasks (Maravelias & Sung, 2009). Scheduling decisions can increase performance "by balancing different objectives such as maximizing service level, and minimizing inventory levels, risks of stockouts, overtime and setup time" (Vieira, 2006). Therefore, production scheduling plays a key role in the operational performance of manufacturing companies.

Process Optimization

Process optimization is the repeated and iterative activity of ensuring that certain processes achieve the specific goal they were designed to achieve (Del Castillo, 2011). Through a set of analytical techniques, processes are scrutinized and constantly evaluated for improvements. Del Castillo emphasizes the importance of empirical methods of testing the performance of processes in a controlled environment before "releasing" them to serial manufacturing (Del Castillo, 2011). With growing competition and increasing customer demands, process optimization is a key tool for manufacturing firms to maintain a competitive advantage (Gröger et al., 2012).

Process Control

Process control relates to the various activities that "exert control over production processes" (Council, 1998). The systems utilized for the control of these manufacturing processes include sensors and actuators, programmable logic controllers, data processing equipment and broader systems such as human–machine interfaces and SCADA systems (Council, 1998; Romagnoli & Palazoglu, 2005). Recent developments in technology have integrated many of these system components into one connected platform allowing for the automation of process execution and control.

Quality Management

Quality management is the approach to manufacturing which aims to "achieve and sustain high quality output, focusing on the maintenance and continuous improvement of processes and defect prevention" (Flynn et al., 1994). Importantly, the management of quality is carried out to meet or exceed the customer's expectations. Continuously delivering higher quality can be a significant driver of competitive advantage.

Maintenance

Maintenance is the set of activities—made necessary by an item's lifecycle—which attempt to "retain it in, or restore it to, a state in which it can perform the required function" (Silvestri et al., 2020, p. 2). As production equipment deteriorates, the integrated planning of production and maintenance is pivotal to ensuring a cost-effective production system (Silvestri et al., 2020). Maintenance can be preventive (planned) or unplanned (Pham & Wang, 1996). While preventive maintenance is based "on industrial or in-plant average-life statistics" to determine the optimal time for intervention, unplanned maintenance is mostly reactive in nature and addresses maintenance needs once the item has started to show signs of failure (Mobley, 2004, p. 5). Within unplanned maintenance, however, is the predictive maintenance approach, which aims to leverage direct item data and sophisticated analytics to implement condition-based monitoring and maintenance (Mobley, 2004).

3.1.2 Maturity Level—Degree of Analytical Foresight

Every use case in Fig. 3.1 has been categorized into the different maturity levels. The following levels indicate the degree of analytical foresight (see also Benfer et al., 2021).

Base Level

The base level of a data analytics framework is, by definition, the acquisition, transmission, and storage of data. These fundamental processes supersede all analytical steps, as they work together to provide the raw information upon which the smart factory vision is based. These foundational activities are supported and enabled by rapidly developing information and communications technologies such as smart sensors, the Internet of Things and cloud computing, to name a few. On top of the three data provisioning activities, the presentation of the data is also a key step in the implementation of data-driven manufacturing.

Descriptive Level

Descriptive analytics is the first step for leveraging data toward understanding and analysing business performance. This kind of analytical insight concerns itself with describing the different dimensions of a given phenomenon and addresses the questions "what happened/what is happening now?" (Banerjee et al., 2013, p. 1). The techniques utilized by descriptive analytics include "data integration, augmentation and reduction", and the results are usually presented as a static overview of past/current performance via a variety of visualization techniques (i.e., dashboard, reports, etc.) (Lustig et al., 2019; Soltanpoor & Sellis, 2016, p. 254). Importantly, descriptive analytics rely on human interpretation and analysis of the data to extract actionable insights (Lustig et al., 2019).

Diagnostics Level

Diagnostic analytics is the first step in analytical autonomy towards actionable data-driven insights. The analytical techniques within this category are focused on data from the past and present, but they also aim to answer the question "why did it happen/why is it happening?" (Soltanpoor & Sellis, 2016, p. 1). Through root cause analysis, diagnostic efforts aim to establish the first patterns and statistical relationships to explain the phenomena as described above. The explanatory analysis of past and current data builds a very solid foundation for the further development of analytical capabilities into the predictive and prescriptive realm (Lustig et al., 2019).

Predictive Level

At the predictive analytics level, the degree of analytical autonomy is increased significantly. Here, sophisticated analytical techniques are applied to the question of "what will happen next?" (Lepenioti et al., 2020). Predictive analytics base themselves on the descriptive analytics output, as well as additional simulation data, to build accurate models to predict future events (Soltanpoor & Sellis, 2016). The analytical techniques used include time series analysis, causal inference and data mining (Lustig et al., 2019). Such tools are implemented with the aim of furthering the understanding of why certain outcomes occur and utilizing these insights for the prediction of future outcomes (Soltanpoor & Sellis, 2016). These analyses deliver a variety of plausible outcomes, each associated with respective probabilities (Soltanpoor & Sellis, 2016). Predictive analytics extend the possibility for data-driven decision making beyond the reactive to the proactive. In this manner, interventions for better performance can be evaluated and considered before they are implemented, lowering costs and risk (Lustig et al., 2019).

Prescriptive Level

Once a solid analytical base has been developed to answer the questions what, why and what next, the subsequent—and highest in terms of analytical autonomy—step is what should be done to improve performance? Prescriptive analytics aim to support the decision-making of employees by applying advanced techniques to optimize given problems (Lustig et al., 2019). Here, "decision alternatives are associated with prediction of outcomes" (Banerjee et al., 2013). Taking into consideration the output of predictive analytics, prescriptive techniques suggest possible courses of action and their respective outcomes in addition to the predictive level. It can go up to autonomous changes, e.g., to a planning board without human interactions. In this context, optimization, simulation and evaluation methods are utilized to arrive at the best decision (Soltanpoor & Sellis, 2016). With prescriptive analytics, the decision-making of human actors is heavily supported or even replaced by analytical techniques. In this way, business interventions can take into consideration the vast complexity presented by a company's operational context and offer reliable and better suggestions.

3.1.3 Technology Mapping

The 25 digital technologies mentioned in Sect. 2.4 make up the toolbox for the 44 identified use cases to design the generic Smart Factory Framework. Each digital technology has a specific application particular to a use case, applied with varying degrees of autonomy and at different process stages (see Table 3.1). This focus is, over time, extended to other use cases. Some technologies are applied to a larger set of functionalities, whereas others have more narrow applications. Figure 3.3 illustrates the distribution of the digital technologies' application frequency. The wide use of the different technologies for the different use cases has emerged over time. As can be seen, the focus of technologies is on IT systems such as ERP and MES in combination with cloud computing and data acquisition technologies, which are more frequently used. Some technologies are more focused on one or two specific use cases. These technologies are not less important than others as the benefit of the smart factory is driven by the specific relevance of the use case rather than the wide usage of a specific technology.

Table 3.1 shows in detail which technologies are appropriate for which use case.

3.2 Base Use Cases

3.2.1 Use Case 01: Data Capturing

Description Along the Framework Categories

Data acquisition refers to the process of "measuring physical world conditions and phenomena such as electricity, sound, temperature and pressure" (Techopedia, 2018). To achieve this, the analogue signals have to be converted into digital, machine-readable signals. This is done by data acquisition systems, which contain three components, namely sensors, signal-condition circuitry as well as analogue-to-digital converters (Techopedia, 2018).

The sensors are responsible for capturing the analogue signals from the real world, which can heavily differ, depending on the area of application as well as the targeted information to be captured. Apart from more common sensors such as cameras, there exist a multitude of different ones, such as electrical, radiation, temperature, or proximity sensors (Tönshoff & Inasaki, 2001). The signal-condition circuity standardizes the captured signals, to make them more feasible to convert. The last component, the analogue-to-digital converter transmits, as its name suggests, the analogue input signals into a digital data stream (Techopedia, 2018). The digital, numeric information then can be transmitted, visualized, stored as well as be further processed with computing-based devices.

Picture of the Model/Framework Explanation

The acquisition of data is probably the most important use case for a smart factory: no data means no reports, no overview, no information, and, in general, no knowledge and learning opportunities. The acquisition of data is required for every well-founded action and decision. This is why this use case is considered a base use case (see Fig. 3.4). In a factory, multiple sources of data are available and generated. This data generation starts with all enterprise data generated in the different enterprise IT systems and is then followed by different process and quality data during the manufacturing process. Machines with sensors & actuators, IoT devices and additional track & trace technologies complement the wide areas of data sources. Understanding and developing this base use case is a key element for the smart factory.

Table 3.1 Mapping of technologies with use cases

	IT Systems / Infrastructure	Enterprise Resource Planning System	Manufacturing Execution System	Product Lifecycle Management	SCADA System	**Data Transmission / Storage**	Cloud Computing	Edge Computing	Long Range Wide Area Network	5G	M2M Communication	Mobile Computing	**Data Capturing**	Internet of Things	Sensors and Actuators	Track & Trace Technologies	Vision Technologies	Cyber-Physical Systems	**Data Analytics**	Big Data Analytics	AI, ML & Simulation	**Production Automation**	Collaborative Robotics	Automated Guided Vehicles	Additive Manufacturing	**New Technologies**	Blockchain	Drone	Wearables	Augmented Reality	Virtual Reality
UC01		●	●	●	●							●		●	●	●		●										●	●	●	●
UC02							●		●	●	●	●		●				●											●		
UC03		●	●	●	●		●	●				●																			
UC04		●	●	●	●		●	●				●		●	●	●													●	●	●
UC05		●	●	●	●		●							●		●				●											
UC06		●	●	●	●		●							●		●				●	●										
UC07		●	●	●	●		●							●		●		●		●	●										
UC08		●	●				●			●				●	●	●				●							●				
UC09		●	●	●			●													●					●		●				
UC10		●	●				●							●	●	●				●							●				
UC11		●	●	●			●			●	●	●		●	●	●	●	●		●			●	●							
UC12		●	●				●			●	●	●		●		●	●	●		●	●								●	●	
UC13		●	●				●							●	●			●		●	●						●				
UC14		●	●	●			●													●	●						●				
UC15		●	●	●			●							●	●	●	●	●		●	●										
UC16		●	●	●			●		●	●				●	●	●	●			●	●								●		
UC17		●	●		●		●					●																			

(continued)

Table 3.1 (continued)

	IT Systems / Infrastructure	Enterprise Resource Planning System	Manufacturing Execution System	Product Lifecycle Management	SCADA System	**Data Transmission / Storage**	Cloud Computing	Edge Computing	Long Range Wide Area Network	5G	M2M Communication	Mobile Computing	**Data Capturing**	Internet of Things	Sensors and Actuators	Track & Trace Technologies	Vision Technologies	Cyber-Physical Systems	**Data Analytics**	Big Data Analytics	AI, ML & Simulation	**Production Automation**	Collaborative Robotics	Automated Guided Vehicles	Additive Manufacturing	**New Technologies**	Blockchain	Drone	Wearables	Augmented Reality	Virtual Reality
UC18		•	•	•	•		•							•						•	•										
UC19		•	•		•		•													•	•										
UC20		•	•		•		•							•				•		•	•										
UC21		•	•		•		•							•				•		•	•										
UC22		•	•		•		•	•								•															
UC23		•	•		•		•	•						•	•					•	•										
UC24		•	•		•		•	•						•	•	•		•		•											
UC25		•	•		•		•	•						•	•			•		•	•										
UC26		•	•	•	•		•	•			•			•	•	•		•		•	•								•		•
UC27		•	•	•	•		•	•						•	•	•	•	•		•	•										
UC28		•	•	•	•		•	•						•	•	•	•	•		•	•										
UC29		•	•	•	•		•	•			•			•	•	•	•	•		•	•										
UC30		•	•		•		•	•				•		•	•	•				•				•					•	•	
UC31		•	•				•							•	•	•				•	•				•				•		
UC32			•				•				•	•		•	•	•	•	•		•	•		•						•		
UC33			•				•	•				•		•	•	•	•	•		•	•									•	•
UC34		•	•	•			•				•			•	•	•	•	•		•	•			•							
UC35		•	•	•	•		•	•	•	•	•	•		•	•	•	•	•		•	•		•								
UC36		•	•	•	•		•							•	•	•	•	•		•	•									•	
UC37		•	•	•	•		•							•	•	•	•	•		•	•										

(continued)

Table 3.1 (continued)

	IT Systems / Infrastructure	Enterprise Resource Planning System	Manufacturing Execution System	Product Lifecycle Management	SCADA System	**Data Transmission / Storage**	Cloud Computing	Edge Computing	Long Range Wide Area Network	5G	M2M Communication	Mobile Computing	**Data Capturing**	Internet of Things	Sensors and Actuators	Track & Trace Technologies	Vision Technologies	Cyber-Physical Systems	**Data Analytics**	Big Data Analytics	AI, ML & Simulation	**Production Automation**	Collaborative Robotics	Automated Guided Vehicles	Additive Manufacturing	**New Technologies**	Blockchain	Drone	Wearables	Augmented Reality	Virtual Reality
UC38		•	•	•	•		•							•	•	•	•	•		•	•										
UC39		•	•	•	•		•	•						•	•	•	•	•		•	•										
UC40		•	•		•		•	•				•		•	•		•	•											•	•	•
UC41		•	•		•		•	•						•	•		•	•		•	•									•	
UC42		•	•				•							•	•		•	•		•	•										
UC43		•	•				•							•	•		•	•		•	•										
UC44		•	•				•				•			•	•		•	•		•	•										

UC01: Data Capturing | UC02: Data Transmission | UC03: Data Storage | UC04: Information Visualization | UC05: Demand Diagnostics | UC06: Demand Prediction | UC07: Autonomous Demand Forecast | UC08: End-To-End Performance Tracking | UC09: Marketplace Platforms | UC10: Sustainability & Carbon Footprint Tracking | UC11: Digital Inventory Management | UC12: Digital Picking Assistance | UC13: Supply Chain Early Alert System | UC14: Supply Chain Collaboration Platforms | UC15: Autonomous Inventory Control | UC16: Identification & Asset Tracking | UC17: Shift Planning | UC18: Production Network Configuration & Coordination Assistance | UC19: Scheduling Assistance | UC20: Scheduling Decision Support | UC21: Autonomous Job Scheduling | UC22: Incident Database | UC23: Incident Patter Recognition | UC24: Intelligent Performance Dashboard | UC25: Process Mining | UC26: Digital Twin | UC27: Process Performance Prediction | UC28: Optimal Process Parameter Identification | UC29: Autonomous Process Optimization | UC30: Digital Shopfloor | UC31: Low Batch Size Manufacturing | UC32: Human-Centered Flexible Automation | UC33: Digitally Guided Operations | UC34: Autonomous Controlled Material Flow | UC35: Autonomous Controlled Manufacturing/Assembly | UC36: Fault Detection | UC37: Quality Diagnostics | UC38: Quality Prediction | UC39: Autonomous Quality Management | UC40: Remote Maintenance Assistance | UC41: Condition Monitoring | UC42: Failure Diagnostics | UC43: Predictive Maintenance | UC44: Autonomous Maintenance

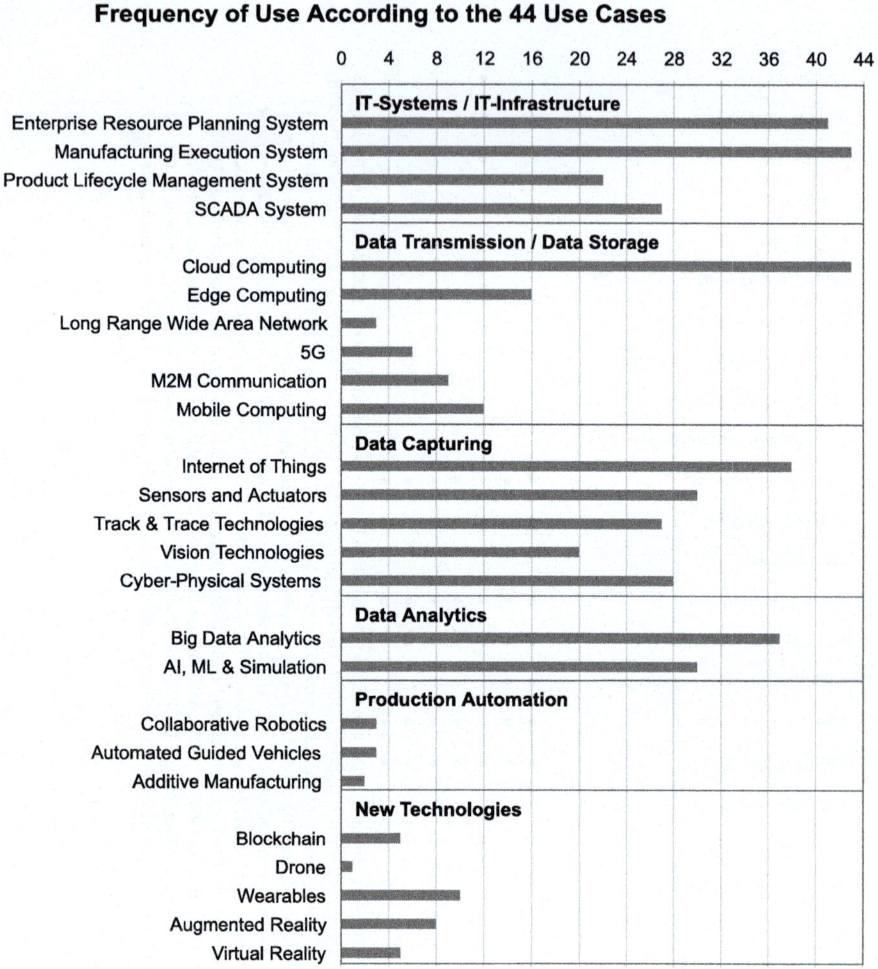

Fig. 3.3 Application frequency of digital technologies in the 44 use cases

Relevant Technologies

The following technologies are considered relevant for this use case:

- Enterprise Resource Planning System
- Manufacturing Execution System
- Product Lifecycle Management
- SCADA System
- Mobile Computing
- Internet of Things
- Sensors and Actuators
- Track & Trace Technologies

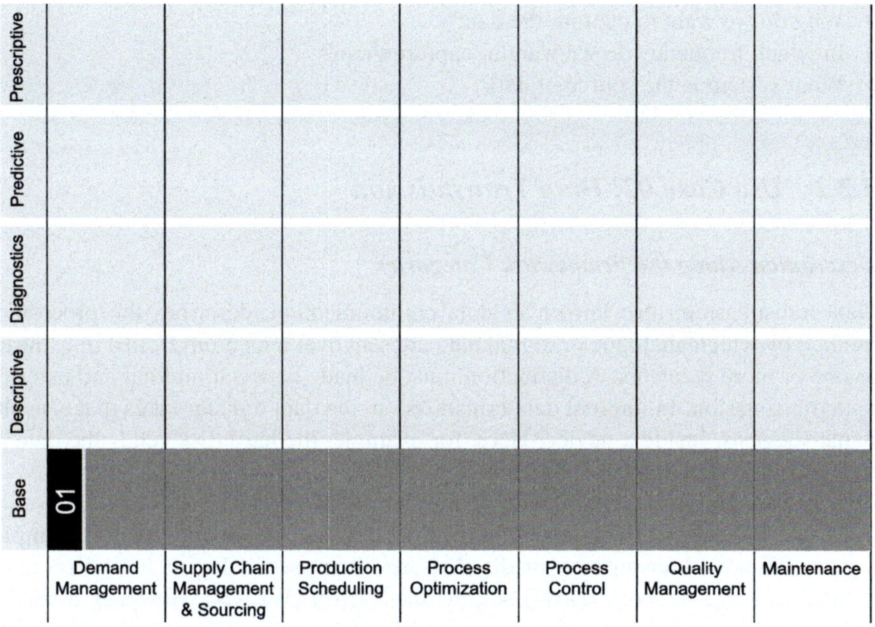

Fig. 3.4 Smart factory framework—use case 01

- Cyber-Physical Systems
- Drone
- Wearables
- Augmented Reality
- Virtual Reality.

Impact on Competitive Priorities

This use case strongly influences all competitive priorities.

Innovators/Exemplary Cases

A globally connected enterprise IT system with different components such as ERP, PLM and MES is relevant for a successful implementation of this base case. These IT systems must be shaped according to the business model and business processes, which are company-specific.

When and How to Apply This Use Case?

An implementation of the smart factory and any other use cases will need to address the topic of data capturing.

Questions to Ask

- What kind of data are important to us?

- Why do we want to capture the data?
- In which frequency do we want to capture them?
- What system is the source of data?

3.2.2 Use Case 02: Data Transmission

Description Along the Framework Categories

Data transmission, also known as data communication, describes the process or method by which analogue or digital data are sent over a medium from a transmitter to one or more receivers. A distinction must be made between internal and external data transmission. In internal data transmission, the data transfer takes place, as the name suggests, inside a device. Here, for example, the hard disk sends the data or data objects to the processor (Techopedia, 2013).

Alternatively, in external data transmission, a device—e.g., a notebook—sends a file or a data object to one or more receiver devices, which may be, for example, servers. Famous transmission media that are often used today include fibre optic cable, local area network (LAN), and wireless LAN (WLAN). However, a variety of other wireless data transmission methods also exist, such as radio-frequency identification (RFID), Bluetooth, or infrared.

Therefore, the exchange of files and data objects can take place within a digital device, or between at least two different entities, which either belong to the same or different companies.

The overall effectiveness of a data link depends heavily on the transmission speed and amplitude. Depending on the type of application, or the amount of data required, stronger or less strong data connections are needed. The data transfer rate is defined by the amount of data sent within a certain period of time, which is measured in bits per second. The bandwidth specifies the maximum amount of data that can theoretically be sent within the period. (CDNetworks, 2021).

Picture of the Model/Framework Explanation

Data transmission refers to the sending of data objects either within a device or from one device to another. This is crucial for every digital, data-based technology, which is why this use case is considered a base use case (see Fig. 3.5). The two key questions to be asked for the data transmission use case are "which data" and "how many data" need to be transferred?

Relevant Technologies

The following technologies are considered relevant for this use case:

- Cloud Computing
- Long Range Wide Area Network
- 5G
- M2M Communication

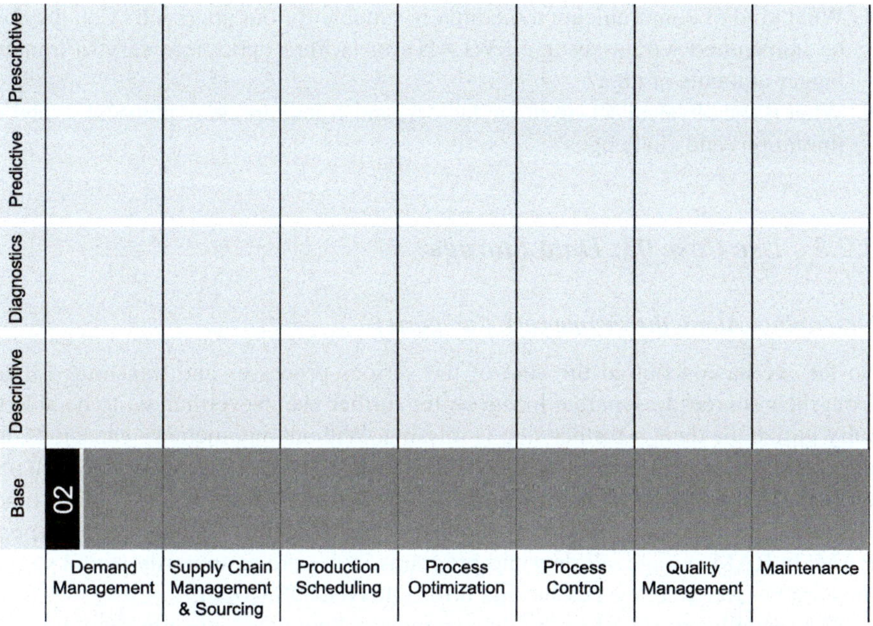

Fig. 3.5 Smart factory framework—use case 02

- Mobile Computing
- Internet of Things
- Cyber-Physical Systems
- Wearables

Impact on Competitive Priorities

This use case strongly influences all competitive priorities.

Innovators/Exemplary Cases

The specific physical set-up of a factory will determine the data transmission path. Therefore, there is no right or wrong approach. It needs to be built according to the particular specifications of each individual factory.

When and How to Apply This Use Case?

For any use case that stores data, there will be the need to transmit it. Therefore, this use case is a vital or even mandatory use case for a smart factory.

Questions to Ask

- Which kind of data do we have to transfer?
- What amount of data do we have to transfer?

- What kind of communication medium is suitable for our approach? Can the data be transmitted wirelessly (e.g., WLAN), or is fibre optic necessary to transfer bigger amounts of data?
- What is the physical set-up in the factory, and where are there physical or layout limitations and challenges?

3.2.3 Use Case 03: Data Storage

Description Along the Framework Categories

So far, we have acquired the data of the various processes and transmitted them from their sources to separate locations for further use. Nevertheless, to be able to fully utilize the data, a further step is missing. Without any memory capacities, all generated data would just decay. Nowadays, almost every electronic device contains a rewritable or non-rewritable memory in some form. Computers, as well as other electronic devices, possess two prominent types of memory storage. One type is RAM. In RAM storage, all data and programs in use are temporarily stored so that they are fast, reliable, executable and utilizable (IONOS, 2020).

The second important form of data storage in computers concerns the long-term storage of data. This is either done by devices that are directly connected to the device in the immediate area (e.g., HDDs) or externally by means of network-based storage (IBM, n.d.). In the context of digital transformation, and the accompanying organizational and structural changes, network-based data storage is becoming increasingly important. The direct and shared access to a variety of different data allows for effective collaboration between the various internal and external players of the company. In general, there are two main methods to store the data on servers, either internally on so-called in-house servers or externally on third-party servers, namely on clouds (Arrow, 2018). Nowadays, cloud-based approaches are becoming increasingly popular for various reasons. These include minimal investment, a reduction of server rooms and IT technicians, the ease of scalability and cross-organizational application, and the possibility of remote collaboration (Arrow, 2018; Axxiv, 2019). Depending on the size and industry of the enterprise, however, one solution might be more promising than the other.

Picture of the Model/Framework Explanation

Data storage, together with data acquisition and data transmission, form the foundation for all other use cases. By using data storage capacities, the collected data is not only recorded for further use but is also made more utilizable, for example by RAMs. The implementation of different use cases of a smart factory poses significant challenges for this base use case (see Fig. 3.6). All data sources can offer an enormous data quantity. Sometimes all this data is needed. Sometimes, however, "less is more". The quantity of data will strongly influence the storage solution due to technical limitations (e.g., upload speed), cost or availability.

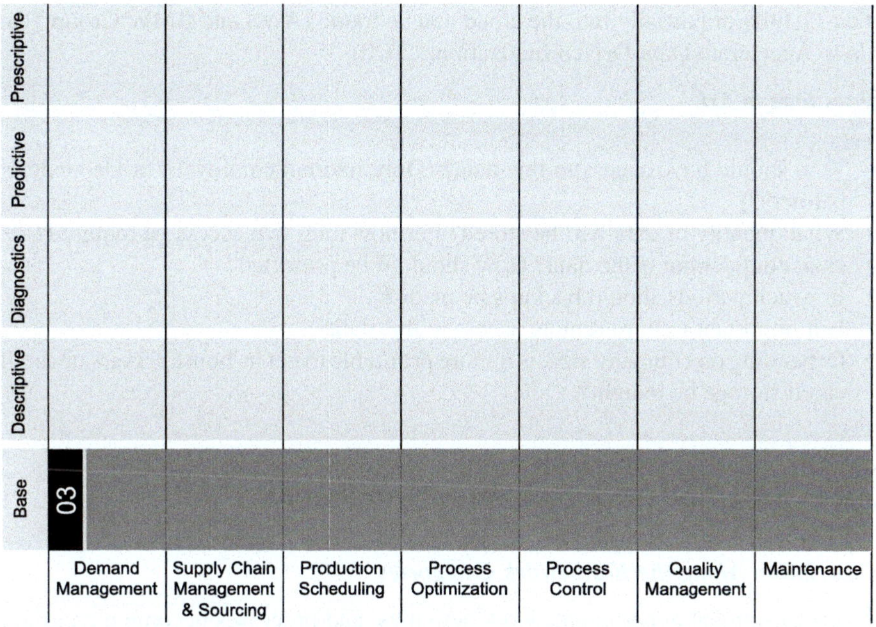

Fig. 3.6 Smart factory framework—use case 03

Relevant Technologies

The following technologies are considered relevant for this use case:

- Enterprise Resource Planning System
- Manufacturing Execution System
- Product Lifecycle Management System
- SCADA System
- Cloud Computing
- Edge Computing
- Mobile Computing

Impact on Competitive Priorities

This use case strongly influences all competitive priorities.

Innovators/Exemplary Cases

Due to the fundamentality of this use case, there are no explicit success stories.

When and How to Apply This Use Case?

Due to the fundamentality of this use case, there is no definitive application. However, cloud computing is playing a key role in data storage today, and this tendency is likely to continue in the future. Therefore, many stories of companies that have transformed their IT fully or partially into the cloud can be found (AWS and BMW Group Team Up to Accelerate Data-Driven Innovation., 2020).

Questions to Ask

- Who should have access to this data? (Only internal employees or also external partners?)
- What quantity of data will be stored? For how long will access be required?
- How confidential is the data? How should it be protected?
- In which periods should backups be made?
- Is it important to have remote access to this data?
- Depending on company size, is it more profitable to get in-house servers or should cloud storage be bought?

3.2.4 Use Case 04: Information Visualization

Description Along the Framework Categories

As organizational departments, work activities, and processes become increasingly networked, the associated complexity within the working environment also increases. Regardless of the role of the employee, this inadvertently increases the complexity of decision-making. To be able to successfully overcome these challenges and make sound decisions, supporting information is pivotal. Information visualization plays a key role in transferring this information to the end-user throughout the organization. The information to be displayed takes a variety of different forms. Static information—e.g., operating manuals or process instructions—is presented to support workers in their assembly and operational tasks and reduce errors. Alternatively, dynamic data—production data, measurements, etc.—are often presented in structured forms such as key performance indicators (KPIs). KPIs need to consider the different dimensions—performance, enabler, and organization—and present different kinds of information according to the needs of the intended end-user. The effective visualization of KPIs can support the decision-making of employees across the organization, from operators to top-level management, turning data into insightful and actionable insights (see Fig. 3.7).

KPIs can cover different operational priorities and different timespans. Each metric is constructed keeping in mind what decisions the end-users need to make and what kind of information will support such decisions. The Fig. 3.8 showcases a variety of performance measures according to different frequencies and needs to act.

There are a lot of relevant KPIs for a factory. One very relevant KPI is overall equipment efficiency (OEE). This metric records the various wastes and relates them directly to the operational performance of the factory. It measures the utilization, performance, and quality of the process in one number. The KPI OEE is often the first step into more advanced KPIs or even use cases (Hänggi et al., 2021).

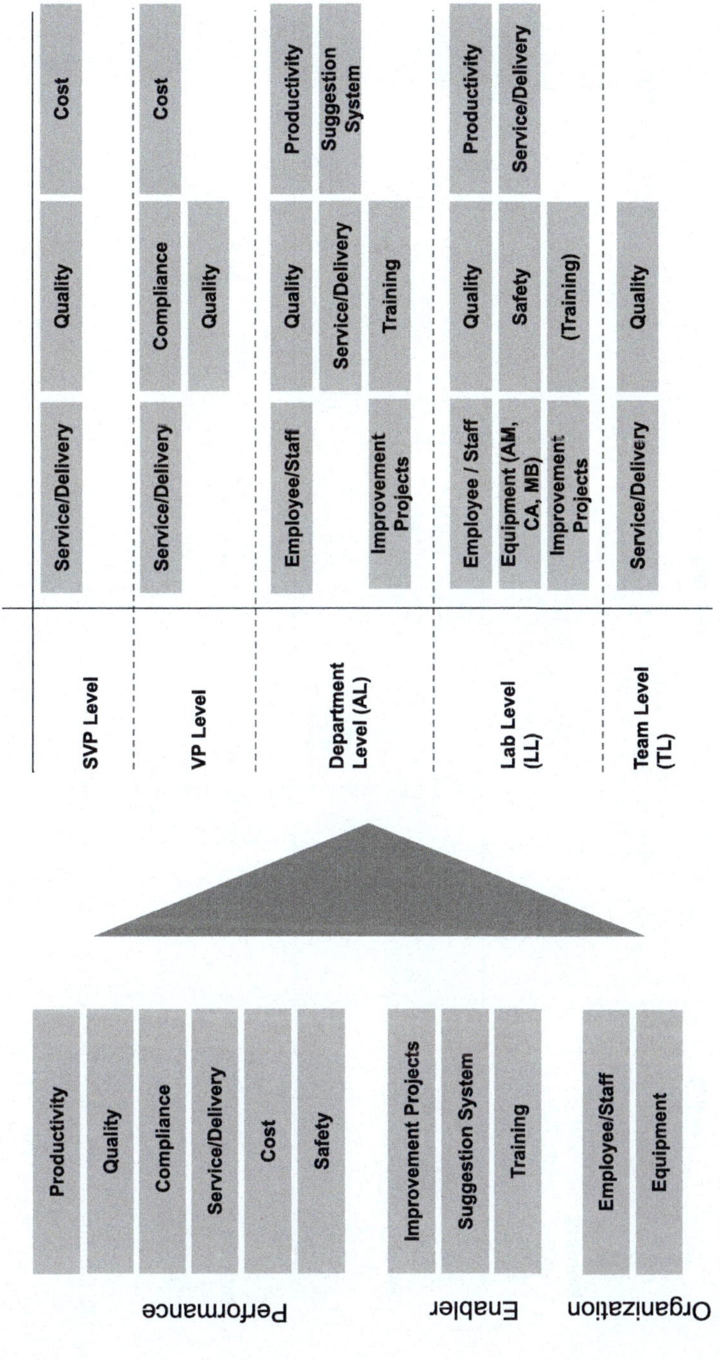

Fig. 3.7 KPIs according to seniority level

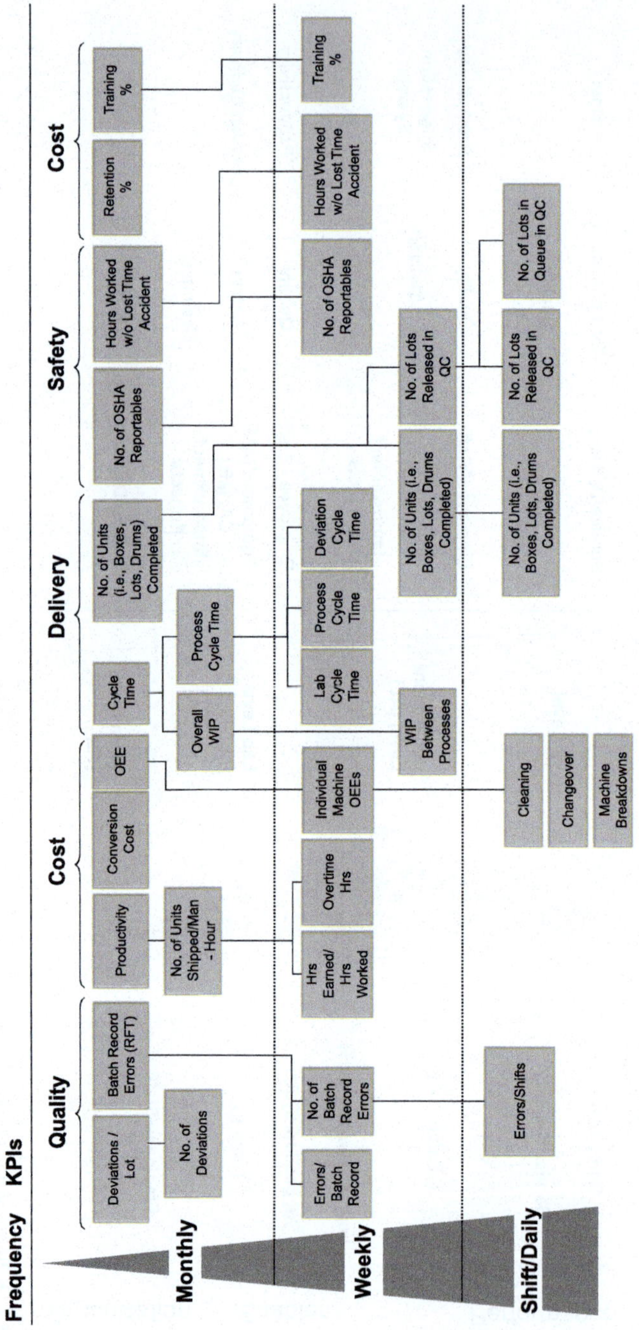

Fig. 3.8 Examples of different KPIs according to different frequencies and needs to act

Ultimately, the use case information visualization aims to facilitate the presentation and utilization of processed and unprocessed data into KPIs. The latter are often communicated through digital solutions ranging from conventional and well-established displays, such as computer monitors, to more modern ones, such as head-mounted displays (HMDs) (e.g., smart glasses). The medium for presenting the data varies according to the functions of the end-users and spans from unprocessed data in an Excel file to a real-time KPI dashboard with interactive visualizations.

Picture of the Model/Framework Explanation

The use case information visualization is somewhat special. It is relevant and respectively applied to practically every other use case presented in this framework. Specifically, every occasion where information in the form of raw data, graphs, and other types is displayed for utilization by human employees. The depiction of unprocessed or slightly aggregated data are contained in the base dimension of our framework. Due to its wide applicability, information visualization is relevant for each process stage (see Fig. 3.9).

Relevant Technologies

The following technologies are considered relevant for this use case:

- Enterprise Resource Planning System
- Manufacturing Execution System

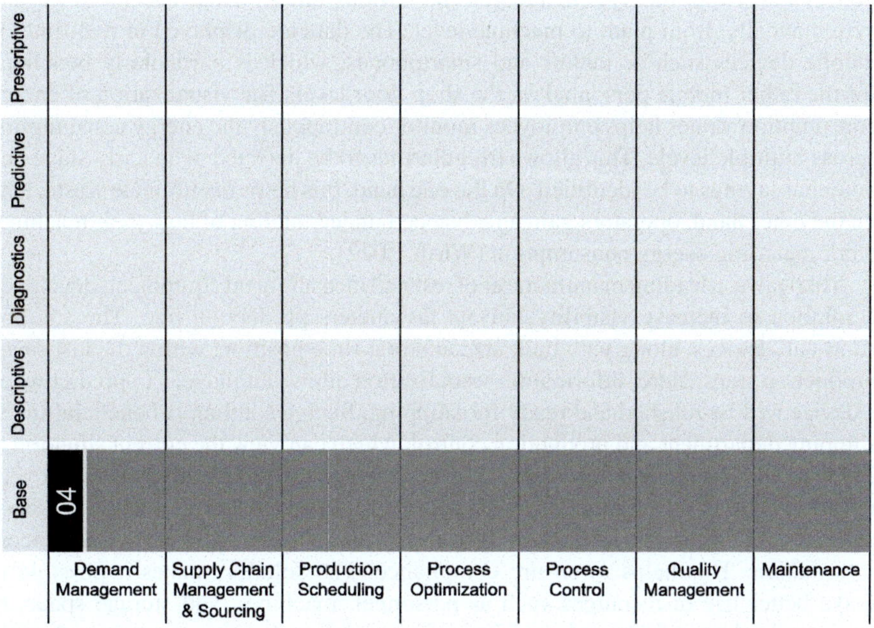

Fig. 3.9 Smart factory framework—use case 04

- Product Lifecycle Management System
- SCADA System
- SCADA System
- Cloud Computing
- Edge Computing
- Mobile Computing
- Internet of Things
- Sensors and Actuators
- Track & Trace Technologies
- Wearables
- Augmented Reality
- Virtual Reality

Impact on Competitive Priorities

This use case strongly influences all competitive priorities.

Innovators/Exemplary Cases

Beside displaying different KPIs in plant dashboard or visualize in a shopfloor, more sophisticated approaches also fall into this use case. In the context of holistic plant and process optimization, the reduction of consumable resources such as electricity is becoming increasingly important. To gain an overview of energy consumption, the company Festo created an energy transparency system for industrial production plants. Here, the system records the various energy data, decentralized and systematically, from plant to machine level. The data are displayed in real-time on mobile devices such as tablets and smartphones, which is particularly beneficial for the rather mobile personnel on the shop floor level. The visualization of energy consumption values helps employees monitor continuously the energy consumption across multiple levels. This allows irregularities to be detected at an early stage and potential savings to be identified. On the one hand, this helps to eliminate waste, and, on the other hand, operators gain an advanced understanding of how their machines work regarding energy consumption (Wirth, 2022).

Yokogawa, a leading manufacturer of test and measurement equipment, developed a solution to increase visibility into its flowmeters production line. The solution shows all devices, along with their size and real-time position, within the last seven production steps. Here, information visualization allows employees to predict when a device will be finished and ready for shipping. Such a solution is beneficial to the shipping department and production control, which can view the current situation at any time and intervene if necessary. To gain access to the production line, employees must log on to the computer with their code card. This is to ensure that the person is authorized to access the line and is up to date with the latest regulations and process information. This helps to ensure compliance. The solution makes it possible to make better use of resources such as personnel, machines, and storage space, to process small production orders more economically, and also to react more flexibly by facilitating the prioritization of individual orders (Trump, 2022).

When and How to Apply This Use Case?

Due to its versatile characteristics, no specific recommendations can be made for the specific focus area of this use case, but every company will need to implement this base use case. Fact based decision making, is driven by information visualization. Wherever a person directly makes situational decisions or performs actions, data visualizations are often required. The visualization supports the employees with decision making, either by giving an overview of the relevant circumstances for the decision or by showing already drawn processes that lead to the required actions.

Questions to Ask

- Who is going to read or use the data?
- For which purpose is the information displayed? Which decisions do they support?
- How should the data be visualized (definition, frequency, graph) so that the user will be able to extract the relevant information?
- Which additional information needs to be added to get a better understanding of the data?

3.3 Demand Management

3.3.1 Use Case 05: Demand Diagnostics

Description Along the Framework Categories

In the context of unpredictable and rapidly changing consumer demands, the effective gathering, analysis and understanding of demand-relevant data can be a highly decisive factor for the success of manufacturing firms (Kehoe & Boughton, 2001). Demand diagnostics can present a useful starting solution in understanding demand data and offering the visibility needed to make key operational considerations. The demand diagnosis analytical process relies heavily on the availability and quality of data fed into the model. The storing and indexing of internal and external data are often done with the help of enterprise resource planning (ERP) systems and cloud computing solutions. A diagnostic approach to demand is rooted in exploring the specific elements of past and current demand data and making initial analyses on possible patterns and causal relationships (Davenport & Harris, 2017). This kind of approach relies on information systems and basic statistics to answer questions such as “what has happened/what is happening and why?” (Mortenson et al., 2015). Typically, these questions are answered with the help of analytical techniques such as data mining and regression analysis (Benfer et al., 2021).

Importantly, diagnostic demand only considers past and current data and thus makes inferences on what has already happened or on what is happening. For this reason, this analytical tool remains a reactive tool, as opposed to a proactive future-oriented solution.

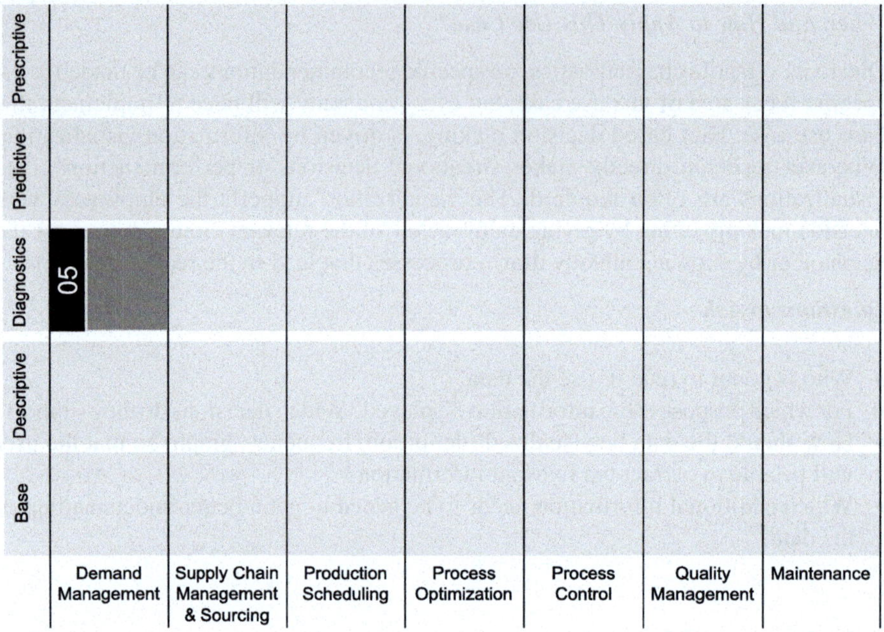

Fig. 3.10 Smart factory framework—use case 05

Picture of the Model/Framework Explanation

As explained above, demand diagnosis considers historical and current data to provide answers to explain demand related questions. While some approaches rely on machine learning algorithms for classification, and the statistical analysis can be rather sophisticated, the scope of the use case is still limited to past and present. For this reason, the use case is classified as diagnostic (see Fig. 3.10).

Relevant Technologies

The following technologies are considered relevant for this use case.

- Enterprise Resource Planning System
- Manufacturing Execution System
- Product Lifecycle Management System
- SCADA System
- Cloud Computing
- Internet of Things
- Track & Trace Technologies
- Big Data Analytics

Impact on Competitive Priorities

The impact of this use case on competitive priorities is shown in generic form in the Fig. 3.11. The specific impact depends on the context of the company.

Impact on flexibility • Enables shorter response time to contextual changes	Impact on costs • Reduces stock-out situations that lead to bottlenecks and idle times • Optimizes the inventory
Impact on delivery • Increases the availability of parts • Reduces out-of-stock situations	Impact on quality • No direct impact

Innovators/Exemplary Cases

The analytics company saviant supported a food and beverage client to understand sales and demand information through a cloud-based analytics platform. With over 9,000 stores and 36 million data records per day, the utilization of sales data to make effective operational decisions becomes challenging. The old solution utilized by the client provided information but was far too slow and individual to each store. Instead, saviant suggested the implementation of a cloud-based analytics platform to eliminate data silos and rapidly make diagnostics upon which to base operational decisions. The new cloud-based business intelligence (BI) solution provided quality data for each store and multiple drill-down possibilities (Food Services Case Study | Sales Analytics Solution, 2022).

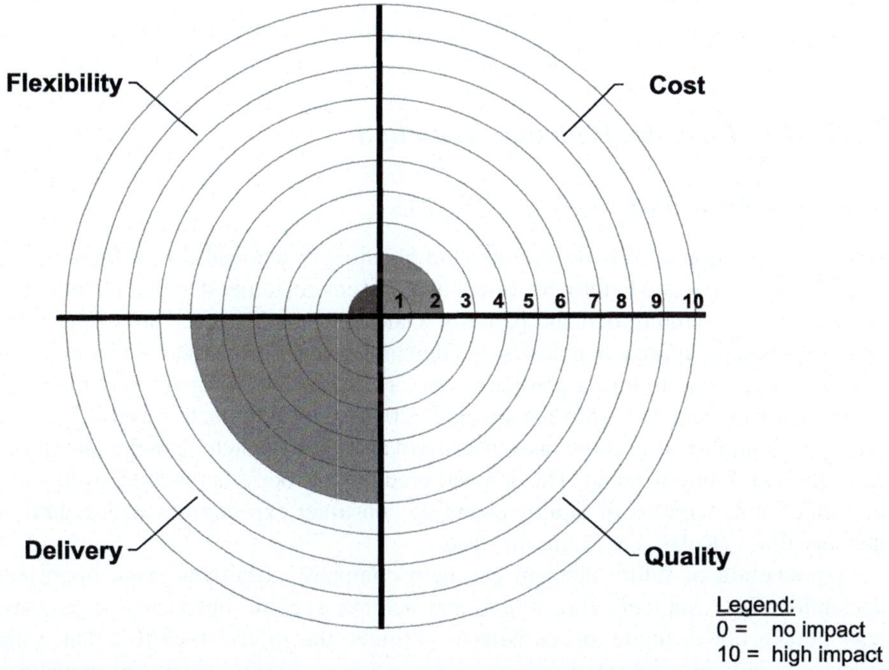

Fig. 3.11 Impact—use case 05

When and How to Apply This Use Case?

Demand diagnostics is especially useful in the following situations:

- **Volatility in customer demand**: A comprehensive data-backed model for diagnosing demand will help companies operating in industries with heavily fluctuating demand.
- **High lead times**: Diagnostics on-demand data can better inform actions—such as processing of inventory orders—and work towards reducing the supplier and delivery lead times of a company.
- **High warehousing costs/limited storage**: A better understanding of current and past demand patterns can help optimize inventory and reduce avoidable warehousing/storage costs
- **Long and complex supply chains**: Demand diagnostics can offer some added visibility for companies with long and complex supply chains.

Questions to Ask

- Is the demand for our product particularly volatile?
- Is the demand for our product different in different geographical regions?
- Which total lead-time do we want to achieve?
- What are the costs of different scenarios? (e.g., inventory costs, profit loss due to stock-out situations, etc.)
- How reliable and fast are our inbound and outbound order lead times?

3.3.2 Use Case 06: Demand Prediction

Description Along the Framework Categories

Demand prediction, also known as demand planning or demand-driven forecasting, describes the process whereby historical and present customer demand information is used to predict future demand patterns (Gartner, n.d.). The predictive approach, or future-looking approach, extends this demand management use case from being simply a diagnostic tool to a predictive one. The question being answered now is, "what will happen next?" (Mortenson et al., 2015). This approach is useful for both retail and manufacturing companies in terms of accurately matching their planning of capacity with future demand. The demand prediction process is rendered even more important in the context of rapidly changing consumer expectations with regard to delivery times (Kolassa & Siemsen, 2016).

A prediction of future demand can help company executives make important decisions (e.g. resources, allocations) and achieve specific objectives (e.g. costs, lead times). For example, it can help to optimize the safety stock to balance the stock-out, delivery, and inventory costs. An average of 90% of demand prediction is driven by process and analytics, and only 10% is driven by enabling technologies

(Chase, 2009). However, although the latter amounts only for a tenth of the whole demand prediction process, the other 90% relies on this technology and cannot be achieved without a sufficient technology infrastructure (Chase, 2009).

To create a reliable and accurate demand forecast, the prediction model should incorporate all demand-related data that is available to the company and its supply chain. These data can be, for example, directly market-related (trends, seasonal consummation patterns, etc.) or non-related such as weather forecasts, etc. The complexity of such prediction techniques ranges from rather simplistic exponential smoothing methods to AI-based approaches that consider many hundred different factors as well as several billion data points (Lemmon, 2017).

Picture of the Model/Framework Explanation

As its name suggests, demand prediction is a predictive use case (see Fig. 3.12). By forecasting future demand trends based on multiple company internal and external datasets, a highly sophisticated approach is used to determine and model the future.

Relevant Technologies

The following technologies are considered relevant for this use case:

- Enterprise Resource Planning System
- Manufacturing Execution System
- Product Lifecycle Management System

Fig. 3.12 Smart factory framework—use case 06

- SCADA System
- Cloud Computing
- Internet of Things
- Track & Trace Technologies
- Big Data Analytics
- AI, ML & Simulation

Impact on Competitive Priorities

The impact of this use case on competitive priorities is shown in generic form in the Fig. 3.13. The specific impact depends on the context of the company.

Impact on flexibility • Enables shorter response times to changes in demand	Impact on costs • Prevents stock-out situations that lead to bottlenecks and idle times • Optimizes the inventory
Impact on delivery • Increases the availability of parts • Avoids of out-of-stock situations • Reduces the bullwhip effect	Impact on quality • No direct impact

Innovators/Exemplary Cases

The utility of such approaches is exemplified in the following cases. A retailer of replacement parts for construction machines wished to predict the demand for its products in several locations to optimize its inventories. By integrating firm company-internal and -external datasets into a machine learning (ML) application, the firm managed to predict the demand in a time horizon of twelve months. This led to higher availability of the single workpieces, as well as to a 50% decrease in profit loss due to stock-outs (Alexander Thamm GmbH, n.d.).

Another example of the effectiveness of demand prediction is the case of a producer of consumer electronics. The company shifted its demand planning from an ill-defined decentralized approach to a more structured approach in the early 2000s. The initiative led to a 50% decrease in on-hand inventory, as well as to the abolishment of practically all excess and obsolete costs (Oliva & Watson, 2009).

When and How to Apply This Use Case?

Demand prediction is especially useful in the following situations:

- **Volatility in customer demand**: A comprehensive data-backed model for forecasting demand will help companies operating in industries with heavily fluctuating demand
- **High lead times**: Actions informed by the demand forecast—such as processing inventory orders—will be useful for reducing the supplier and delivery lead times of a company.
- **High warehousing costs/limited storage**: A better understanding of current and past demand patterns can help optimize inventory and reduce avoidable warehousing/storage costs

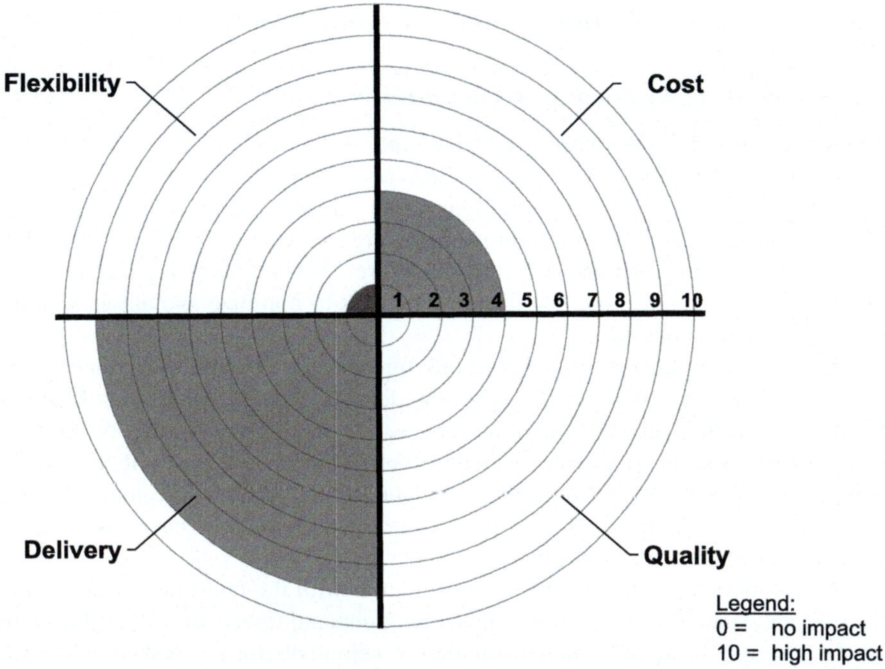

Fig. 3.13 Impact—use case 06

- **Long and complex supply chains**: Demand diagnostics can offer some added visibility for companies with long and complex supply chains.
- **To promote a lean approach**: The prediction of future demand can support a broader implementation of a lean manufacturing approach. Smoothing of production can be implemented on a wider range.

Questions to Ask

- How important is demand prediction for our operations model? Some companies, for example, companies with a make-to-orders system do not need to forecast demand.
- Which total lead time do we want to achieve?
- What are the costs of the different scenarios? (e.g., inventory costs, profit loss due to stock-out situations, etc.)
- How reliable and fast are our inbound and outbound order lead times?

3.3.3 Use Case 07: Autonomous Demand Forecast

Description Along the Framework Categories

The importance of accurately predicting and understanding the demand for a company's products or services has already been established in the previous use cases. Similarly, the exploratory questions of the demand management process have also been addressed (what/why is (this) happening and what will happen next?). The use case autonomous demand forecast, therefore, builds upon and is informed by these previous diagnostic and predictive use cases to optimize and automatize the decision-making processes.

The techniques utilized in autonomous forecasting are similar to some of the more advanced approaches described in the demand prediction use case. Machine learning algorithms are applied to the forecasting process as a variety of different data sources—weather data, calendar data, other historical data—are utilized to build a model on the interdependencies between data and actual demand volume (Kloosterman, 2019). Once the algorithms produce reliable enough forecasts, they can be applied to predict future demand.

The autonomous aspect of the use case comes with the real-time connection of the relevant data inputs to the machine learning demand forecasting algorithm. This can be achieved through the implementation of a cloud platform which allows for the seamless integration of autonomously generated databases with the demand forecast application. Achieving this level of data integration and process autonomy allows for a more effective forecast as data is fed into the model continuously and can be utilized to prescribe operational decisions such as processing new inventory orders.

Picture of the Model/Framework Explanation

This use case consists of the autonomous forecasting of demand and the subsequent automatic triggering of certain decisions. As this application not only forecasts demand data but also uses them to execute tasks autonomously, it is classified as prescriptive (see Fig. 3.14).

Relevant Technologies

The following technologies are considered relevant for this use case:

- Enterprise Resource Planning System
- Manufacturing Execution System
- Product Lifecycle Management System
- SCADA System
- Cloud Computing
- Internet of Things
- Track & Trace Technologies
- Cyber-Physical Systems
- Big Data Analytics
- AI, ML & Simulation

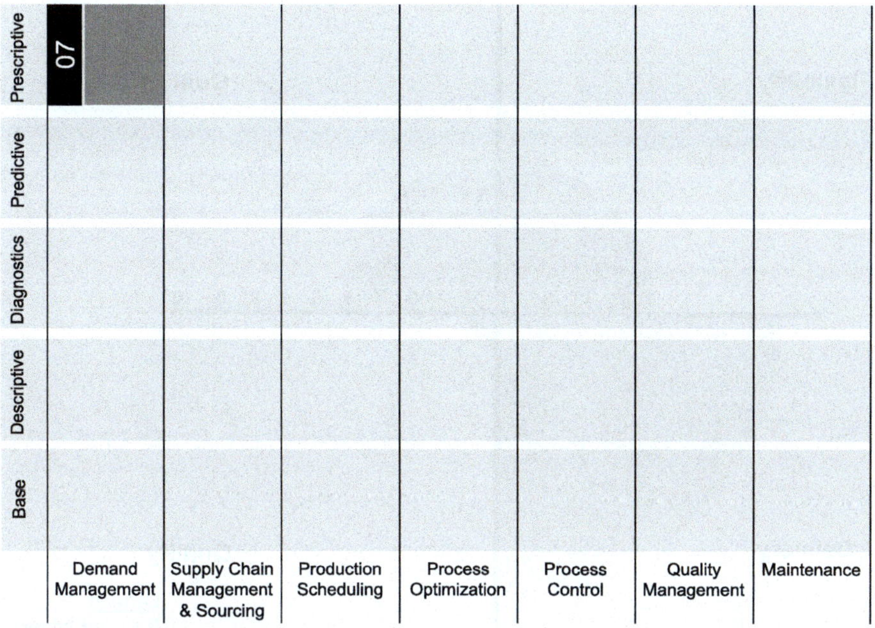

Fig. 3.14 Smart factory framework—use case 07

Impact on Competitive Priorities

The impact of this use case on competitive priorities is shown in generic form in the Fig. 3.15. The specific impact depends on the context of the company.

Impact on flexibility • Provides continuous response to changes in demand	Impact on costs • Prevents stock-out situations that lead to bottlenecks and idle times • Optimizes the inventory • Reduces labour costs—optimizes scheduling and planning activities
Impact on delivery • Reduces lead times—cuts ordering and processing times	Impact on quality • Data-driven planning allows for more regional specification of products

Innovators/Exemplary Cases

FleetPride, a truck and trailer parts supplier, turned to IBM to implement a series of analytical solutions to improve their customer-oriented spare parts service. As the business is built on the fast and dynamic response to customer demand, Fleet-Pride was thinking seriously of a solution to improve the efficiency of their supply chain. To this end, the company adopted a variety of analytics solutions one of which was a demand modelling application. The model was built on three years of shipping data to accurately forecast the amount of in and out shipping orders for their

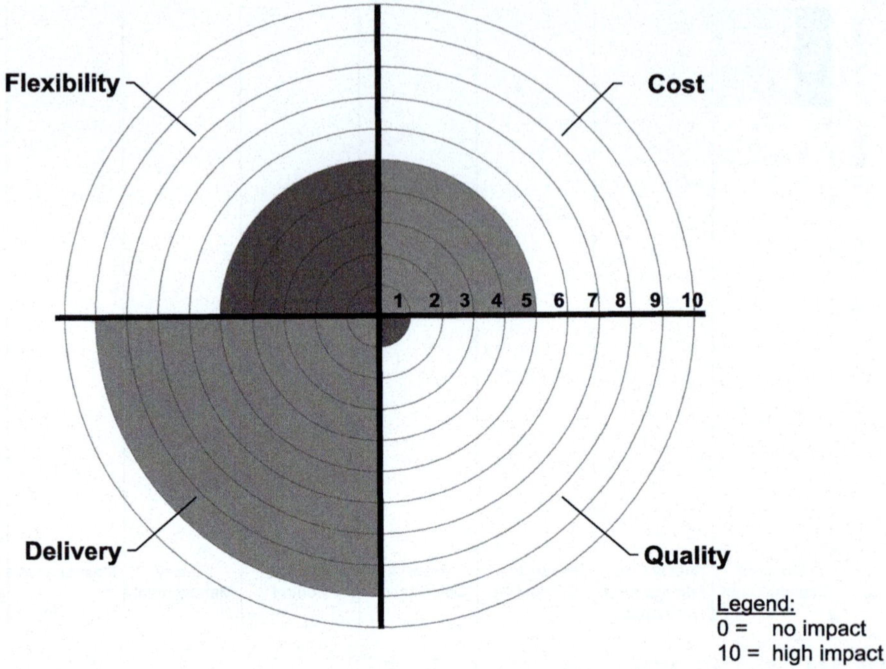

Fig. 3.15 Impact—use case 07

respective warehouses (FleetPride Keeps the Wheels of Commerce Turning with Seamless Supply Chain Management, 2021; Why Prescriptive Analytics & Decision Optimization Are Crucial, 2020). In addition, decision optimization tools were applied to the model and optimal actions were determined for handling customer demand—i.e., staff scheduling and inventory location (FleetPride Keeps the Wheels of Commerce Turning with Seamless Supply Chain Management, 2021; Why Prescriptive Analytics & Decision Optimization Are Crucial, 2020). With these data-driven insights, the company was able to significantly reduce overtime costs caused by poor planning (FleetPride Keeps the Wheels of Commerce Turning with Seamless Supply Chain Management, 2021).

In terms of software offering, Anadot has emerged as a promising solution when it comes to autonomous demand forecasting. The solution relies on machine learning algorithms to create accurate forecasting models, tapping into a wide variety of data sources to improve performance compared to traditional approaches. The Anadot autonomous demand forecast has been applied to inventory capacity management for example, factoring in seasonal, weather, and geographical data to better plan the stocking of individual stores across the country (Cohen, 2022). Such a solution allows retailers to better meet consumer expectations and increase revenues (Cohen, 2022).

When and How to Apply This Use Case?

This use case is especially useful in the following situations:

- **Volatility in customer demand**: A comprehensive data-backed model for forecasting demand autonomously will help companies operating in industries with heavily fluctuating demand
- **High lead times**: Autonomous actions informed by the demand forecast—such as processing of inventory orders—will be useful for reducing the supplier and delivery lead times of a company.
- **High overtime/scheduling costs**: Autonomous demand forecast can increase visibility into future need for labour and will reduce unplanned overtime costs by optimizing scheduling activities.
- **Products are regionally differentiated**: Companies offering products in different geographical regions will face differing demands. Autonomous demand forecasting can manage this complexity and optimize demand planning.
- **Enabling a lean approach**: The prediction of future demand can support a broader implementation of a lean manufacturing approach.

Questions to Ask

- What is the most pertinent demand-relevant data?
- Is this data easily accessible?
- What processes and operations could benefit from data-driven autonomous demand forecasts?

3.4 Supply Chain Management and Sourcing

3.4.1 Use Case 08: End-To-End Performance Tracking

Description Along the Framework Categories

In the age of globalized value chains and distributed production, supply chains are becoming ever more complex (R. Oliveira et al., 2019; Stefanovic & Stefanovic, 2011). The information available for analysis is much larger than in the past, and companies are struggling to keep up their visibility and performance optimization (Stefanovic & Stefanovic, 2011). End-to-end performance tracking can offer a simple yet effective solution for a more holistic data-informed supply chain management. The greater visibility provided by a well-constructed KPI dashboard can offer useful insights in terms of the specific positive or negative determinants influencing performance.

Generally, throughout the supply chain processes, large volumes of diverse data are generated at different supply chain partners. The supply chain dashboard requires the collection of all relevant data, ranging from ERP data to supply chain management (SCM) and transactional systems data (Stefanovic & Stefanovic, 2011). Here, the key step for the implementation of an effective dashboard is the identification and isolation of relevant information. For each process step in the supply chain, the most relevant information in terms of optimization potential must be identified and subsequently isolated. Once this information base is ready, balanced and expressive performance measures and metrics (KPIs) can be constructed to drive improvements.

A major obstacle to this holistic supply chain approach is the heterogeneity of data across different regions and facilities in the supply network ("Staying on Top of Your Supply Chain," 2022). Once the right KPIs have been defined, companies need to harmonize the data collection across all blocks within the supply chain and decide how and where this data is stored. A successful solution will standardize information gathered across facilities and store it in a central cloud-based repository. This allows companies to draw insightful and actionable insights from the dashboard.

Picture of the Model/Framework Explanation

End-to-end performance tracking can be useful for any manufacturing company and provides an additional degree of visibility into the company's supply chain performance. An effective dashboard provides important information in terms of the specific levers influencing performance and offers actionable insights in terms of where to direct optimization efforts. For this reason, the use case is classified as descriptive (see Fig. 3.16).

Relevant Technologies

The following technologies are considered relevant for this use case:

- Enterprise Resource Planning System
- Manufacturing Execution System
- Cloud Computing
- 5G
- Internet of Things
- Sensors and Actuators
- Track & Trace Technologies
- Big Data Analytics
- Blockchain

Impact on Competitive Priorities

The impact of this use case on competitive priorities is shown in generic form in the Fig. 3.17. The specific impact depends on the context of the company.

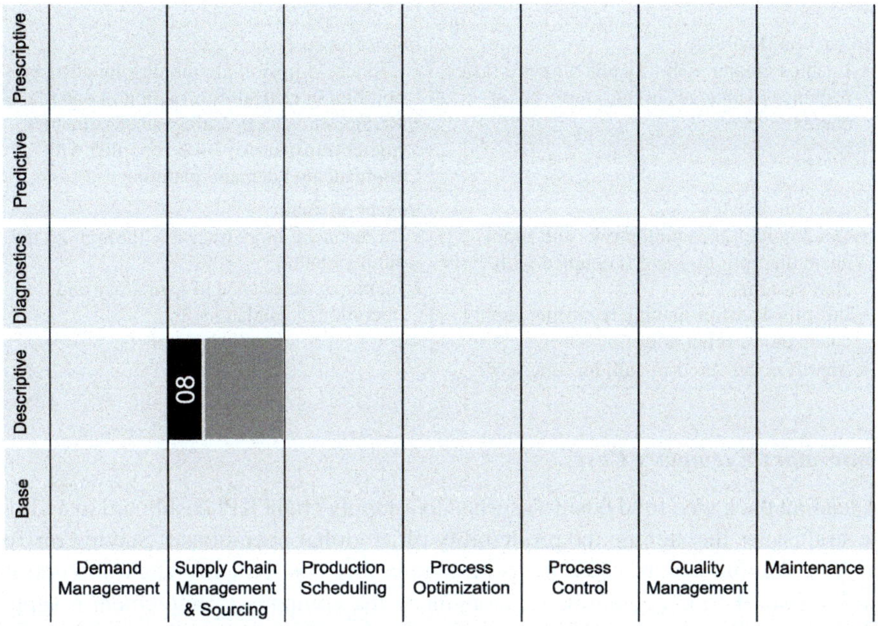

Fig. 3.16 Smart factory framework—use case 08

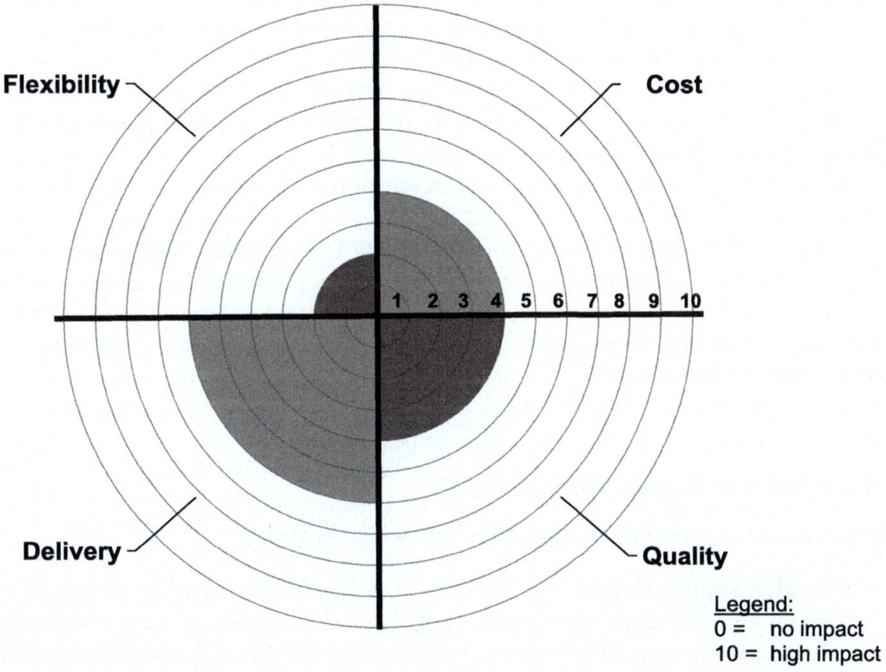

Fig. 3.17 Impact—use case 08

Impact on flexibility • Enables greater visibility and allows a faster response to changes in the supply chain context	Impact on costs • Reduces supply chain monitoring costs by combining critical information in one place • Reduces inventory (safety stock) due to higher transparency—i.e., coupled with sophisticated demand planning use cases
Impact on delivery • Quickly identifies bottlenecks and supply chain disruptions—and is coupled with early alert systems • Informs decision-making regarding supply chain and/or order adjustments • Improves delivered-in-full, on-time score	Impact on quality • Products are more traceable throughout the supply chain • Increase compliance to regulatory and operational standards

Innovators/Exemplary Cases

A leading packaged food business turned to a supply chain KPI dashboard to address several issues threatening the profitability of its global operations ("Staying on Top of Your Supply Chain," 2022). Rising commodity costs, supply chain disruptions and a case of food contamination convinced the company to implement a supply chain management dashboard to remain a leading player and maintain its quality commitments to consumers. First, relevant information and KPIs were constructed focusing on three main areas of improvement. Second, employees were instructed on how to best interpret the information and turn insights into action. Finally, the data streams were fully integrated into the cloud. The system allowed the company's employees to make more informed decisions. Information sharing across sites and the increased identification of quality issues positively affected both product quality and the number of regulations violations.

TESCO, a groceries and general merchandiser company, offers a second successful case study of supply chain KPI dashboard implementation ("Supply Chain Analytics," 2022). In order to keep up with data from roughly 3000 stores in the United Kingdom, with an average of 15000 products per store, the company opted for a dashboard solution to improve its delivery performance. Drawing from a cloud database, containing historic data of up to 5 years, the company implemented a supply-chain dashboard with a particular focus on meeting the fluctuating demands characterizing the industry, relying on effective demand planning systems to match the expectations of customers and offer the best possible service.

When and How to Apply This Use Case?

This use case is especially useful in the following situations:

- **Complex supply chains**: The benefits of a supply chain dashboard are most relevant for companies with complex supply chains containing many geographically dispersed steps and partners. A dashboard can help reduce uncertainty and maintain visibility.

- **Supply chain inefficiencies**: Mismatches between supply and demand, rigid forecasts, etc. might lead to inefficiencies within the supply chain. Supply chain dashboards can help to gain visibility and thus help to avoid misconceptions and organize more efficient supply chain operations.
- **Disruption-rich environment**: The more disruptions in the environment of a company occur, the more it might be in favour of creating a collaboration network to mitigate the issues.
- **Offering services in a highly scrutinized industry**: A supply chain dashboard can offer the needed visibility into a company's sustainability and good governance practices. Presenting this information can ease consumer and regulatory pressures.

Questions to Ask

- Does our supply chain involve many different and geographically dispersed steps?
- Does our supply chain involve many different and geographically dispersed partners?
- Is the company exposed to increased supply chain risk?
- Are the end products subject to increased regulation and scrutiny with regards to their origin?

3.4.2 Use Case 09: Marketplace Platforms

Description Along the Framework Categories

The COVID-19 crisis caused a ripple effect in many industries around the world (Queiroz et al., 2020). Due to failures in companies' suppliers (e.g. production shutdowns), many industrial companies could not adequately supply their own customers, some of which are industrial companies themselves. As a countermeasure, many companies looked for alternative, often regional sources, to reduce the risks and to ensure delivery reliability (World Economic Forum (WEF), 2020). One approach to facilitate the networking of companies with new suppliers is the participation in mediation platforms or so-called marketplaces. Here, the companies can register themselves as customers or suppliers and can request or offer products and services.

However, it would be wrong to assume that such marketplace platforms were driven exclusively by the pandemic. The contemporary high fluctuations in demand (Holtewert & Bauernhansl, 2016) and increasingly shorter product life cycles (Sorostinean et al., 2021) as well as not utilized manufacturing equipment have led to a more dynamic manufacturing environment. Marketplace platforms can be used to turn those challenges into opportunities and leverage them effectively. For example, a manufacturer who is challenged with fewer orders in the medium-term can offer its additional manufacturing capacities on the marketplace platform. From a demand side, companies might be able to design products more quickly and easily with rapid prototyping or to have special parts produced at short notice for customized orders by outsourcing the different tasks to specialized enterprises. Nowadays, a wide variety of different manufacturing-as-a-service-related marketplaces exist—for example, for contract production (OrderFox, n.d.), prototyping services (Pahwa et al., 2018), and other services (Startups Against Corona, 2020).

Marketplace platforms can be distinguished into two types, namely centralized and decentralized platforms. In decentralized platforms, the customer is provided with a list of the suppliers whose offer matches the requirements of the client. The comparison and selection of the right suppliers are done by the customers themselves. In centralized manufacturing marketplaces, customers determine the providers as well as the price of the job themselves. This makes them much more actively involved in the matchmaking process, whereas decentralized marketplaces only provide the platform (Pahwa et al., 2018).

Picture of the Model/Framework Explanation

Marketplace platforms help to bring together providers and customers of all types of industrial services. Those platforms often consist of matchmaking mechanics, which themselves do not contain any advanced analytical functions. Only centralized platforms, which involve analytics to determine the price of the industrial services can be considered diagnostic. This results in the use case being classified as descriptive (see Fig. 3.18).

Relevant Technologies

The following technologies are considered relevant for this use case:

- Enterprise Resource Planning System
- Manufacturing Execution System
- Product Lifecycle Management System
- Cloud Computing
- Big Data Analytics
- Additive Manufacturing
- Blockchain

Impact on Competitive Priorities

The impact of this use case on competitive priorities is shown in generic form in the Fig. 3.19. The specific impact depends on the context of the company.

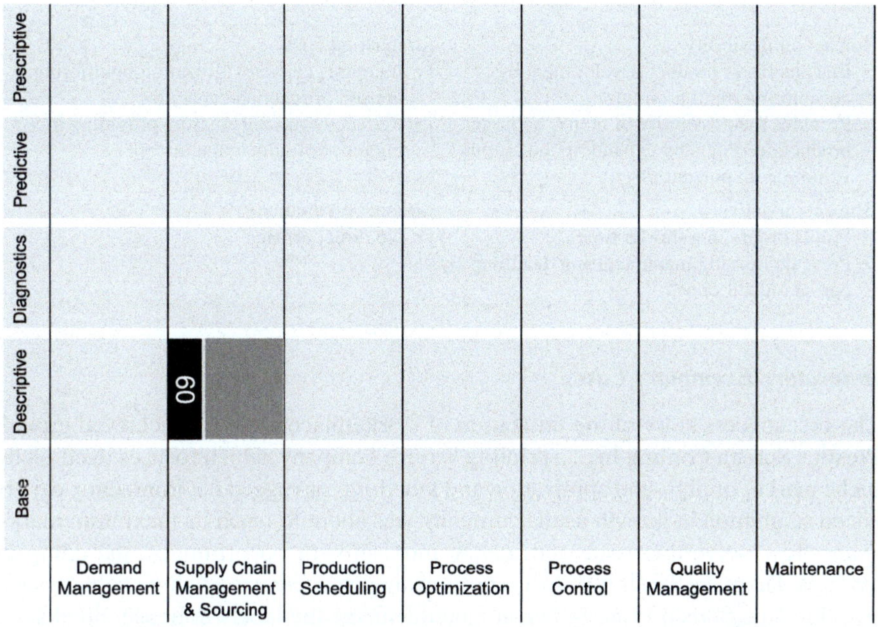

Fig. 3.18 Smart factory framework—use case 09

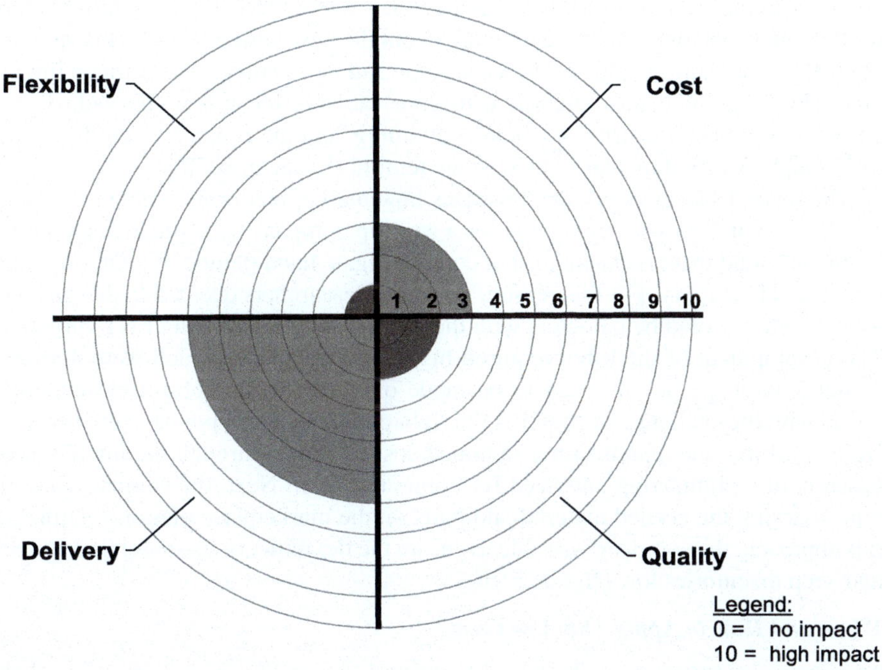

Fig. 3.19 Impact—use case 09

Impact on flexibility • Enables faster product development by outsourcing rapid prototyping • Provides the possibility of taking on larger production orders by expanding the output volume with outsourcing	Impact on costs • Increases capacity utilization by offering unused production capacities • Reduces sourcing costs by providing a bigger pool of manufacturers
Impact on delivery • Fulfils orders in a shorter time • Prevents missing manufacturing deadlines by partial outsourcing	Impact on quality • No direct impact

Innovators/Exemplary Cases

The first success story of the utilization of marketplace platforms achieved growth. Prestige Screen Printing Inc., a printing service company, which focuses itself on the niche market of high-end application and labelling for rugged environments, experienced stagnation in growth as the company was about to reach its maximum market share. To resolve this issue and realize more growth opportunities, the company attempted to make the shift from a supply-chain perspective to a value-chain perspective. Prestige Screen Printing began manufacturing the labels in-house, thanks to a newly purchased 3,000-W laser and was able to reduce costs and production time as well as offer new services. At the start of its business expansion journey, the company relied on Xometry, a manufacturing-as-a-service marketplace, to obtain more orders for their new machines and to learn more about the new business area. This strategy helped the company to achieve higher revenue and thus achieve the return on investment (ROI) faster. It also enabled it to gain valuable know-how. Nowadays, the growth of the manufacturing segment is booming, and the company went from two to 12 employees and increased its revenue tenfold (Krueger, 2020).

The second success story demonstrates how market platforms enabled the rapid prototyping of highly specific parts for a start-up company. A medicinal survey by a student found that no existing robot could mitigate hand tremors for ear, nose and throat (ENT) surgery. For this reason, a team of researchers decided to design such a robot, which could help doctors with their trembling hands during ENT surgeries. The development of the robot required highly specific parts, which were not only complex in shape but also had to be made of a specific biocompatible material. Faced with the challenge of sourcing the right parts, the development team turned to Xometry. First, the manufacturer obtained cost estimates through the intelligence-based portal, eliminating any need for communication. Next, the company moved onto sourcing the needed material and parts on the marketplace platform. Utilizing this approach, the company was able to ensure the flexibility of its sourcing activities and keep inventories low (Ngoh, 2020).

When and How to Apply This Use Case?

The situation differs depending on whether a company wants to act as a service provider or a customer.

As a service provider

- **Excess capacities**: If a company experiences occasionally excess capacities, it might be reasonable to offer them on market platforms to get additional production orders. This might be especially helpful for start-ups or new manufacturing segments within a company.
- **New sale channels**: Another sales-related reason to offer manufacturing services on market platforms is to add a new sale channel to the company.

As a service receiver

- **Need for highly specific parts**: A bigger pool of companies facilitates the search for the right manufacturer of specific parts (in terms of materials, properties, delivery time, etc.).
- **Need for highly specific services**: A bigger pool of companies facilitates the search for the right manufacturer of specific services.
- **Optimization of the supply chain**: Marketplace platforms with bigger pools of manufacturers may help to optimize the supply chain configurations regarding specific objectives (short lead time, cost, quality, etc.).
- **Finding alternative suppliers in a short time**: If an existing supplier cancels the delivery, online marketplaces can help identify an alternative supplier quickly.

Questions to Ask

- What is our reason for engaging in marketplace platforms?
- Are there existing marketplaces in our field? How many relevant companies (regarding proximity, offered services, etc.) are on this platform?
- What data do we want to share over a platform?

3.4.3 Use Case 10: Sustainability and Carbon Footprint Tracking

Description Along the Framework Categories

The Intergovernmental Panel on Climate Change recently issued a report declaring the current state of the climate crisis as a "code red for humanity" ("IPCC Report: 'Code Red' for Human Driven Global Heating, Warns UN Chief," 2021). For industrial players, therefore, sustainability initiatives and carbon footprint tracking efforts are gaining momentum, as key stakeholders—investors, customers, and employees—are starting to raise their sustainability expectations (Arviem Cargo Monitoring, 2017). To face this increasing pressure and minimize negative climate externalities, a selection of smart carbon footprint tracking solutions can be of use to manufacturing firms. The carbon emissions of manufacturing firms are typically broken down into three distinct categories. Scope one emissions are defined as the "direct carbon emissions from sources owned or controlled by the company" (Lee, 2011). Scope two emissions are defined as the "indirect carbon emissions associated with the purchase of utilities used by the company" (Lee, 2011). Scope three emissions include "all other indirect emissions including the total supply chain up to the production gate" (Lee, 2011). Generally, scope one and two emissions are operational emissions (occurring at the production site), and scope three emissions are all other emissions caused by the different links in the supply chain. While all three scopes of emissions should be considered in sustainability efforts, scope three emissions are—conceptually—the most relevant in the context of supply chain management. In addition, regardless of the supply chain management focus, greenhouse gas (GHG) supply chain emissions (scope three) are "on average 11.4 times higher than operational (scope one and two) emissions" (Climate Disclosure Project, 2021).

In terms of scope three GHG emissions, there are two further categorizations: suppliers' operational emissions and logistics' emissions. Tracking the individual carbon footprint of every supplier's operations is a highly complex task. The lack of an agreed standardized methodology, scarce primary data, and the sheer complexity of value chains, renders the collection of actionable data highly difficult (World Business Council for Sustainable Development, 2021). To this end, cloud solutions can prove useful in managing this complexity. Buying firms—or other willing players in the chain—can open and run a cloud platform to facilitate the collection, storage, and analysis of emission data (Singh et al., 2015; SupplHi, 2021). A variety of software as a service (SaaS) solution—in the form of footprint calculators or even supplier carbon trackers—can be utilized to increase carbon footprint visibility. Nevertheless, cooperation and harmonization of commitments across the supply chain are still fundamental. With regards to logistics, tracking a manufacturer's transport emissions is easier. IoT enabled tracking solutions allow companies to accurately monitor logistic operations and collect granular data on transport emissions. This way, companies can increase visibility and make the needed adjustments to lower environmental impact as well as operational costs (Arviem Cargo Monitoring, 2017).

Picture of the Model/Framework Explanation

The sustainability and carbon footprint tracking use case revolves around understanding more about a company's own environmental impact and drawing actionable insights. The technologies mentioned here rely on diagnostic analytic capabilities, and, thus, the use case is categorized accordingly (see Fig. 3.20).

Relevant Technologies

The following technologies are considered relevant for this use case:

- Enterprise Resource Planning System
- Manufacturing Execution System
- Cloud Computing
- Internet of Things
- Sensors and Actuators
- Track & Trace Technologies

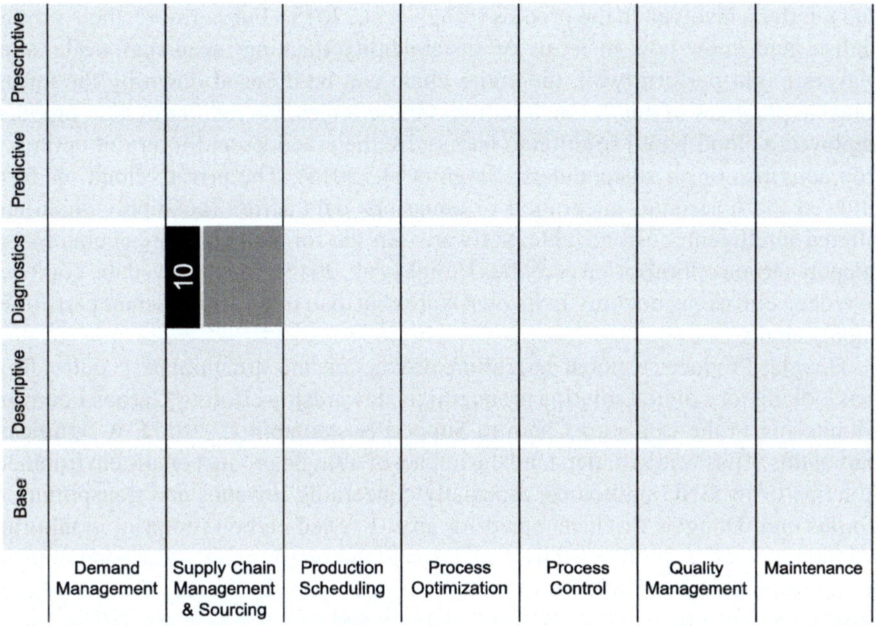

Fig. 3.20 Smart factory framework—use case 10

- Big Data Analytics
- Blockchain

Impact on Competitive Priorities

The impact of this use case on competitive priorities is shown in generic form in the Fig. 3.21. The specific impact depends on the context of the company.

Impact on flexibility • A horizontally integrated green supply chain can adapt quickly to sudden changes in market demand	Impact on costs • Greater visibility into logistics operations can reduce redundancies and inefficiencies • Lowering the carbon footprint can lead to a decrease in costs incurred by regulatory entities—e.g., carbon tax
Impact on delivery • Better integration and coordination of supply chain actors can reduce lead time	Impact on quality • Greater customer satisfaction by meeting climate expectations of consumers

Innovators/Exemplary Cases

In response to rising government pressure to reduce its carbon footprint, a retail store turned to a digital solution to reduce the GHG emissions across its entire beef supply chain (Singh et al., 2015). The management of beef supply chains is notoriously complex due to the number of players such as farmers, abattoirs, processors and retailers, involved in the process (Singh et al., 2015). Furthermore, their varying skillset and know-how in terms of sustainability tracking mean that while some players might perform well, the entire chain can be dragged down by the unsustainable practices of one actor. To address these issues, the retailer of beef products deployed a cloud-based solution to harmonize the practices and footprint optimization activities of all stakeholders (Singh et al., 2015). The private cloud platform allowed the horizontal integration of emissions data across the supply chain and offered intelligent, customizable, software—in the form of carbon calculators—to support carbon optimization activities (Singh et al., 2015). This allowed the company to reduce emissions not only in its own stores but also in the most distant parts of its supply chain.

Douglas Products, a global agriculture production and structural pest control firm, was looking for a digital solution to support its stewardship efforts ("Carbon Footprint Monitoring in the Logistics Chain to Support Sustainability," 2017). A significant part of this effort was to understand the impact of a single product on the environment in terms of its GHG emissions, especially concerning logistics and transportation. To this end, Douglas Products opted for an IoT based carbon footprint monitoring solution to track its product all the way through the supply chain ("Carbon Footprint Monitoring in the Logistics Chain to Support Sustainability," 2017). Such a solution provided the company with increased visibility and concrete primary data to "make

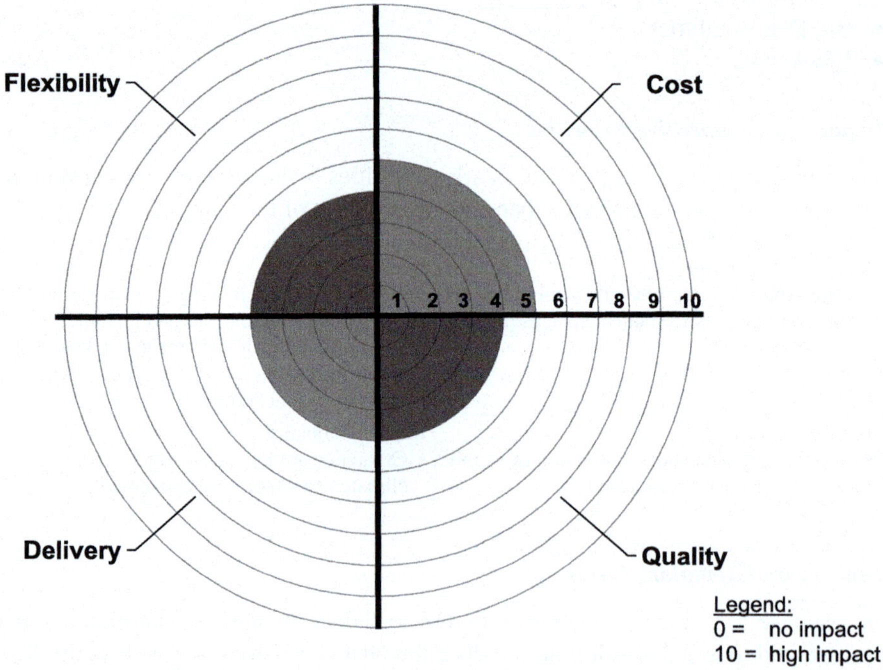

Fig. 3.21 Impact—use case 10

key business decisions, to reduce energy use and to lower their carbon footprint" (Arviem Cargo Monitoring, 2017).

When and How to Apply This Use Case?

While all manufacturing firms should consider sustainability and carbon footprint tracking solutions, implementation can bring the most benefits in the following cases:

- **Operating in a carbon-intensive industry**: If operating in a carbon-intensive industry, greater visibility into current emissions might reveal some low-hanging fruits in terms of reducing environmental impact and operational costs—for instance, the optimization of delivery routes for freight/transportation companies.
- **Supply chain composed of diverse, smaller actors**: Small and mid-size enterprises (SMEs) or local suppliers might have difficulty tracking and reducing their own emissions, leading to higher scope three emissions. In this case, companies can benefit from integrating the supply chain actors and providing the missing knowledge or capabilities (e.g., by implementing a SaaS carbon calculator).
- **Offering services in a highly scrutinized industry**: Transparency regarding carbon footprint can ease consumer and regulatory pressures regarding sustainability concerns. Tracking and reporting this data can help serve as a differentiating factor to beat competition.

Questions to Ask

- Do we operate in a carbon-intensive industry?
- Where are the carbon hotspots in our supply chain?
- What are more significant, our scope one and two or our scope three emissions?
- What are the costs of more sustainable supply chain practices?
- Is it possible to harmonize data collection and practice across our supply chain?
- How much control do we have over suppliers and partners?

3.4.4 Use Case 11: Digital Inventory Management

Description Along the Framework Categories

For many companies, complexity has increased rapidly in recent years. This can be attributed to various causes, such as the increasing number of product variants and the high variance in the size and product variety of individual orders (Lee et al., 2018). To cope with these challenges, effective management of different materials and finished products in the warehouses is crucial. Managing and tracking inventory allows for a constant overview of the materials readily available for production, and this information can be used to make sure plants produce goods on time and meet shipping requirements.

Currently, individual warehouse stocks are still captured, adjusted, and reviewed manually. This means that the entry and exit of finished products, semi-finished products, and other materials are still partly booked with scanners or by hand into their warehouse management systems (WMS) or paper-based documents (Schmeltzpfenning & Krämer, 2019). As a result, human errors may lead to incorrect stock levels being recorded, which in turn can lead to losses in productivity due to bottlenecks as well as delays in the delivery of products.

To tackle these challenges, digital technologies such as IoT, radio-frequency identification (RFID) chips, and analytics have started to become increasingly utilized by the warehouse management of manufacturing companies (Lim et al., 2013; Schmeltzpfenning & Krämer, 2019). This use case represents the entirety of all technologies that aim to improve the processes within the warehouse. IoT seems to be especially promising with an estimated impact of 1.9 trillion USD on overall logistic operations (DHL Trend Research, 2018). Schmeltzpfenning and Krämer, (2019) mention present and future application fields of AI within warehouse operations, such as the control of outgoing goods by visual recognition, the organization (and optimization) of inventories in warehouses, real-time optimization of routes and machine vision-based booking systems, among others.

Picture of the Model/Framework Explanation

The technologies presented in this generic use case are very diverse. Nevertheless, they all have something in common: their application creates transparency within the warehouse. The addition of analysis tools further improves the warehousing processes. For this reason, the use case can be considered both descriptive and diagnostic (see Fig. 3.22).

Relevant Technologies

The following technologies are considered relevant for this use case:

- Enterprise Resource Planning System
- Manufacturing Execution System
- Product Lifecycle Management System
- Cloud Computing
- 5G
- M2M Communication
- Big Data Analytics
- Collaborative Robotics
- Mobile Computing
- Internet of Things
- Sensors and Actuators
- Track & Trace Technologies
- Vision Technologies
- Cyber-Physical Systems
- Automated Guided Vehicles

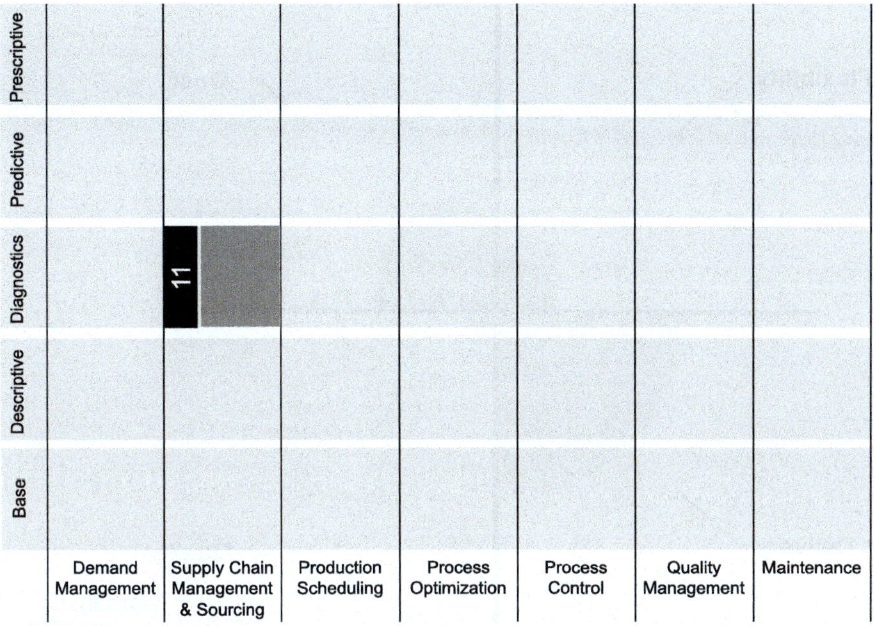

Fig. 3.22 Smart factory framework—use case 11

Impact on Competitive Priorities

The impact of this use case on competitive priorities is shown in generic form in the Fig. 3.23. The specific impact depends on the context of the company.

Impact on flexibility • No direct impact	Impact on costs • Reduces inventory due to greater transparency • Improves warehouse organization and efficiency • Prevents bottlenecks
Impact on delivery • Enables faster order fulfilment • Improves delivered-in-full, on-time score	Impact on quality • Improves transparency and traceability of the products during the operational part of their lifecycle

Innovators/Exemplary Cases

A pan-European producer of upholstery and upholstery-related products managed its warehouse in a very traditional way: all bookings and the processing of data were done manually, and the commissioning process was performed using a paper list. Due to the lack of information about the warehouse, the company employed several commission agents at each location who were responsible for monitoring and optimizing stock levels. This meant that the company was performing all the

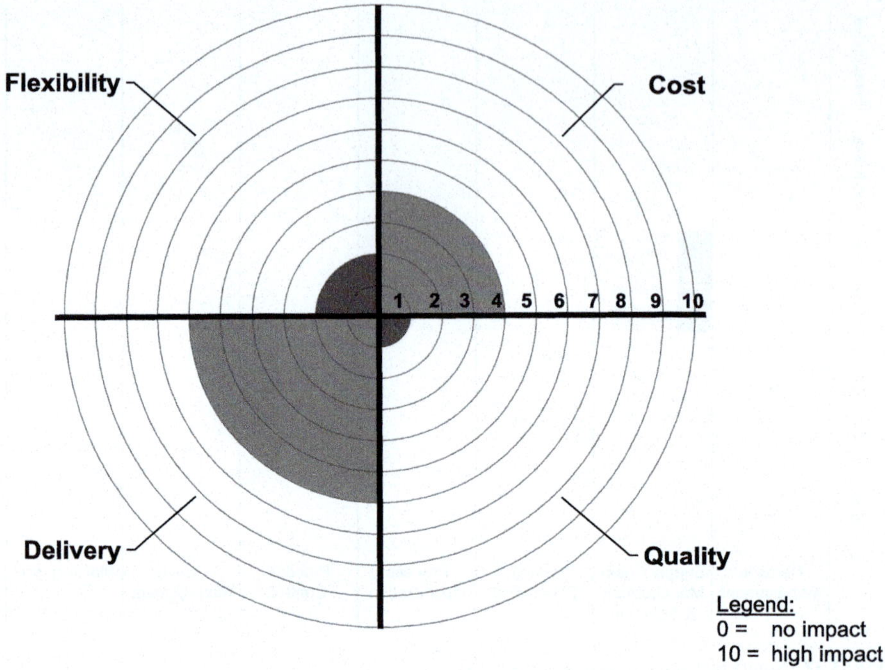

Fig. 3.23 Impact—use case 11

tasks of a WMS manually, which proved very time-consuming and extremely cost-intensive. As a solution, the company implemented an advanced cloud-based WMS. A fast and error-free implementation process meant that the system could be changed with almost no noticeable interruptions. Benefits of the implemented system included a significant reduction in administrative overhead, halving of the error rate due to picking, a two-thirds reduction in order lead time, and 80% less manual processing. In the future, the system will also be able to provide a real-time overview of stock levels (SnapFulfil, n.d.).

Often hundreds, if not thousands, of different materials and products are stored in a warehouse. This increases the complexity within the warehouse, which itself leads to operational inefficiencies (e.g., empty transport aisles, untraceable products, etc.). This might not only lead to a delay in the goods shipped but can also have a negative effect on the production, for example, when a specific part is untraceable. To overcome these challenges, Infsoft created a solution that consisted of the application of two digital technologies in the inventory management system, namely ultra-wideband (UWB) and RFID. UWB locates all vehicles and reports the position in real-time to a dashboard. Furthermore, the system detects and reports all (possible) delays, which allows for early intervention. On top of the tracking, the system calculates the best routes for the individual vehicles to optimize their runways. The vehicles were furthermore equipped with an RFID reader, which allowed them to automatically read the RFID labels attached to the goods. The combination of vehicle and inventory locations provided not only a real-time overview of the ongoing operations in the warehouse but also allowed tracking back to the different items. In addition, the analysis modules helped to optimize the warehouse processes (Wurms, 2017).

When and How to Apply This Use Case?

As mentioned earlier, this use case contains a variety of different hardware and software technologies that help to gain a better overview as well as to manage the warehouse in a more efficient manner. Consequently, this use case is especially useful for those warehouses where errors and other inefficiencies often occur. Specifically, digital inventory management can be applied in the following circumstances:

- **Short lifetime Goods**: Perishable goods should be tracked in the warehouse to a) guarantee the customer the promised quality and b) gain transparency over the stock, enabling the utilization of the soon-to-be-perished products first.
- **Reduction of inventory**: The transparency created enables the reduction of safety stocks, which might go in hand with the companies' strategy.
- **Varied Stocks**: The more different products, materials, etc., a company has in its inventory, the harder it becomes to manually track the different products stored.

Questions to Ask

- Which challenges are we facing regarding the warehouse operations, or which processes do we want to optimize?
- Which hardware and software technologies help us to achieve the desired object?

3.4.5 Use Case 12: Digital Picking Assistance

Description Along the Framework Categories

The commissioning process in warehouses has become increasingly difficult. This can be traced back to the fact that many companies have increased the number of their product variants sharply in recent years (Schwerdtfeger et al., 2009). Simultaneously, the pickers are challenged with the trend towards smaller batches and smaller order sizes (Hammerstein, 2019). The increase of variants is usually accompanied by an increase in the number of production materials required. In specific production strategies and processes (e.g., make-to-order), this means that the worker has to prepare a wider array of material configurations, choosing from a broader variety of resources. The setup and organization of assembly with labels become difficult due to increasing numbers of parts. Quality problems in assembly due to wrong parts being assembled are a risk. A picker who has to select the right items for shipping faces the same problem. Many small orders with a selection of many different products increase the complexity significantly. To successfully overcome these challenges, reduce the error rate in the picking process and increase efficiency, digital solutions exist that support the warehouse worker actively in selecting the materials for production. These technologies, mostly defined as pick-by-x or pick-to-x applications, are summarized in this generic use case under the umbrella term digital picking assistance. In the following paragraphs, the most prominent methods are briefly described:

The pick-by-light, also known as pick-to-light, applications are among the more prominent assistance systems. In this approach, displays together with lights are attached to each item's position on the shelf. The employee receives the information about the number of items they have to collect from the display on the shelf, and the lights indicate the relevant items by lighting up. After the retrieval, the operator pushes the button next to the light to confirm the removal (Bächler et al., 2016).

Pick-by-voice is a method where the picker wears a headset with an integrated microphone. Both are connected to a terminal, which itself communicates with the WMS. In the warehouse, the picker is acoustically guided to the location of the item. By speaking a specific code as well as the number of items retrieved into the microphone, the worker confirms the removal (De Vries et al., 2015).

A more novel approach to guide the employee through the picking process is pick-by-vision. This solution uses smart glasses that allow information in the form of text and/or images to be superimposed on the users' field of vision. This allows the worker to be informed, for example, of the location and number of items to be removed while simultaneously keeping both hands free (Werning et al., 2020).

A similar approach is the pick-by-display method. Here, however, the information is not displayed with smart glasses. Instead, a display is attached to the warehouse worker and/or their picking cart as a wearable and/or tablet. The required instructions are sent wirelessly to the display in form of texts (and images) (Bächler et al., 2016).

Picture of the Model/Framework Explanation

The technologies in this use case range from rather simple visualization and validation approaches to advanced solutions to optimize the picking route, voice recognition systems, etc. Because of the analytical components that many of the tools contain, the use case is considered diagnostic (see Fig. 3.24).

Relevant Technologies

The following technologies are considered relevant for this use case:

- Enterprise Resource Planning System
- Manufacturing Execution System
- Cloud Computing
- 5G
- M2M Communication
- Mobile Computing

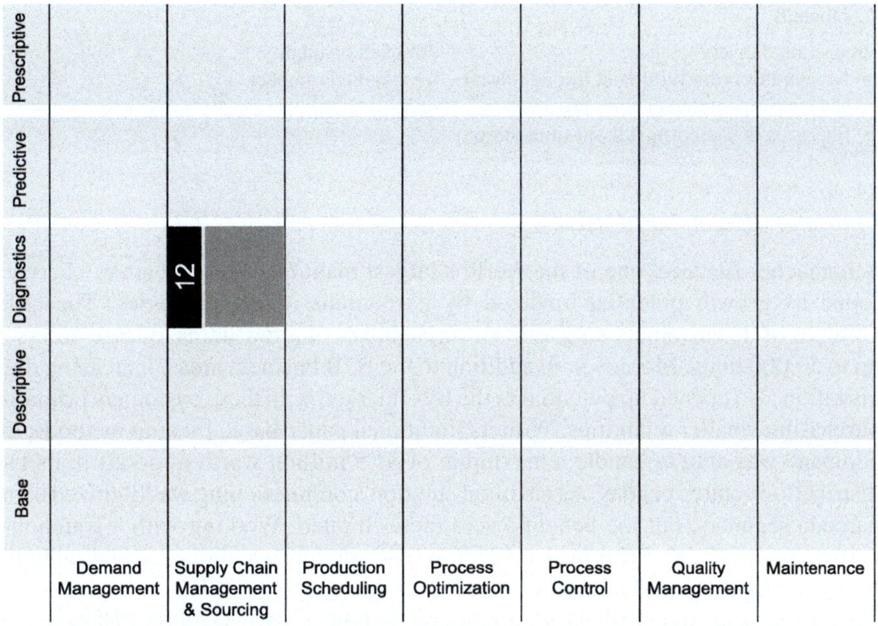

Fig. 3.24 Smart factory framework—use case 12

- Internet of Things
- Track & Trace Technologies
- Vision Technologies
- Cyber-Physical Systems
- Big Data Analytics
- AI, ML & Simulation
- Wearables
- Augmented Reality

Impact on Competitive Priorities

The impact of this use case on competitive priorities is shown in generic form in the Fig. 3.25. The specific impact depends on the context of the company.

Impact on flexibility	Impact on costs
• Improves flexibility regarding the commissioning of product variants and different products	• Reduces personnel costs by utilizing low-skilled workers • Increases labour productivity and reduces picking errors • Reduced quality problems due to reduced amount of wrong assembled parts

(continued)

(continued)

Impact on delivery	Impact on quality
• Accelerates order fulfilment through faster and less error-prone picking processes • Improves delivered-in-full, on-time score	• No direct impact

Innovators/Exemplary Cases

Schumacher Electric, one of the world's largest manufacturers of battery chargers, found its growth potential hindered by its commissioning processes. The order volumes of its customers ranged from single boxes of a few items to large orders of up to 50,000 units. Moreover, in addition to the B2B business areas, increasing digitalization also opened up potential in the B2C market, with these customers primarily purchasing smaller quantities. With its traditional paper-based picking methods, the company was able to handle a maximum of \$1.5 million worth of goods in its two distribution centres per day. As an initial solution, commissioning was digitized using barcode scanners, but the benefits were rather limited. Working with a warehouse automation provider, Schumacher Electric innovated its warehouse operation holistically. Firstly, the most commonly used stock-keeping units (SKUs) were relocated close to the conveyor belt at an ergonomic height to save walking distances and reduce employee inconvenience. Furthermore, picking processes were significantly improved through digital technologies such as pick-by-voice solutions and order sortation systems. The pick-by-voice solution guides the employee to the location of the relevant SKUs. Once there, they scan the code with their wearable device or speak directly into the microphone. The system then informs the worker of the number of specific SKUs to pick from the container. Once the step is completed, the worker informs the system, after which they are directed to the next SKU. The pick-by-voice solution combined with path optimization increased process reliability to 99.99% and efficiency by 70%. This not only generated a higher throughput but also helped to avoid incorrect deliveries and resulting penalties. Furthermore, the pick-by-voice system also eliminated most of the effort involved in training new workers. The maximum handling capacity increased from 1.5 million to over five million per day. The overall productivity gain resulted in a payback of the \$1.4 million investment within one year (The Numina Group, 2016).

The logistics group Schnellecke optimized the picking and sequencing processes in its subsidiary, which produces automotive parts, using digital technologies. The company produces in just-in-time / just-in-sequence, which makes error-free and optimized production imperative. Components were traditionally picked using handheld scanners, but this often led to errors and inefficiencies. To increase process speed while reducing errors, the company introduced smart glasses for pick-by-vision and a multi-sensor RFID wristband. The data glasses continuously show workers all the information they need to pick. The wristband automatically confirms all picking steps, eliminating the time-consuming scanning process. The benefits are accelerated picking processes, a reduction of the error rate to 0, increased ergonomics for the employees as they now have both hands available, a shortened training time for

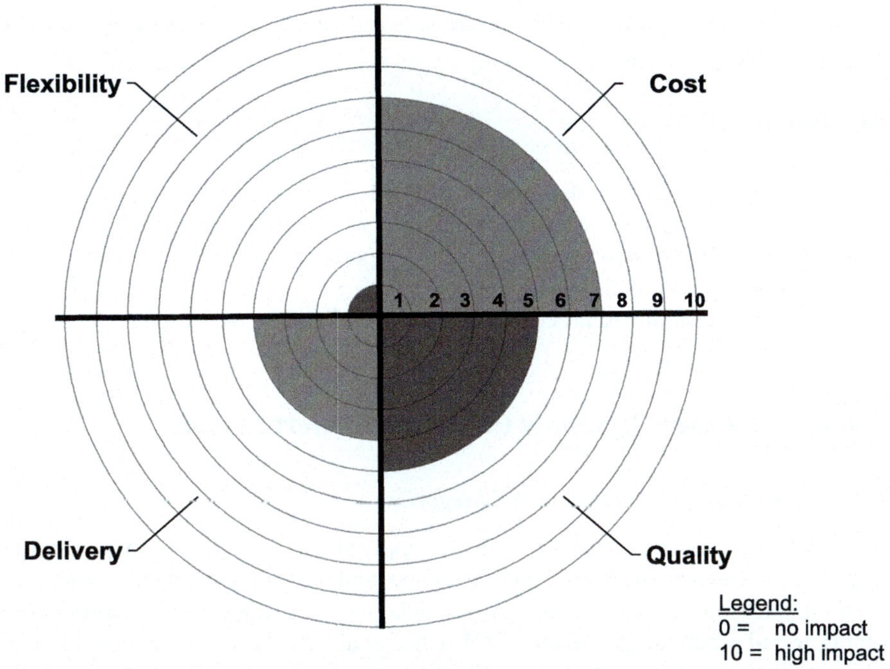

Fig. 3.25 Impact—use case 12

new employees as well as a more flexible arrangement of the employee deployment (Schnellecke Logistics, n.d.).

When and How to Apply This Use Case?

This use case is especially useful in the following situations:

- **Lack of qualified personnel**: A shortage of skilled personnel can be counteracted with the employment of semi-skilled personnel who are then equipped with digital technologies to compensate for their lack of skills.
- **High employee turnover**: Regardless of their skills, new employees must each be introduced to a new plant, familiarizing them with the respective storage locations and work practices. A digital picking assistance system helps workers achieve productivity without the need for major induction.
- **Many different SKUs**: The greater the variety of SKUs that exists within a warehouse, the more complicated the picking processes are. This is due to the size of the warehouse on the one hand and the higher risk of picking the wrong items on the other. Digital technologies can help to organize the best route and show the locations of the single SKUs while also reducing the error rate of picking.
- **A multitude of small orders**: Many orders of small sizes but with a great variety of different items can quickly become a challenge for the warehouse workers.

Digital picking assistance technologies can mitigate these issues by ensuring the right items are picked.

Questions to Ask

- What is the maximum daily turnover of goods in our warehouse?
- How many dispatched orders have contained wrong or missing items in the past?
- Do we have the necessary amount of skilled personnel?
- How is the international logistic organized?
- Do we have to optimize processes as well (reducing lot size, Kanban)?

3.4.6 Use Case 13: Supply Chain Early Alert System

Description Along the Framework Categories

Sudden disruptions and disturbances in the supply chain—e.g., the recent port bottleneck in Los Angeles, the Suez Canal blockade, and the global pandemic—can have costly impacts on a company's operations, delaying the shipment of vital production inputs or end products (Anguiano, 2021; Chae et al., 2014). The operational impact of these disruptions is intensified by the horizontal nature of global supply chains as well as the increased prominence of lean management approaches (Baryannis et al., 2019). In addition, events in supply chains "tend to propagate across partners" meaning one unplanned event in the chain can lead to serious disruption across the entire network (Liu et al., 2007). Given the potential impacts of disruptions, early alert systems have a high potential to positively affect operational outcomes such as cost, delivery, and flexibility.

Standard risk-based warning systems lack the dynamism to effectively manage the unpredictable circumstances companies are currently operating under (KPMG, 2020). This use case, therefore, aims to significantly improve the early detection and resolution of unplanned events with the use of proactive and dynamic analytical solutions. The process of early alert systems can be divided into several steps. Firstly, the monitoring areas for the individual steps in the supply chain must be defined and mapped out (Bodendorf & Zimmermann, 2005). Next, the data from each of these monitoring blocks need to be obtained and processed. Effective early alert systems will supplement internal data—process data describing qualitative features of orders and products—with external data—meteorological, news feeds, social media (Genc et al., 2014; KPMG, 2020). Once relevant data are collected and integrated, they are automatically monitored along "specific indicators, trends, pattern and experiences to identify critical events" (Genc et al., 2014). This process relies heavily on artificial intelligence and machine learning tools. Finally, simulations are carried out to determine the potential effects of the disruptive event on performance and strategic (alternative) decisions can be taken if the risk is deemed critical enough (Bodendorf & Zimmermann, 2005). Importantly, the early alert system aims to work at the "indicative" stage of a disruption timeline, where patterns and signals indicate the possible occurrence of a disturbance (Genc et al., 2014).

The early alert system is usually embedded in a broader supply chain management platform, often deployed through a real-time dashboard illustrating the entire supply chain network and the predetermined monitoring hubs.

Picture of the Model/Framework Explanation

The technologies required for the effective implementation of an early alert system range from descriptive tools such as data gathering and visualization technologies to more complex analytical tools such as artificial intelligence, machine learning and simulation techniques. Ultimately, the scope is to predict potential supply chain disruptions dynamically. For this reason, the use case is classified as predictive (Fig. 3.26).

Relevant Technologies

The following technologies are considered relevant for this use case:

- Enterprise Resource Planning System
- Manufacturing Execution System
- Cloud Computing

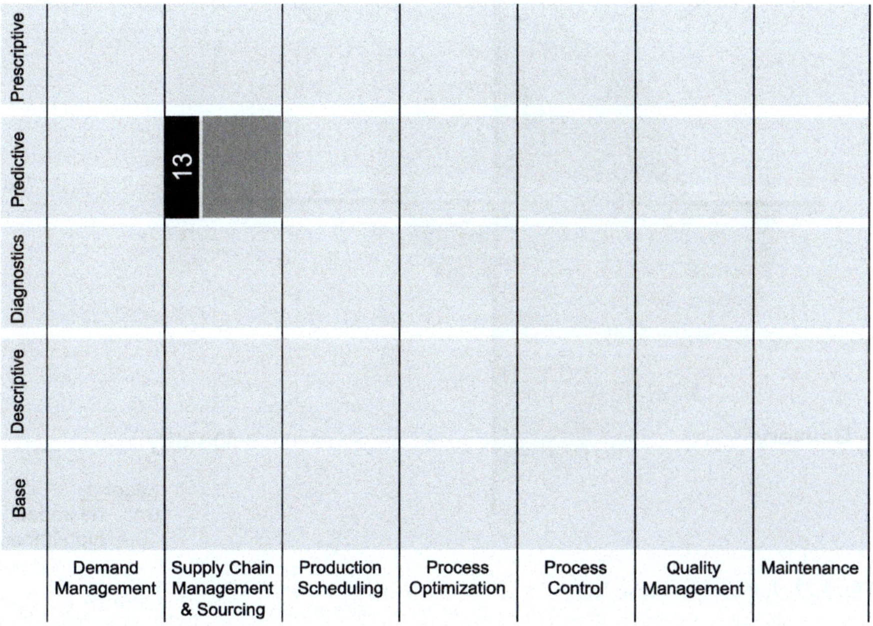

Fig. 3.26 Smart factory framework—use case 13

- Internet of Things
- Sensors and Actuators
- Cyber-Physical Systems
- Big Data Analytics
- AI, ML & Simulation
- Blockchain

Impact on Competitive Priorities

The impact of this use case on competitive priorities is shown in generic form in the Fig. 3.27. The specific impact depends on the context of the company.

Impact on flexibility	Impact on costs
• Improves reaction to unexpected events, both positive (sudden surge in demand) and negative (environment-related disruption)	• Reduces monitoring costs • Reduces unexpected incident-related costs • Reduces the likelihood of a stockout

(continued)

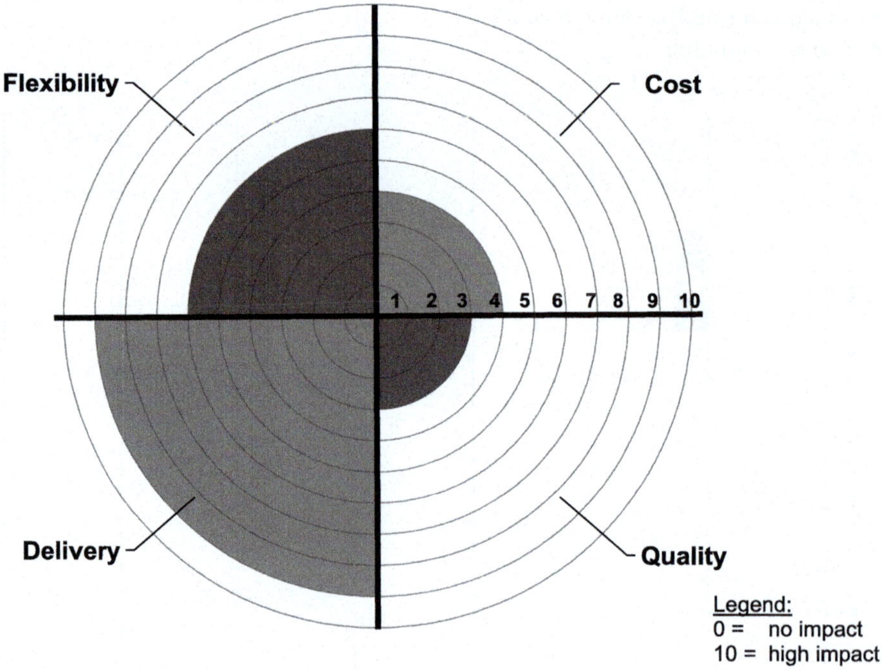

Fig. 3.27 Impact—use case 13

(continued)

Impact on delivery	Impact on quality
• Increases contingency planning time in the case of lead time disruption • Maintains on-time delivery	• Prevents the damage/waste of raw materials or end products

The Innovators/Exemplary Cases

Given the recent disruptions to traditional supply and demand cycles, a specialized European retailer opted for the Supply Chain Early Alert Dashboard provided by IT Company Infosys. The solution involved the visualization of the retailer's supply chain on a map, with the different data from the network hubs organized in a central database. Early alerts were administered thanks to the close analysis of external datasets—social media and news outlets—and impact probabilities were calculated utilizing artificial intelligence—AI utilized historical data of comparable events to estimate impact factors. Overall, the implementation of this novel event management solution led to the increase in prediction accuracy of shortages and likely delays (Infosys, 2021).

The second exemplary case concerns the early alert solution implemented by Lexis Nexis and State of Flux to tackle the issue of supplier failure in supply chains. Relying on Lexis Nexis's Smartindexing technology, the companies analyzed 90,000 newspaper articles concerning a sample of 28 suppliers which had gone bankrupt. The indexing solution allowed the companies to identify "early warning signs of failure in more than 80% of the cases analyzed" (State of Flux, n.d.). Such a solution offers a more mid-term alarm system, inferring and predicting the financial vulnerability of supply chain partners up to six months in advance (State of Flux, n.d.).

When and How to Apply This Use Case?

This use case is especially useful in the following situations:

- **Complex supply chains**: The benefits of a supply chain early alert systems are most relevant for companies with complex supply chains containing many geographically dispersed steps and partners. Complexity increases the risk exposure of supply chains, and early intervention can prove pivotal.
- **Long lead times**: The longer the lead time for a specific resource is, and the more important it is, the more it will benefit the manufacturer to have early alerts if the shipment is delayed.
- **Disruption-rich environment**: The more disruptions in the environment of a company occur, the higher the incentives to develop an early alert system to mitigate the issues.
- **Dynamic environment**: Apart from disruptions, there might be other dynamics within and outside of the company (e.g., demand fluctuations, etc.) which result in difficulties. Such challenges can also be tackled with the implementation of early alert systems.

Questions to Ask

- How exposed is our supply chain to unpredictable disturbances—e.g., weather-related events?
- What are the critical steps in our supply chain?
- How consistent are the deliveries of our suppliers?
- What is the lead time for certain components/resources we use?
- How much do we have to rely on them?
- How well can current risk management solutions cope with supply chain disturbances?
- Is our current approach reactive or proactive?

3.4.7 Use Case 14: Supply Chain Collaboration Platforms

Description Along the Framework Categories

The global COVID-19 pandemic has had a serious impact on the global supply chains (SCs) of many industrial enterprises. Among the challenges are issues such as order cancellations from customers, decreased production volumes due to material and personnel shortages, inconsistent deliveries, as well as slow shipments (Taqi et al., 2020). Such challenges have led to the need for an increase in supply chain resilience (Koch et al., 2020). To achieve this, most companies are opting to increase visibility along their supply chain through enhanced collaborations and close relationships with their customers and suppliers (Harapko, 2020, 2021). However, it is important to note that the COVID-19 pandemic simply accelerated the demands for and implementation of a more transparent supply chain. (Harapko, 2021; Steinberg, 2019).

Cloud-based supply chain collaboration platforms, which connect the manufacturing company with the relevant players along the value chain, have the potential to meet the changing expectations described above (Alicke et al., 2016; Pfohl et al., 2016). Supply chain platforms compile technologies, and information within databases, which enable the players to collaborate by sharing their resources (Global eTrade Services (GeTS), 2020). By enabling a real-time exchange of information about production capabilities, the location of shipped orders, and specific customer demands—between different stakeholders along the supply chain—a general and insightful overview can be gained (Pfohl et al., 2016). This knowledge can either be utilized unprocessed or processed (with data analytics approaches) to improve supply chain performance. In their literature review, Duong and Chong (2020) identified several measures to promote collaboration that lead to competitive advantages for a company and its partners. These include, for example, collaborative planning, replenishment, better resource allocation, and identifying performance measures.

Four different generic types of supply chain-related collaboration platform configurations exist: collaborations between the company and its supplier, between the

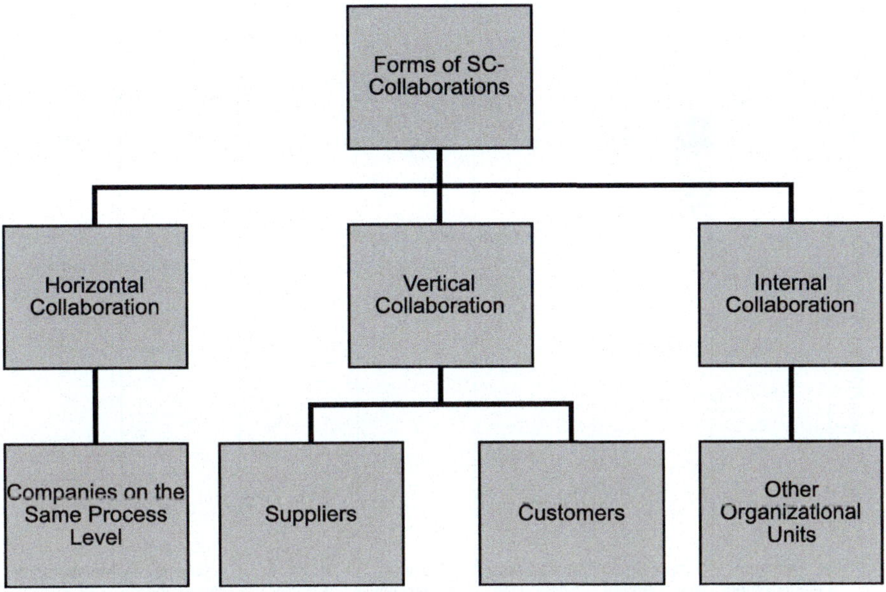

Fig. 3.28 Different Types of Supply Chain Collaborations. Own Illustration According to the Findings of Chan and Prakash (2012) and Frank et al., (2019)

company and its customers, between multiple companies of the same value-creation level, and between different organizational units (Chan & Prakash, 2012; Frank et al., 2019). The different types of collaboration along the value chain are depicted in Fig. 3.28. It should be noted that also a mix of different collaboration types can exist, such as a combination of horizontal and vertical collaboration forms (e.g., for the synchronization of shipments to maximize capacities) (Chan & Prakash, 2012).

Picture of the Model/Framework Explanation

As the name suggests, the collaboration with players along the supply chain is at the centre of this use case. By connecting, for example, an automotive manufacturer together with its original equipment manufacturer (OEM) supplier, increased visibility can be achieved. The cloud platforms often contain analytical elements, which allow the forecasting and modelling of different scenarios. Due to these analytical and forecasting capabilities, the use case is considered predictive (see Fig. 3.29).

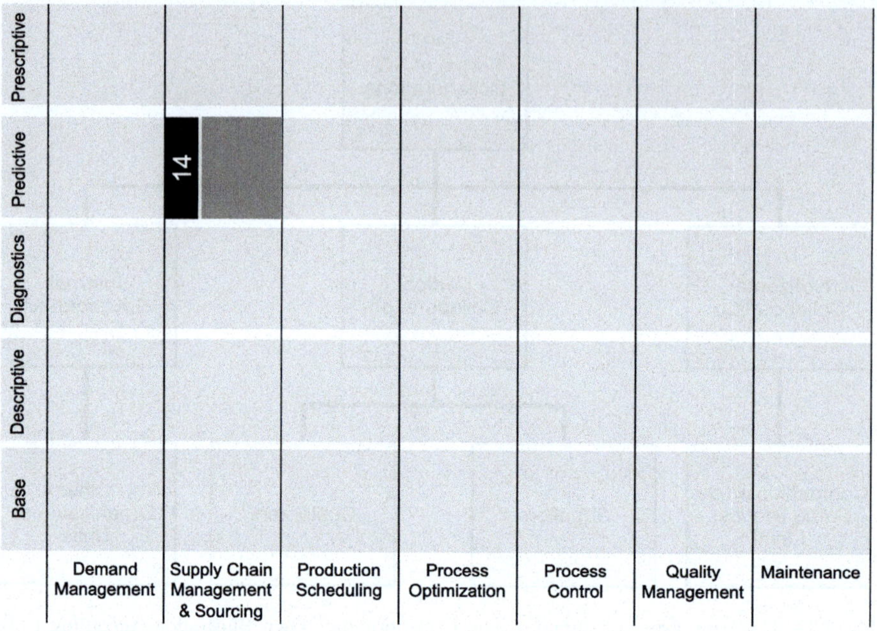

Fig. 3.29 Smart factory framework—use case 14

Relevant Technologies

The following technologies are considered relevant for this use case:

- Enterprise Resource Planning System
- Manufacturing Execution System
- Product Lifecycle Management System
- Cloud Computing
- Big Data Analytics
- AI, ML & Simulation
- Blockchain

Impact on Competitive Priorities

The impact of this use case on competitive priorities is shown in generic form in the Fig. 3.30. The specific impact depends on the context of the company.

Impact on flexibility	Impact on costs
• Enables flexible sourcing • Enables flexible delivery	• Reduces idle time due to prevention of stock-outs in the production • Reduces inventory (safety stock) due to higher transparency and more agile reordering processes

(continued)

(continued)

Impact on delivery • Facilitates the early identification of bottlenecks, supply chain disconnects, etc • Facilitates better and faster decision-making for supply chain and/or order adjustments • Improves delivered-in-full, on-time score	Impact on quality • Improves traceability of the products throughout the supply chain

Innovators/Exemplary Cases

Due to an increase in demand, Airbus Group had to rely on a perfectly working, visible supply chain. To achieve that, Airbus established in collaboration with other players of the European aerospace industry the supply chain collaboration platform AirSupply. The platform enables effective collaboration between suppliers and customers of different tiers throughout the whole supply chain with one single tool. Its high customizability ensures the provision of relevant data to each manager. The software allows the interactive fine planning of delivery quantities and dates, future demand planning, as well as an enhanced visibility for supply chain planning, ranging from short-term to long-term horizons. The order approach, for example, works as follows: Firstly, the demand for the upcoming week is calculated by the ERP of the customer. After the validation of the demand planner, the request is automatically sent to the supplier, who then can accept the inquiry, propose a different date or reject the request. Another service that is offered, is the vendor-managed inventory, where a supplier can plan their deliveries, based on the inventory as well as the forecasted demands of their customers in order to guarantee the desired stock level. Airbus themselves has integrated over 600 of their suppliers into the platform to execute the order and shipment processes more efficiently. Apart from the more efficient processes, possible bottlenecks can be detected early, which helps to achieve a more stable supply chain. Furthermore, the communication between the different organizations is fostered, which helps to create long-term synergetic relationships. (Platform Industrie 4.0, 2022; SupplyOn—The Supply Chain Business Network, 2013).

Exostar, a cloud platform provider, follows a similar approach. Its supply chain platform enables manufacturing- and supply chain transparency throughout multiple customer and supplier levels. The company operates with the data taken by the network organizations. The platform includes functions such as the tracking of deliveries, a real-time overview of supply and demand, the autonomous triggering of orders, the early detection of possible shortages and bottlenecks, and a reduction in the inventory. Furthermore, the software can detect mismatches between forecasts and commitments, late shipments, non-compliance, and inventory imbalances. The different applications within the supplier platform help to realize globalization strategies by decreasing risk while simultaneously increasing agility with enhanced visibility. (Opitz, 2013).

When and How to Apply This Use Case?

This use case is especially useful in the following situations:

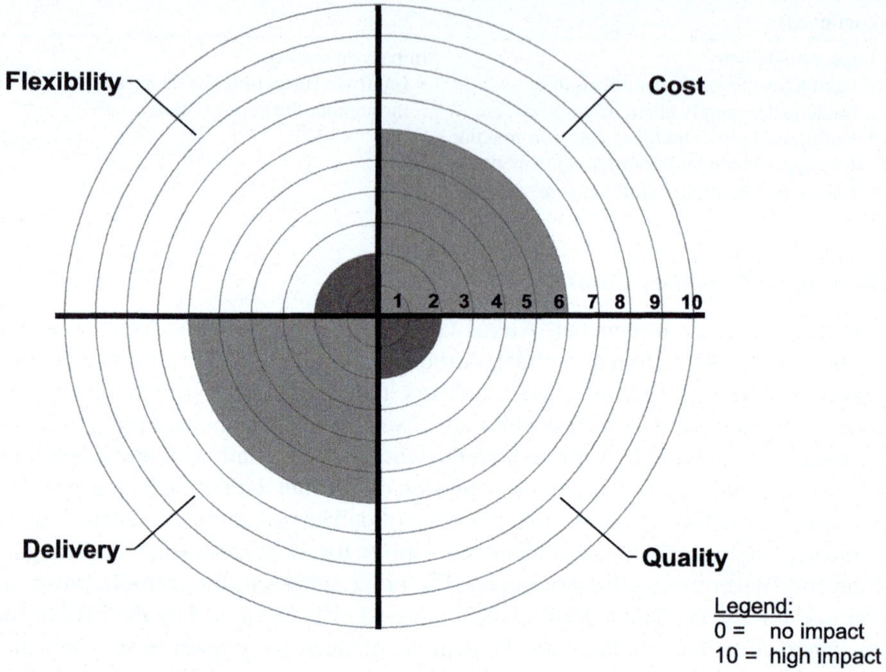

Fig. 3.30 Impact—use case 14

- **Supply chain inefficiencies**: Mismatches between supply and demand, rigid forecasts, etc., might lead to inefficiencies within the supply chain. Supply chain platforms can help to gain visibility and thus helps to avoid misconceptions and organize more efficient supply chain operations.
- **Disruption-rich environments**: The more disruptions in the environment of a company occur, the more it might be in favor of creating a collaboration network to mitigate the issues.
- **Dynamic environments**: Apart from disruptions, there might be other dynamics within and outside of the company (e.g., demand fluctuations, etc.) which result in difficulties. These challenges can also be tackled with the implementation of supply chain platforms.
- **Low supply chain resilience**: A low supply chain resilience means that internal and external disruptions have a grave impact on the performance of a supply chain. Creating visibility is the first step to enhancing resilience.
- **Long lead times**: The longer the lead time for a specific resource is (and the more important it is), the more it will benefit the manufacturer to have a constant overview of its shipment to enact early countermeasures if a delivery is delayed.

Questions to Ask

- What is the lead time for certain components/resources we use?
- How much do we have to rely on them?
- Do we have any secondary supplier to undertake any measures in case of any disruptions in deliveries?
- How consistent are the deliveries of our suppliers?
- How reliable are our demand forecasts? Do they have to be adjusted throughout the week?
- Is there already any existing supply chain collaboration platform for our industry? Are any of our customers/ suppliers already using such a platform?

3.4.8 Use Case 15: Autonomous Inventory Control

Description Along the Framework Categories

The objective of this use case is to automatize not only the monitoring of the stock but also the reorder/refill process. Every manufacturer faces the same problem when it comes to maintaining optimal stock. On the one hand, a company opts to keep sourcing and inventory management costs as low as possible. On the other hand, the stock should be high enough to deal with events such as unexpected orders or high scrap rates. This is especially important with C-parts, which are used in high volume and whose procurement costs are relatively large in comparison to their value (Thomas Warburton, n.d.). While keeping the right inventory is crucial, it is also important to monitor and resupply the production machines with the relevant raw and/or auxiliary material on the shop floor to ensure production flow. In this regard, two areas of application of this use case are emerging. The first is its application in the warehouse. Information about the stock is used together with historical and future data to calculate and model different scenarios. As soon as a certain minimum is reached, an autonomous order is placed with the supplier. The second application is its application directly on the shop floor. Sensors constantly measure the remaining amount of the material and autonomously trigger a refill order in the warehouse as soon as a certain threshold is reached.

Picture of the Model/Framework Explanation

Autonomous inventory control is used to automatically control the (future) quantity of certain stocks. Due to its high analytical properties as well as the autonomous execution of refill/resupply orders, this use case is classified as prescriptive (Fig. 3.31).

Relevant Technologies

The following technologies are considered relevant for this use case:

- Enterprise Resource Planning System

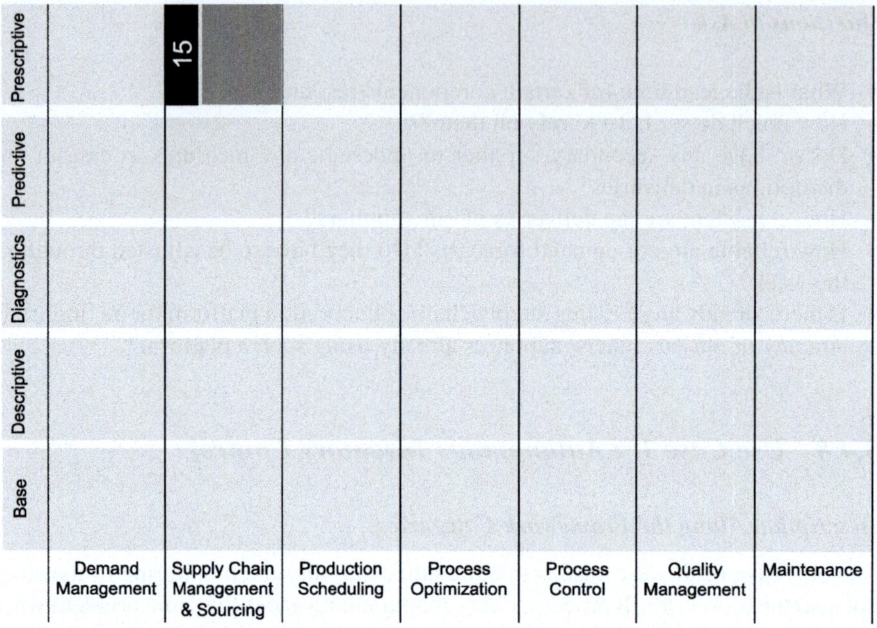

Fig. 3.31 Smart factory framework—use case 15

- Manufacturing Execution System
- Product Lifecycle Management System
- Cloud Computing
- Internet of Things
- Sensors and Actuators
- Track & Trace Technologies
- Vision Technologies
- Cyber-Physical Systems
- Big Data Analytics
- AI, ML & Simulation

Impact on Competitive Priorities

The impact of this use case on competitive priorities is shown in generic form in the Fig. 3.32. The specific impact depends on the context of the company.

Impact on flexibility • No direct impact	Impact on costs • Reduces monitoring and sourcing expenses • Reduces inventory due to better planning
Impact on delivery • Prevents stockouts and downtimes	Impact on quality • Improves traceability of the products throughout the supply chain

Innovators/Exemplary Cases

The global wholesaler Würth introduced a novel, digital Kanban container to prevent stockouts of C-parts. The first time the container is filled with the C-parts, the camera module records a picture to create a reference for a filled container. From then on, the camera constantly monitors the content of the containers, calculates the rough amount of the remaining C-parts, and sends the result to the inventory control system. As soon as the stock falls below a certain minimum limit, a reorder is automatically triggered at the supplier's side. The benefits include a real-time overview of the remaining parts, immediate recognition of unplanned requirements, a lower capital commitment, and direct workplace supply (Würth Industrie Service GmbH & Co. KG, 2022).

The logistics company Bossard implemented a similar solution. The company introduced containers with embedded weight sensors. These record the weight of each container, ascertain the remaining amount of the specific B- and C-parts and sends the data to the respective supplier. The supplier then sends, according to the specific demands, a resupply to the manufacturer. Furthermore, a digital label can be put in the container that visualizes the status of each container and when it needs to be resupplied. The solution leads to overall better handling of B- and C-parts which can lead to cost savings of up to 40%. This results in decreased storage and

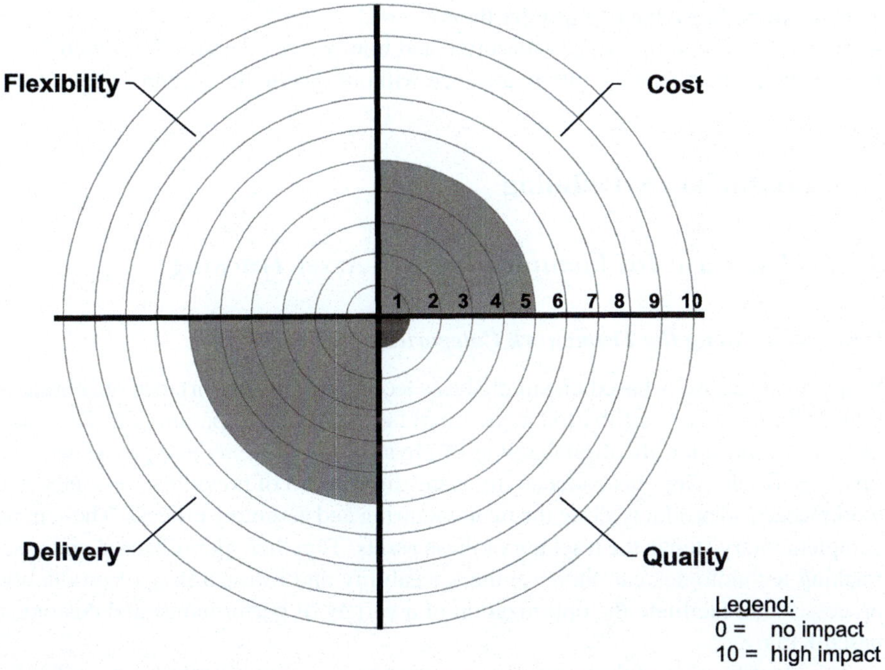

Fig. 3.32 Impact—use case 15

sourcing expenditure while simultaneously increasing the security of supply and creating higher transparency (van der Velden, 2022).

When and How to Apply This Use Case?

This use case is especially useful in the following situations:

- **Little room for additional stock**: A small warehouse with not much space for additional stock prevents bulk buying of additional material.
- **A multitude of different parts**: The greater the variety of parts a company uses for its products, the harder it is to maintain a clear overview of the stock of the different parts.
- **Hardship in monitoring and tracking the usage of certain materials**: The usage/consumption of some materials is hard to monitor, especially for smaller and more commonly used parts (C-parts), such as bolts, etc.
- **High sourcing expenses**: The ratio between sourcing expenses and their value is for some parts significantly higher than for other parts. To reduce these costs, an autonomous reorder/monitor system can help.

Questions to Ask

- Which parts do we use, and how often?
- How often do we have to reorder them?
- How is the remaining stock measured, and how are reorders/refills triggered?
- How many parts can we buy in advance without raising the inventory costs?

3.5 Production Scheduling

3.5.1 Use Case 16: Identification and Asset Tracking

Description Along the Framework Categories

While this use case is based on simple basis technology, its importance on a broader scale, in terms of its contribution to the smart factory vision, is paramount. The smart factory is built upon the digitalization of physical components and processes. The first step in achieving this connectivity is the digitization of previously disconnected workpieces and products along the manufacturing and assembly process. The second complementary step is the tracking of these assets. Together, identification and asset tracking technologies can increase the traceability and transparency of production processes and facilitate the optimization of a variety of performance and customer-based KPIs.

The identification of the various workpieces is made possible thanks to technologies such as QR-CODES and RFID chips. These labels allow the reading and storing of information, either in the system or on a chip, and facilitate the dynamic information exchange between the workpiece and its environment. If the workpiece does not already have a sensor embedded, the QR-Code, data matrix code or RFID chip can be attached either onto the workpiece itself (Werner, 2022), onto the skid (Wickel-Bajak, 2022), or onto the packaging (Spotsee, 2020). In this regard, RFID technology can be considered an enabler for the (industrial) Internet of Things (Wang, 2014), as it enables the ground level information exchange needed for vertical integration (Frank et al., 2019).

Once workpieces and other elements have been identified and integrated into the digital factory, they can be located and tracked for further optimization possibilities. The term "locating" describes the concept of tracking and recording the locations and movements of the subjects and objects within the manufacturing floor, such as the workers, tools, or products. The systems that aggregate and store these data are often referred to as identification and traceability systems. These automated systems enable the acquisition, control, and analysis of data about the operations, employees, and other resources (Nosenko et al., 2019).

Identification and tracking can be used as a foundation for various process improvements. The identification and automatic exchange of various information is an important prerequisite for the automation of processes. The tracking of the movements of subjects and objects involved in production allows for improved transparency. The recording and storage of this information can facilitate process optimization and identify waste in the context of lean management. Traceability can also be useful in the context of product defects and product recalls. Increased traceability makes it easier to carry out product recalls, as the faulty products can be traced to the individual consumers reducing expense and the likelihood of ensuing lawsuits (IQMS, 2019). The trend towards traceability is increasing in multiples sectors such as automotive or medical products.

Picture of the Model/Framework Explanation

The identification aspect of this use case involves the usage of a chip or similar object which communicates with its environment. The tracking aspect relates to the recording, storing and visualization of product related data. Neither of the applications relies on sophisticated analytics. For this reason, the use case is classified as descriptive (see Fig. 3.33).

Relevant Technologies

The following technologies are considered relevant for this use case:

- Enterprise Resource Planning System
- Manufacturing Execution System
- Product Lifecycle Management System
- Cloud Computing
- Long Range Wide Area Network
- 5G
- Internet of Things

- Sensors and Actuators
- Track & Trace Technologies
- Vision Technologies
- Big Data Analytics
- AI, ML & Simulation
- Wearables

Impact on Competitive Priorities

The impact of this use case on competitive priorities is shown in generic form in the Fig. 3.34. The specific impact depends on the context of the company.

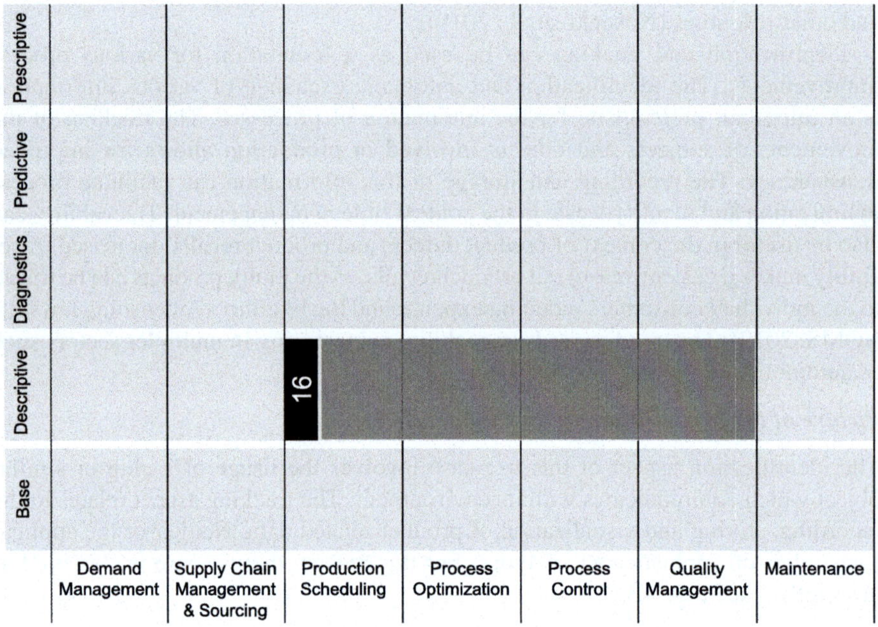

Fig. 3.33 Smart factory framework—use case 16

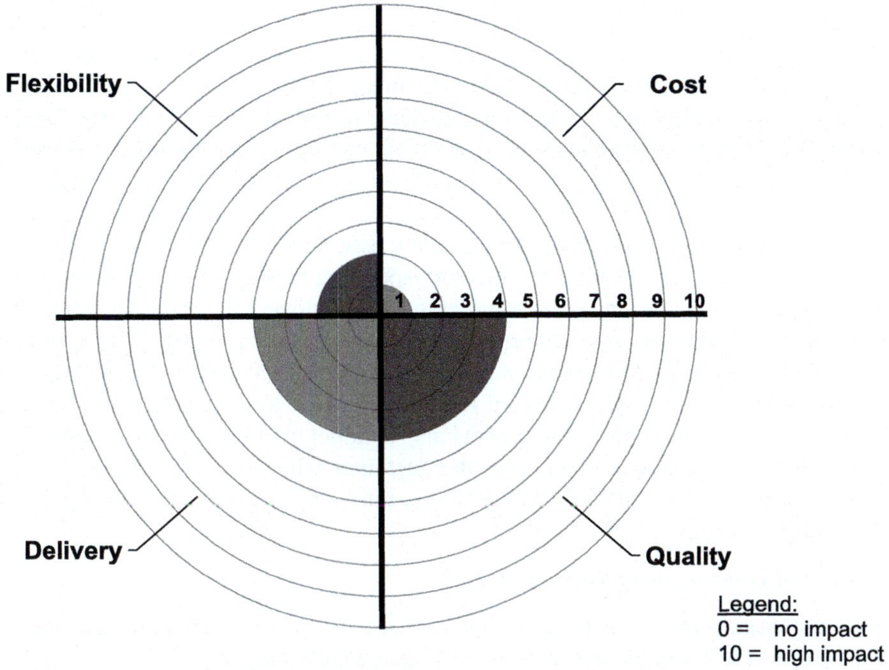

Fig. 3.34 Impact—use case 16

Impact on flexibility • Improves handling of multiple product variants	Impact on costs • Improves inventory accuracy and, therefore, better planning and scheduling • Increases the process efficiency • Avoids fines and lawsuits for non-compliance issues
Impact on delivery • Improves delivery-in-full on-time score due to better quality assurance and process control • Improves orders on time due to real-time tracking of workpieces	Impact on quality • Provides an efficient and reliable quality assurance instrument • Reduces scrap and reworking by screening-out • Facilitates traceability

Innovators/Exemplary Cases

The first success story stems from the German manufacturer Bosch, which introduced skids with RFID chips on an assembly line in one of its plants. The RFID labels enabled the connection of the skids with the process data, which resulted in multiple benefits. Thanks to the tracking capabilities, missing skids could be identified easily and inspected for defaults, preventing undetected faults from causing defects in the products. Additional data, such as part and tool ID, the characteristics of the screwdriving process, and the ID of the workpiece carrier used, could all be documented for each screwdriving process automatically, leading to a significant increase in transparency and traceability (Etemadi, 2016).

Lenze, a manufacturer of electronic devices, introduced novel software for the holistic tracking of each of its products and their assembly parts. By fully integrating and developing its manufacturing execution system (MES), the company was able to network all process steps. Every process step for each specific product was holistically displayed and stored. Every device, every assembly part, and every component could be identified and assigned to the corresponding process, test, and repair data. The real-time visualization function supported the employees in the monitoring of the processes, enabled early intervention and allowed for greater control of the product variations. The increased transparency helped to generate new knowledge about the various processes and thus increased reliability. Furthermore, once quality-related requirements were met, the system could also conduct bookings autonomously. The stored data was also used to support the customer when using the product. Ultimately, the increased traceability helped Lenze improve its quality and overall process performance (Riemann, 2022).

When and How to Apply This Use Case?

The reasons and conditions for applying this use case can be divided into reasons for identification technology and reasons for tracking technology.

Identification technologies:

- **High customization**: A high customization rate leads to higher complexity. Digital labels can help to decrease the process complexity by showing the required working steps and materials.
- **Large, expensive goods**: The larger and/or more expensive the goods are, the more it is in the interest of the manufacturer to document the working piece throughout its process.
- **Complex products/processes**: Complex products usually require multiple process steps, which all have to be done properly. Digital labels not only can help to guide the operator/assembler but also helps to document and validate the working steps done at each station.
- **Dangerous goods**: Goods that can become dangerous if they are defective, such as parts for aeroplane engines, must be fully traceable during the manufacturing process as well as their respective quality tests. Digital workpieces can facilitate the documentation processes.
- **Decrease of documentation costs**: Regular scanning of the labels of working pieces automates the documentation process and therefore decreases the documentation costs.

- **Automation of customizable products**: Digital labels, such as RFID, can help to achieve a manufacturing system where (more complex) customizable products are produced automatically.

Tracking technologies:

- **High product error rate**: A high product error rate can be traced back by comparing the products with their respective process parameters to gain knowledge and resolve the issues.
- **Significant negative impact if a product does not work as intended or breaks**: Products whose malfunctions can have significant negative consequences, such as profit setbacks or even the endangerment of lives, should be precisely documented on the shop floor so that the company is protected in the case of possible lawsuits.

Questions to Ask

- What kind of label should we use? RFID, QR-code, or something different?
- How should we mark the item without causing any later issues?
- Where should the label be placed? On the workpiece, the packaging, or the skid?
- In which size should we define our traceable resource unit (TRU)? Single products, batches, lots, or shipments?
- Which data do we want to record?
- What can we learn from these data?
- What do we want to use these data later for?
- How should the unique identification number be created?
- Which IT-system is the provider of this number?

3.5.2 Use Case 17: Shift Planning

Description Along the Framework Categories

The following use case, shift planning, describes the process of creating a schedule for the working times and responsibilities of the individual workers.

Shift planning has experienced multiple innovations since the more simplistic approaches of Excel spreadsheets. Nowadays, the planning process is supported in a variety of ways by different technologies. These enhancements range from the optional setting of individual preferences regarding shift times, through automated reports, to the possibility of operators organizing the shifts themselves according to the required manpower and working times for different tasks (Gerlach, 2022; Snap Schedule, 2020). Different processes require different competences on human side as well as on equipment and machines. To plan shifts, production lines need to be aligned with machines and worker competences. As more product variance and processes accrue the more complex the planning of the shifts will be. This use case provides guidelines on this topic.

Picture of the Model/Framework Explanation

Shift planning tools help the planner/scheduler by allocating the right workforce to the different tasks. In order to perform this act, only a rather simple software without any real analytical properties is required. For this reason, the use case is categorized as descriptive (see Fig. 3.35).

Relevant Technologies

The following technologies are considered relevant for this use case:

- Enterprise Resource Planning System
- Manufacturing Execution System
- SCADA System
- Cloud Computing
- Mobile Computing

Impact on Competitive Priorities

The impact of this use case on competitive priorities is shown in generic form in the Fig. 3.36. The specific impact depends on the context of the company.

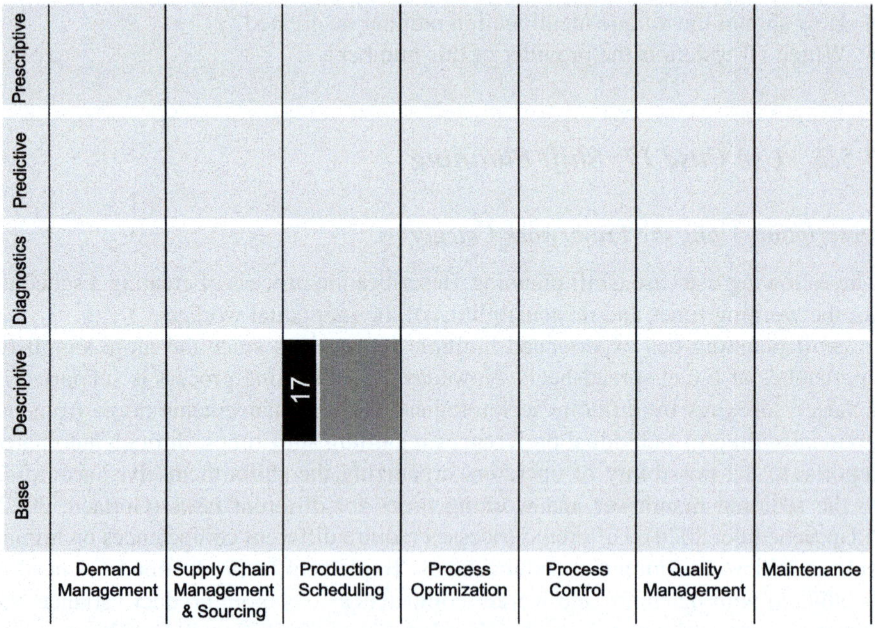

Fig. 3.35 Smart factory framework—use case 17

Impact on flexibility • Enables increased transparency and flexibility regarding changes in the production schedule	Impact on costs • Deploys the right number of workers with the right skills at the right time, increasing efficiency
Impact on delivery • Increases the efficient deployment of the workforce, helping to fulfil the orders in time • Reduces idle periods	Impact on quality • No direct impact

Innovators/Exemplary Cases

The project KapaflexCy is a good example of the utility of more interconnected and sophisticated shift planning tools, which let the operators plan their shifts, in due consideration of the required manpower and skills, by themselves (Gerlach, 2022). This helps to achieve higher responsiveness and flexibility regarding fluctuations in demand and therefore to prevent non-productive periods (Gerlach, 2022).

Other shift planning tools come with integrated dashboards which are connected to other devices (e.g., the time clock). Such setups create transparency and help to provide an overview about specific labour-related metrics such as the accruing labour costs (Homebase, 2022).

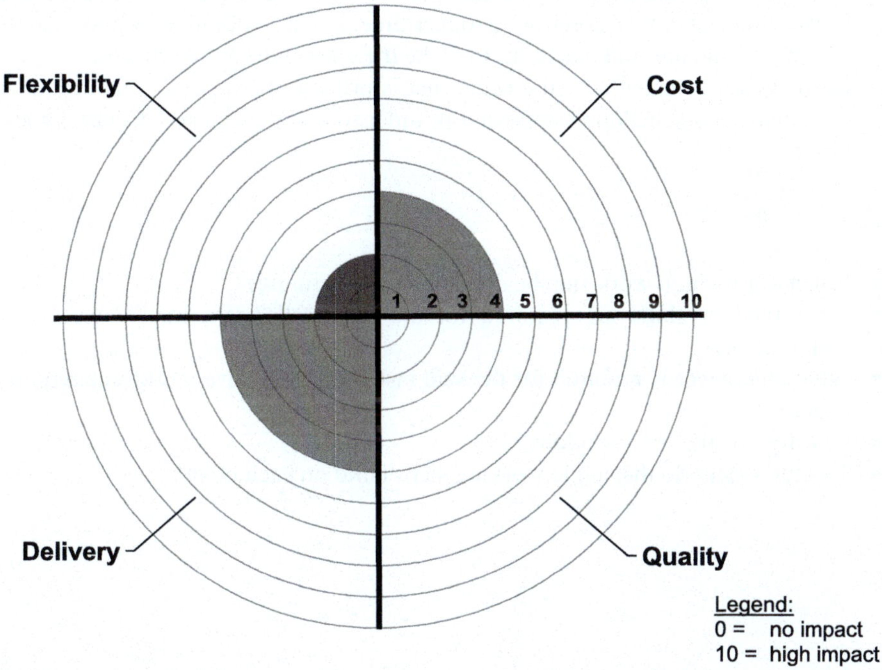

Fig. 3.36 Impact—use case 17

When and How to Apply This Use Case?

Although the determination of more or less flexible working times is conducted nowadays in practically every industry, there are a few factors that make the shift planning of employees especially important.

- **Tight time window**: Within a tight timeframe, planning errors which lead to bottlenecks are especially fatal.
- **Many Sequential and intertwined working steps**: The more process steps for a single product are required, and the more the different steps are intertwined and rely on each other, the more likely bottlenecks may occur if the coordination of humans and machines is not done properly.
- **Volatile markets and low storage possibilities**: The combination of volatile markets (which are, at least to a certain extent, predictable) and low storage options result in the need for time-efficient manufacturing. Efficient scheduling is the foundation enabling this.
- **Great variety of products**: A great variety of products is sometimes accompanied by smaller batch sizes, which leads to increased operational complexity. Therefore, the need for more efficient schedules, to manufacture the different product types, emerges.
- **24/7 operations**: In a permanent ongoing manufacturing facility, there is usually little room to work beyond a machine's allotted time.
- **Bad storage possibilities**: Raw materials and semi-finished goods (such as fresh fruits) often cannot be stored for a longer time. Because of this, it is important to coordinate humans and machines to make the operations more efficient.
- **Bottlenecks**: A manufacturing setup that is prone to bottlenecks often requires a sophisticated scheduling approach in order to avoid or make bottlenecks less likely.

Questions to Ask

- When must which work step be done (in the near future)?
- What kind of skills are required for the said tasks? (Qualitative workforce requirements)
- What manpower is required for the said tasks? (Quantitative workforce requirements)
- How tight is the time window?
- To what extent do the single working steps build on each other?

3.5.3 Use Case 18: Production Network Configuration and Coordination Assistance

Description Along the Framework Categories

The increased economic interconnectedness brought about by globalizing forces has pushed manufacturing enterprises to shift their production horizon from the site level to the network level. Accordingly, the operational management of the value chain has also shifted along these lines to consider the macro end-to-end perspective. A successful production network will possess a variety of different capabilities: accessibility of strategic objectives, thriftiness ability, manufacturing mobility and learning ability. These capabilities are best achieved through the active configuration and coordination of the network. Network configuration can be understood as "network dispersion" or, how "operations are distributed within a network" (Wenking et al., 2021). The focus revolves around strategic considerations as to how to structure and manage the individual entities of a network—i.e., a site might be established in or near a growing market to exploit its full potential. On the other hand, network coordination refers to the "internal management" of the various sites within the network (Wenking et al., 2021). In this aspect, the focus is on maximizing the collective value of the network—as opposed to the individual sites. Coordination efforts include network cost, resource efficiency, manufacturing flexibility and knowledge transfer.

This use case aims to assist these production network management processes with novel, digital solutions. The management of global networks should be understood as a key pillar of Industry 4.0 as it allows for greater flexibility and optimization activities in a firm's production (Veza et al., 2015). With regards to production network configuration, digital solutions have been developed to support companies in the dynamic evaluation and re-configuration of their global networks with simulation and modelling tools (Young et al., 2015). Current solutions, on the one hand, rely on analytical techniques to "intelligently evaluate the likely business and technical integration issues" associated with potential reconfigurations of the production network (Young et al., 2015). On the other hand, network coordination solutions focus on supporting operational decisions—i.e., the allocation of production capacity—to strike the right balance "between resilience and business efficacy" (Ogura et al., 2021). The technological approach here consists of aggregating and comparing the plant-level data (ERP, MES, WMS, etc.) of the different sites within the network (Ogura et al., 2021). This data is exploited through the use of simulation to determine the optimal conditions for specific orders and inform network planning decisions (Ogura et al., 2021).

Picture of the Model/Framework Explanation

The active coordination and configuration of a company's production network is a vital part of production scheduling and can lead to significant optimization opportunities. The technological solutions used in this use case focus on the statistical analysis of available plant and macro data to inform decisions. Accordingly, it is classified as diagnostic (see Fig. 3.37).

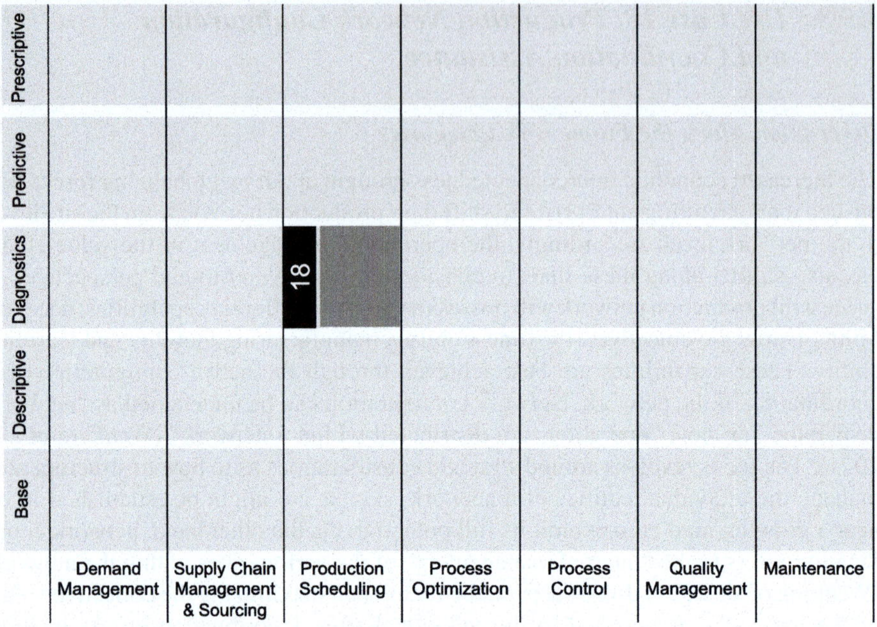

Fig. 3.37 Smart factory framework—use case 18

Relevant Technologies

The following technologies are considered relevant for this use case:

- Enterprise Resource Planning System
- Manufacturing Execution System
- Product Lifecycle Management System
- SCADA System
- Cloud Computing
- Internet of Things
- Big Data Analytics
- AI, ML & Simulation

Impact on Competitive Priorities

The impact of this use case on competitive priorities is shown in generic form in the Fig. 3.38. The specific impact depends on the context of the company.

Impact on flexibility • Facilitates the shifting of manufacturing activities—product lines/individual processes—across the network • Facilitates the shifting of production volume and resources across the network to meet changing demands	Impact on costs • Reduces operational and overhead costs—greater resource efficiency (leveraging economies of scale and of scope) • Reduces shipping costs • Reduces costs caused by international trade restrictions (tariffs) • Reduces costs associated with network expansion/investment • Reduces costs by identifying outsourcing potential
Impact on delivery • Decreases lead times in markets across the world • Ensures on-time delivery in markets across the world	Impact on quality • Enables knowledge transfer between sites and improves the quality of products being produced across the network

Innovators/Exemplary Cases

The FLEXINET initiative, developed by the European Commission in collaboration with the University of Loughborough, aims to provide a series of network configuration tools to help firms deal with the "complex, dynamic, and transient nature of global production networks" (Young et al., 2015). The initiative delivered the one-step platform FLEXINET, whereby manufacturing firms could make use of "business modelling tools, production network configuration tools and cost and risk analysis tools" to exploit potential business and investment opportunities (Young et al., 2015). Additionally, the platform enabled further network configuration support by providing compliance information. The platform was designed in such a way to integrate a variety of different applications and is expected to guarantee users significant cost savings and better business performance (Young et al., 2015).

With regards to network coordination, the Japanese conglomerate Hitachi developed a "value chain coordination service" to support the orchestration of the different resources at the disposal of a firm with a complex global production network (Ogura et al., 2021). The proposed solution is heavily customer driven. Given certain customization requests, the platform generates "draft plans and execution instructions" and matches these to specific plant equipment and capabilities (Ogura et al., 2021). Subsequently, simulations are run to determine the ideal planning parameters for a specific order—i.e., where to produce what. Data from all the individual sites within the network are analyzed, and firms with the greatest potential for successful delivery are allocated different production functions. The process is enabled by the network-wide integration of data and applications. In this way, the platform aims to facilitate network coordination by enabling business efficacy while maintaining resilience (Ogura et al., 2021).

A Swiss mid-size industrial company Baumann developed a simulation-based tool to constantly optimize the network configuration and drives operation decisions (Liao et al., 2021). The solution is based on an initial model used to understand the

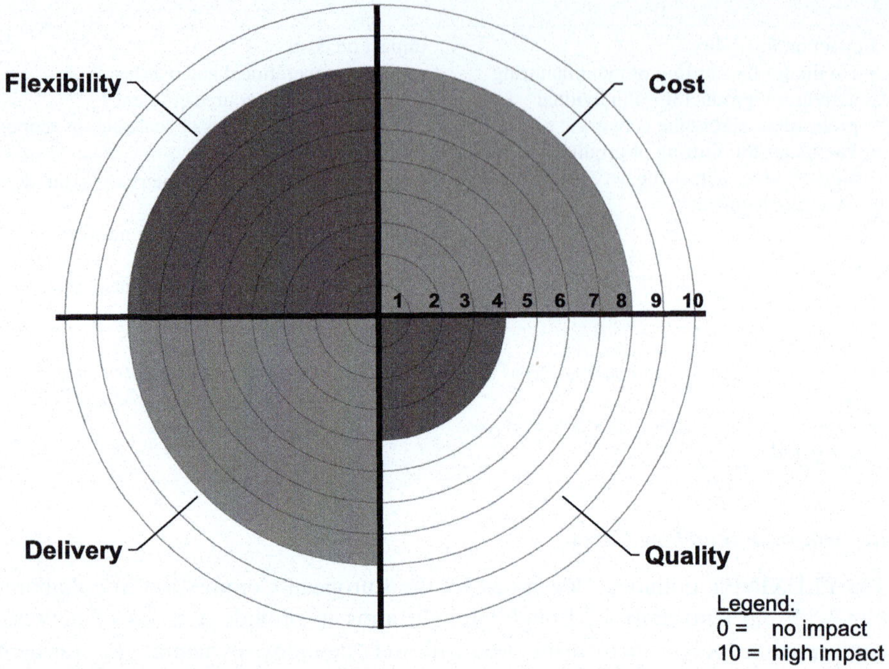

Fig. 3.38 Impact—use case 18

production network. All the relevant production data such as production, historical demand, cost, material cost, lead time, overall equipment efficiency, fulfilment rate or machine utilization create a strong analytical model with discrete event simulation. This simulation complements the actual demand forecasts to determine operational decisions on delivery time and plant utilization. It also helps support strategic decisions on production network decisions for the different manufactured products.

When and How to Apply This Use Case?

Wenking et al. (2021) emphasize that not all companies can benefit from production network optimization outright. The positive impacts of this use case on competitive priorities depend upon the competitive environment a company is operating in.

- **Hostile competitive environment**: Companies operating in markets characterized by hostile competition and falling margins benefit from the added benefits of network configuration and coordination, particularly in leveraging broader network resources to perform in local markets (Wenking et al., 2021).
- **Complexity of market**: Network optimization support will benefit companies supplying particularly complex markets in which there are many players and where sourcing and exchange occur on a global scale (Wenking et al., 2021).

Questions to Ask

- What is the size of our network? How many sites are currently in operation?
- Is there the potential and desire to expand the production network?
- Is there enough information available to inform strategic decisions regarding global production?
- Are the individual elements of the network disconnected or performing differently?
- Is the company exploiting the full potential of the network in terms of resource allocation efficiency and knowledge transfer?

3.5.4 Use Case 19: Scheduling Assistance

Description Along the Framework Categories

The following use case builds on the previous shift planning framework. Whereas shift planning focuses on the optimal deployment of the workforce regarding the pending and upcoming tasks, this use case also integrates the physical means of production such as conveyor bands, welding stations, and forklifts into the planning process. This additional dimension leads to higher complexity due to the higher administrative effort and therefore requires more sophisticated analytical tools to coordinate the human and physical means of production.

Production planning is associated with a number of challenges. For example, production environments are dynamic and uncertain (e.g., order arrivals, order cancellations, machine breakdowns), which causes computational difficulties for most optimization methods. The complexity of production systems caused by heterogeneous production processes (e.g., batching, sequence-dependent setup times, assembly) also lead to scheduling tasks being particularly difficult. In addition, production scheduling must take multiple competing objectives into account to ensure that the schedules produced are approved and applicable (Nguyen et al., 2017). Production planning has become necessary for manufacturing companies looking to take their production facility to the next level. Production planning refers to the allocation of resources, operations and processes required to produce goods and services. Companies can adjust their production schedules based on the availability of resources and customer orders. The goal of a production plan is to match customer needs with available resources while operating cost-effectively. If the production plan is not accurate and feasible given the available resources, there will be problems producing and delivering items to customers on time (PLANETTOGETHER, 2020).

Picture of the Model/Framework Explanation

In this use case, smart applications help the employee plan the schedule by showing the quantitative requirements for the desired output. On top of this, the use case helps

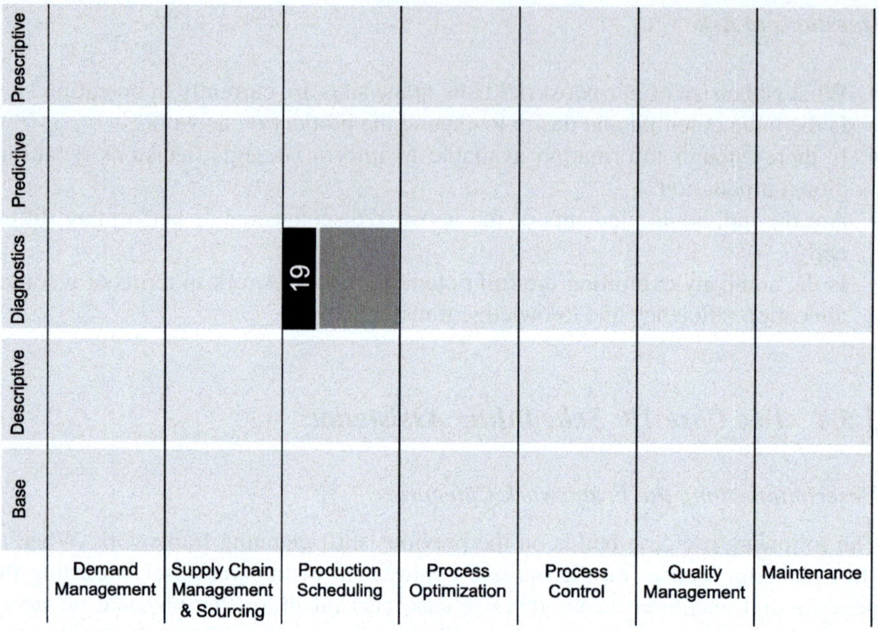

Fig. 3.39 Smart factory framework—use case 19

the optimization of production by tracking goods and showing real-time processing of different orders. As such, scheduling assistance is categorized as a diagnostic tool (see Fig. 3.39).

Relevant Technologies

The following technologies are considered relevant for this use case:

- Enterprise Resource Planning System
- Manufacturing Execution System
- SCADA System
- Cloud Computing
- Big Data Analytics
- AI, ML & Simulation

Impact on Competitive Priorities

The impact of this use case on competitive priorities is shown in generic form in the Fig. 3.40. The specific impact depends on the context of the company.

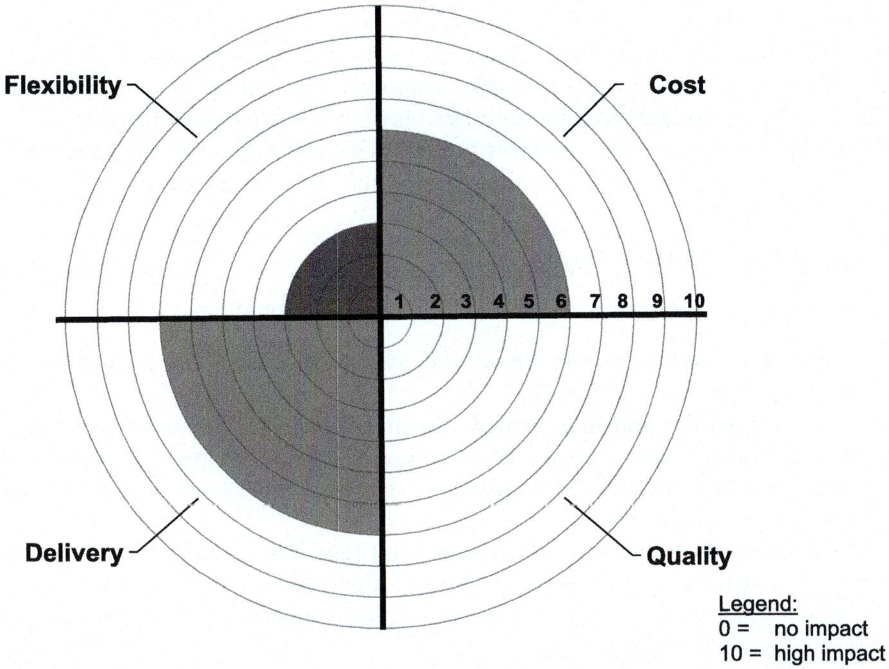

Fig. 3.40 Impact—use case 19

Impact on flexibility • Shortens response time to fluctuating order situation	Impact on costs • Increases productivity • Reduces idle periods due to the increased avoidance of bottlenecks, stock-outs etc
Impact on delivery • Shortens cycle times due to a better combination of workforce and machines • Provides a constant overview of the pending orders as well as their progress	Impact on quality • No direct impact

Innovators/Exemplary Cases

The planning system of a small handbag manufacturer can be used as an excellent example of an efficient planning system. Due to a sudden increase in demand, it became increasingly difficult to keep track of production and inventory. To solve this problem, the company decided to switch from Excel spreadsheets to a suitable planning software. This helped to improve the planning process by analyzing whether the materials needed for a specific order were available and by identifying future risks related to material-related shortages (Katana, n.d.). Such systems use underlying production planning data to help manufacturing plan, route, schedule, dispatch and execute production plans. These systems provide a comprehensive insight into production operations. Advanced planning and scheduling software has become a must for modern manufacturing operations as customers demand higher flexibility in a wider range of products with quick turnaround times. These systems help planners save time while providing more flexibility in updating ever-changing priorities. Such systems provide interfaces to ERP/MRP software to fill the gaps where these systems lack flexibility, accuracy and efficiency in planning and scheduling (PLANETTOGETHER, n.d.).

When and How to Apply This Use Case?

This use case is especially useful in the following situations:

- **Tight time window**: Within a tight time window, planning errors which lead to bottlenecks are especially fatal.
- **Many sequential and intertwined working steps**: The more process a single product requires, and the more these different steps are intertwined and rely on each other, the more likely bottlenecks may occur if the coordination between humans and machines fails.
- **Volatile markets and low storage possibilities**: The combination of volatile markets (which are, at least to a certain extent, predictable) and low storage options result in the need for time-efficient manufacturing. Efficient scheduling is the foundation of this.
- **Great variety of products**: A great variety of products is sometimes accompanied by smaller batch sizes, which leads to higher operational complexity. Therefore, the need for more efficient schedules to manufacture the different product types emerges.
- **24/7 operations**: In a permanent ongoing manufacturing facility, there is usually little room to work beyond a machine's allotted time.
- **Bad storage possibilities**: Raw materials and semi-finished goods (such as fresh fruits) often cannot be successfully stored for long periods. Because of this, it is important to coordinate humans and machines to make the operations more efficient.
- **Bottlenecks**: A manufacturing setup that is prone to bottlenecks often requires a sophisticated scheduling approach to avoid them, or at least make them less likely.

Questions to Ask

- When must which work step be done (in the near future)?
- What kind of skills are required for the said tasks? (Qualitative workforce requirements)
- What manpower is required for the said tasks? (Quantitative workforce requirements)

- How tight is the time window?
- To what extent do the single working steps build on each other?
- What machines/tools are required for which process steps?
- Are there any maintenance times for the machine? When?
- What other constraints and objectives exist?
- Is the duration of certain activities, lead times, and other factors constant or do they vary?

3.5.5 Use Case 20: Scheduling Decision Support

Description Along the Framework Categories

The use case scheduling decision support can be interpreted as an elaboration of the above-described scheduling applications into a more sophisticated approach. Scheduling decision support systems can be defined as "the application of available and suitable computer-based technology to help improve the effectiveness of managerial decision-making in semi-structured tasks" (Keen, 1978, cited in Ardjmand et al., 2016, p. 1). In the context of production scheduling, decision support systems can be classified into the following categories: communication-driven, data-driven, document-driven, knowledge-driven, and model-driven (Ardjmand et al., 2016). These different approaches may overlap within a specific solution system. Through the integration of several variables, for example, lead times, order execution time forecasts or required third-party input, the different orders are prioritized with the aim of meeting the required time frame and optimizing processes against a set of pre-defined objectives. A scheduling problem consists of three factors, which are as follows (Pinedo, 2016):

- Machine environment
- Processing characteristics
- The objective (group) which is/are to be optimized

The complexity of the decision problem to be resolved is determined by the three above-mentioned factors (Ardjmand et al., 2016).

Picture of the Model/Framework Explanation

To optimize a production schedule, a high degree of analytical insight is required. This is because the solution to a given issue requires the integration of various real and simulated datasets and scenarios, such as potential bottlenecks, possible post-processing requirements, scrap, etc. For this reason, this use case is classified as predictive (see Fig. 3.41).

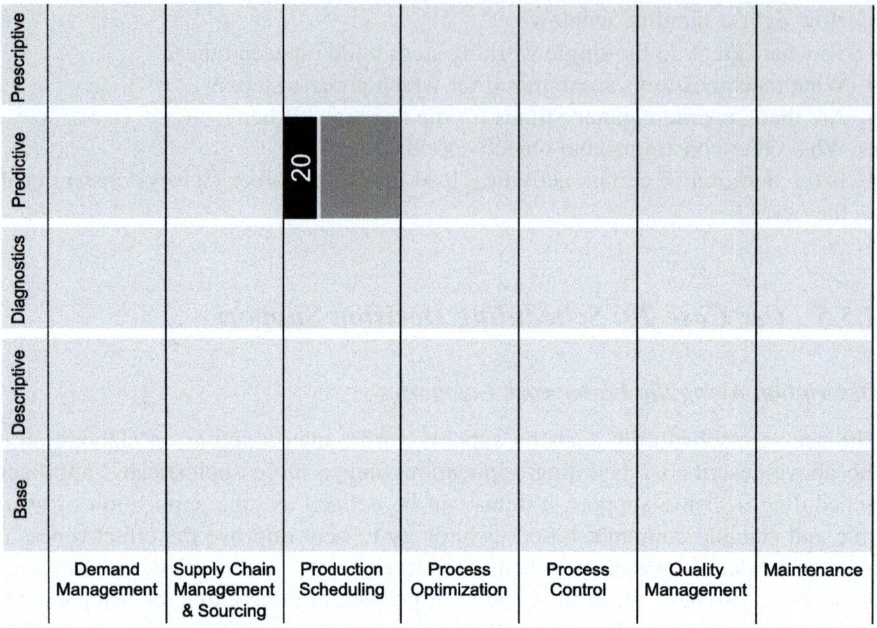

Fig. 3.41 Smart factory framework—use case 20

Relevant Technologies

The following technologies are considered relevant for this use case:

- Enterprise Resource Planning System
- Manufacturing Execution System
- SCADA System
- Cloud Computing
- Internet of Things
- Cyber-Physical Systems
- Big Data Analytics
- AI, ML & Simulation

Impact on Competitive Priorities

The impact of this use case on competitive priorities is shown in generic form in the Fig. 3.42. The specific impact depends on the context of the company.

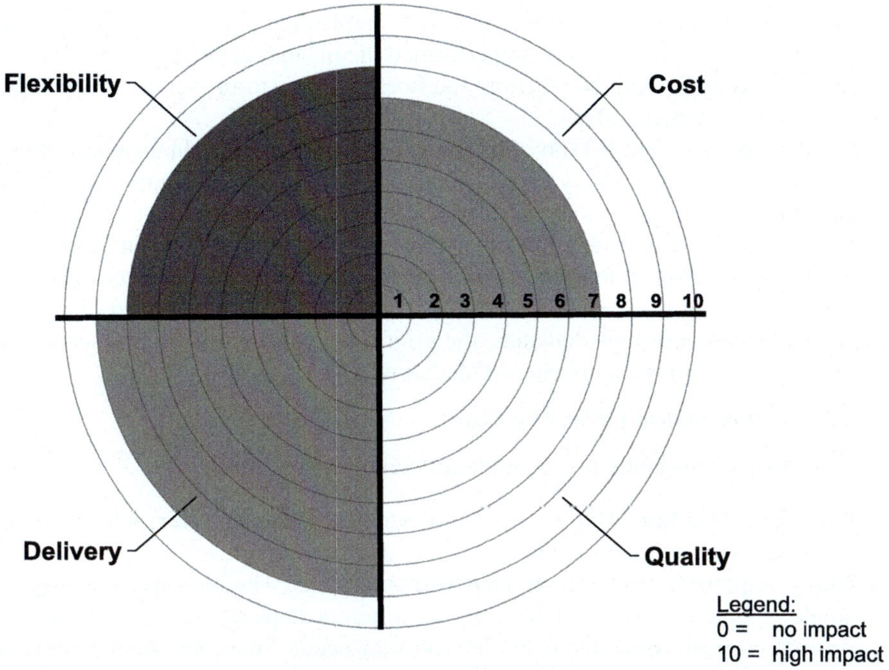

Fig. 3.42 Impact—use case 20

Impact on flexibility • Determines the best sequence for a number of (smaller) orders • Minimizes changeover times	Impact on costs • Provides new insights for optimizing planning and scheduling • Maximizes production capacity • Reduces cycle and idle times • Facilitates the introduction of standard operating times in order to optimize processes
Impact on delivery • Reduces production lead times • Improves on time in full (OTIF) • Facilitates the recalling and managing of production capacities	Impact on quality • No direct impact

Innovators/Exemplary Cases

A manufacturer in the pulp and paper sector was faced with issues such as multiple and shifting bottlenecks due to the configuration of the shop floor and the different process requirements of its products. Its make-to-order system, the variation in resource quality and general disturbances, led to increased complexity. To tackle these challenges while also maximizing manufacturing capacity, a model- and data-driven decision support system was proposed. After implementation, the results showed that the decision support system outperformed the manual planning approach in all main KPIs (backlog, setup, production, and stock), leading to an overall improvement of 7.4% (Figueira et al., 2015).

Goedert and Sekpe (2013) provides a second exemplary case. In their case study, a medium-sized electric utility company implemented a decision support system to schedule multiple projects with rapidly changing and sometimes conflicting objectives and multiple restrictions. The algorithm was also able to integrate and weigh the different decisions of board members, work managers, work planners, and work initiators according to a predefined decision hierarchy. The results showed significant improvements in cash flow demands, the prioritization according to political risks, and the allocation of the employees (Goedert & Sekpe, 2013).

When and How to Apply This Use Case?

This use case is especially useful in the following situations:

- **Tight time window**: Within a tight time window, planning errors which lead to bottlenecks are especially fatal.
- **Many sequential and intertwined working steps**: The more process steps a single product requires, and the more the different steps are intertwined and rely on each other, the more likely bottlenecks may occur if the coordination between humans and machines fails.
- **Volatile markets and low storage possibilities**: The combination of volatile markets (which are, at least to a certain extent, predictable) and low storage options result in the need for time-efficient manufacturing. Efficient scheduling is the foundation for this.
- **Great variety of products**: A great variety of products is sometimes accompanied by smaller batch sizes, which leads to higher operational complexity. Therefore, the need for more efficient schedules to manufacture the different product types emerges.
- **24/7 operations**: In a permanent ongoing manufacturing facility, there is usually little room to work beyond a machine's allotted time.
- **Bad storage possibilities**: Raw materials and semi-finished goods (such as fresh fruits) often cannot be successfully stored for long periods. Because of this, it is important to coordinate humans and machines to make the operations more efficient.
- **Bottlenecks**: A manufacturing setup that is prone to bottlenecks often requires a sophisticated scheduling approach in order to avoid them, or at least make them less likely.

Questions to Ask

- What process step needs to be done and when (in the near future)?
- What kind of skills are required for the said tasks? (Qualitative workforce requirements)

- What manpower is required for the said tasks? (Quantitative workforce requirements)
- How tight is the time window?
- To what extent do the single process steps build on each other?
- Which machine/tool is required for which process step?
- Are there any maintenance times for the machine? When?
- What other constraints and objectives exist?
- Are the durations of certain activities, lead times, and other factors constant or do they vary?

3.5.6 Use Case 21: Autonomous Job Scheduling

Description Along the Framework Categories

The previously presented use cases offer solutions regarding how intelligent systems can support production planning with scheduling activities. The use case of autonomous job scheduling goes a step further and describes the automatization of planning and scheduling tasks. With this approach, the role of the human employee shifts from a scheduling function to a supervisory and monitoring function.

To ensure truly agile production planning and scheduling, all relevant data have to be fed constantly into the tool so that it can react as fast as possible to disruptions. These disruptions can either be internal (e.g. unexpected breakdowns of machines) or external (e.g. product characteristics or order requirements) (Martins et al., 2020).

Picture of the Model/Framework Explanation

To realize an automated scheduling/system approach, highly analytical software is required. Such a tool needs to both monitor the processes continuously and simulate future scenarios to deploy schedules accordingly. Therefore, this use case falls under the prescriptive category (see Fig. 3.43).

Relevant Technologies

The following technologies are considered relevant for this use case:

- Enterprise Resource Planning System
- Manufacturing Execution System
- SCADA System
- Cloud Computing
- Internet of Things
- Cyber-Physical Systems
- Big Data Analytics
- AI, ML & Simulation

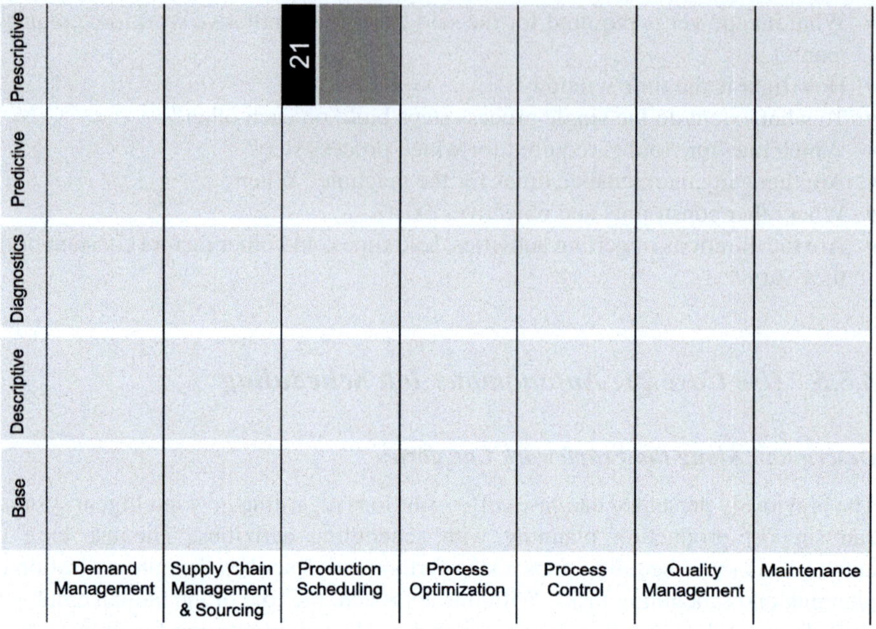

Fig. 3.43 Smart factory framework—use case 21

Impact on Competitive Priorities

The impact of this use case on competitive priorities is shown in generic form in the Fig. 3.44. The specific impact depends on the context of the company.

Impact on flexibility • Improves control over process complexity • Determines the best sequence for several (smaller) orders • Minimizes changeover times	Impact on costs • Optimizes planning and scheduling approach • Shortens cycle times • Optimizes the inventory • Lowers planning and coordination effort
Impact on delivery • Improves compliance with the delivery deadlines • Provides reliable statements about the time of the completion of orders • Reduces production lead times and cycle times automatically	Impact on quality • Enables consistent decision-making, leading to higher traceability • Automatically triggers production orders

Innovators/Exemplary Cases

A printing company implemented a new production control system as part of a larger digitalization project. With this new system, orders are directly sent to the internal system where an algorithm optimizes the production queue according to time requirements and other important KPIs. The orders are subsequently sent directly

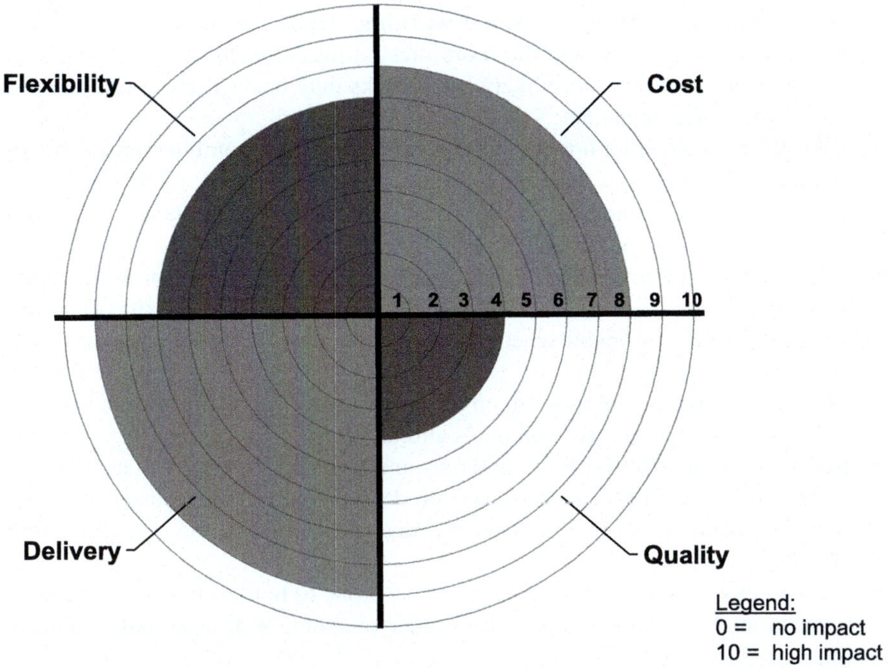

Fig. 3.44 Impact—use case 21

to the connected machines, where they are automatically executed. As soon as the production is finished, a label is printed for the packaging. With the integration of this autonomous scheduling system, the company increased its process quality by 25% and revenue by 30% while also decreasing costs by 15% and cycle times by 30% (BAIRLE Druck & Medien GmbH, 2016).

A provider of system solutions for in-house logistics planned to innovate its previously manually performed production planning and scheduling processes. The decision to innovate was brought about by increases in the complexity of the production and scheduling processes, stagnations that resulted in high inventory, and non-synchronized assembly processes, which led to long cycle times. By implementing an APS software, the production yield was increased by 100% within four years, processes were synchronized, and it became possible to automatically update and optimize the sequence for over 30,000 orders for the following two years (Asprova AG, 2016).

When and How to Apply This Use Case?

This use case is especially useful in the following situations:

- **Tight time window**: Within a tight time window, planning errors which lead to bottlenecks are especially fatal.

- **Many sequential and intertwined working steps**: The more process steps a single product requires, and the more these different steps are intertwined and rely on each other, the more likely bottlenecks may occur if the coordination of humans and machines fails.
- **Volatile markets and low storage possibilities**: The combination of volatile markets (which are, at least to a certain extent, predictable) and low storage options result in the need for time-efficient manufacturing. Efficient scheduling is the foundation for this.
- **Great variety of products**: A great variety of products is often accompanied by smaller batch sizes, which leads to higher operational complexity. Therefore, the need for more efficient schedules to manufacture the different product types emerges.
- **24/7 operations**: In a permanent ongoing manufacturing facility, there is usually little free room to work beyond one's allotted time.
- **Bad storage possibility**: Raw materials and semi-finished goods (such as fresh fruits) often cannot be successfully stored for long periods. Because of this, it is important to coordinate humans and machines to make the operations more efficient.
- **Bottlenecks**: A manufacturing setup that is prone to bottlenecks often requires a sophisticated scheduling approach in order to avoid, or at least make them less likely.

Questions to Ask

- What process step needs to be done and when (in the near future)?
- What kind of skills are required for the said tasks? (Qualitative workforce requirements)
- How much manpower is required for the said tasks? (Quantitative workforce requirements)
- How tight is the time window?
- To what extent do the single process steps build on each other?
- Which machine/tool is required for which process step?
- Are there any maintenance times for the machine? When?
- What other constraints and objectives exist?
- Are the durations of certain activities, lead times, and other factors constant or do they vary?

3.6 Process Optimization

3.6.1 Use Case 22: Incident Database

Description Along the Framework Categories

Manufacturing plants face disturbances, anomalies, and incidents, such as employee accidents or spontaneous machine breakdowns, daily. To find the root causes of such incidents, it is important to record and store related data in a central system for later analysis (Campos Pinto et al., 2013). This is where the incident database comes into the picture. An incident database is used to store various (incident) data concerning products, accidents, processes, machines, or even whole systems, on a central server. These data can then be used for further analyses, which eventually facilitate the prevention of those incidents or other product- or process-related issues in the future. An incident database is used to transform incident management into a more proactive activity. Incident management can be defined as the process to restore the processes of an organizational unit after a disruption in order to minimize its impact on the business' operations (Gartner, n.d.). The processed or unprocessed data of the incidents are gathered, stored, and shared at plant level, cross-plant level, or even cross-company level, for example, within an association of manufacturers (American Institute of Chemical Engineers, 1998). Such an approach helps other lines, plants, or operations to prevent similar issues.

Picture of the Model/Framework Explanation

Incident databases are the foundation for further analysis of historical incidents and other process data. This not only helps to prevent future incidents and accidents but also gives insight into various processes and further enables their optimization. Mostly, these databases do not analyze the gathered data by themselves. Therefore, they are considered descriptive (see Fig. 3.45).

Relevant Technologies

The following technologies are considered relevant for this use case:

- Enterprise Resource Planning System
- Manufacturing Execution System
- SCADA System
- Cloud Computing
- Edge Computing
- Track & Trace Technologies

Impact on Competitive Priorities

The impact of this use case on competitive priorities is shown in generic form in the Fig. 3.46. The specific impact depends on the context of the company.

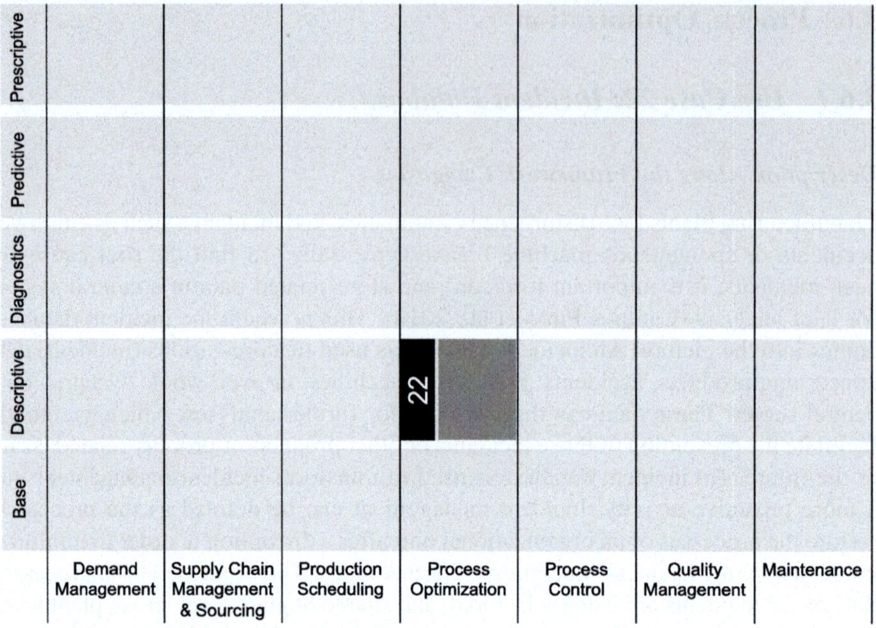

Fig. 3.45 Smart factory framework—use case 22

Impact on flexibility • No direct impact	Impact on costs • Prevents similar scenarios in the future which might be cost-related
Impact on delivery • Prevents similar scenarios in the future which might be delivery related	Impact on quality • prevents similar scenarios in the future which might be quality related

Innovators/Exemplary Cases

The following success story concerns the development and implementation of an incident database in the grid of a Portuguese transmission system operator. The main task of the application was the organization and management of a large set of data, which were needed for future memory record production. To successfully perform the incident analysis, information such as SCADA records, incident cause, identification records, and feeder fault records of the involved feeders were required to be fed into the tool. The benefits of this approach include a faster detection of anomalies within the system, easier and quicker access to incidents and interruptions, as well as better maintenance support (Campos Pinto et al., 2013).

When and How to Apply This Use Case?

This use case is especially useful in the following situations:

- **Complex systems**: Errors are often harder to identify in complex systems. Therefore, the recording, and later analysis of incidents and other error data, help to detect the underlying issues.
- **Error-prone processes**: The more often errors/incidents occur, and the greater the impact they have, the more advisable it is to record and analyze the relevant datasets to resolve those issues.
- **Dangerous working environment**: If the health and safety of the workers are endangered while working, the manufacturer should consider recording and analyzing the data in order to detect and minimize all future risks.

Questions to Ask

- Where do we believe there are correlations between different incidents and/or other product and process metrics?
- Is it possible to quantify these metrics?
- What data are important?
- How will the data be used/analyzed in the future?
- What do we want to learn from these data?
- How are these data reported?

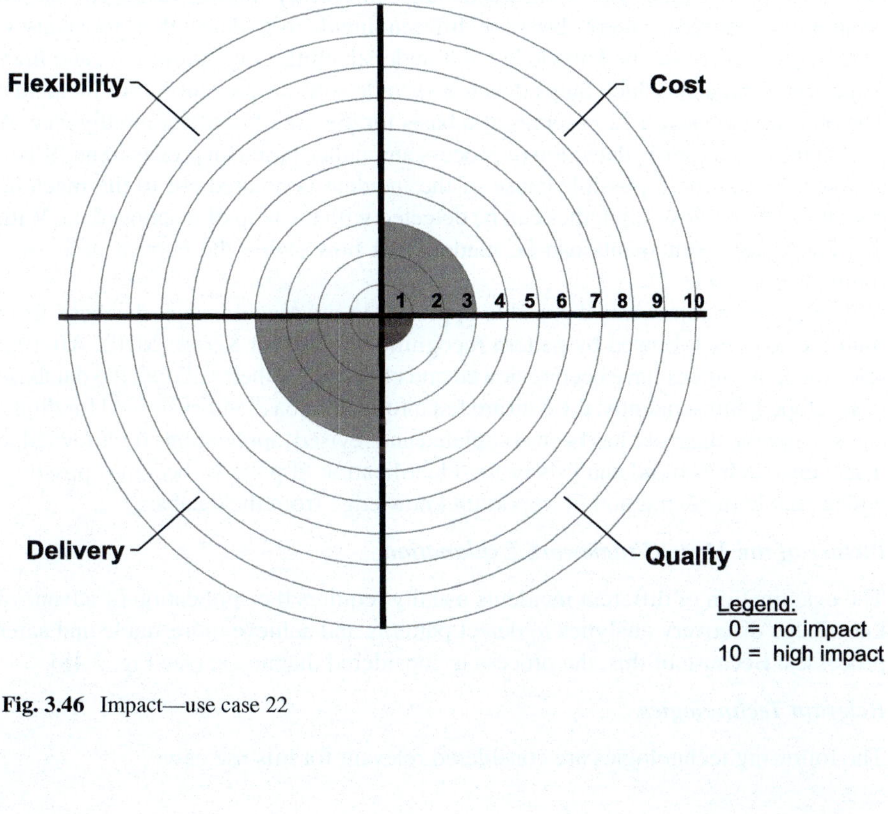

Fig. 3.46 Impact—use case 22

3.6.2 *Use Case 23: Incident Pattern Recognition*

Description Along the Framework Categories

According to the International Labour Organization (ILO), roughly 2.8 million people die every year due to occupational incidents, with most deaths caused by work-related diseases and accidents (SAFETY4SEA, 2019). Similarly, the number of yearly work-related accidents and deaths from work-related diseases amounts to 360 million and 160 million respectively (International Labour Organization, 2011). Apart from this human tragedy, such incidents also have a serious impact on a firm's performance and the economy as a whole. When looking at fatal working accidents, the manufacturing and the warehousing sector belong to the most affected industries (National Safety Council, 2022). Aside from the above-mentioned accidents, there are other non-dangerous incidents and disturbances that might occur in a plant and have a negative impact on the operational performance of a company. To resolve these issues, many firms have implemented an operational incident management system, which alerts the responsible people as soon as an incident occurs and further documents all relevant details of the issue.

As mentioned in the previous use case, an incident database can help to prevent operational incidents by enabling the storage of historical data about past incidents to build a knowledge database. To further enhance this utility, intelligent algorithms can be applied to identify patterns between different incidents to detect their root causes. This helps to increase the knowledge and understanding of the problems (and their respective systems), which helps develop specific rules to prevent future incidents. The built-up data acquisition serves as a basis for the use of artificial intelligence. A central topic is the early detection of process anomalies based on process data. When an anomaly occurs, a possible cause of the incident is pointed out to the machine operator. The process anomalies can be detected with the help of machine data. With this knowledge, statements can be made about how stable the current process is running (Fig. 3.47).

The use case represents a possible approach to creating an event database from multiple sources followed by pattern recognition (based on Keren, 2010). After all relevant data sources have been reviewed and checked for their validity, the database is developed. Subsequently, the data are fed into the database and examined for duplicates. After the database has been completed and revised, applications for knowledge discovery, such as trend analysis or the identification of patterns becomes possible, which enable the extraction of important knowledge from the database.

Picture of the Model/Framework Explanation

The examination of different incidents usually requires the application of advanced knowledge discovery analytics to detect patterns and achieve more stable and safer processes. Because of this, the process is considered diagnostic (see Fig. 3.48).

Relevant Technologies

The following technologies are considered relevant for this use case:

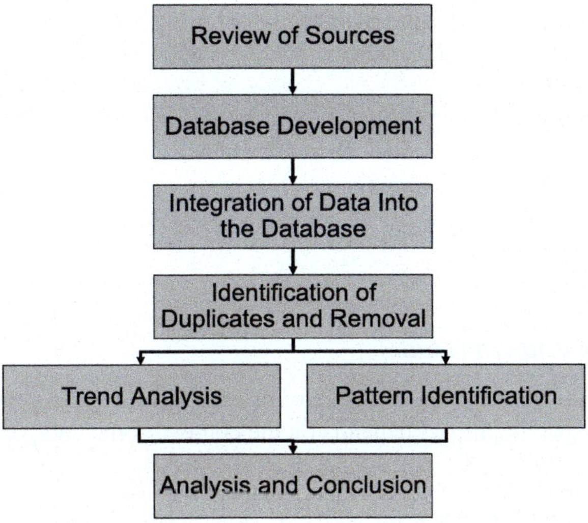

Fig. 3.47 Methodology for creating an incident database with pattern recognition (Keren, 2010)

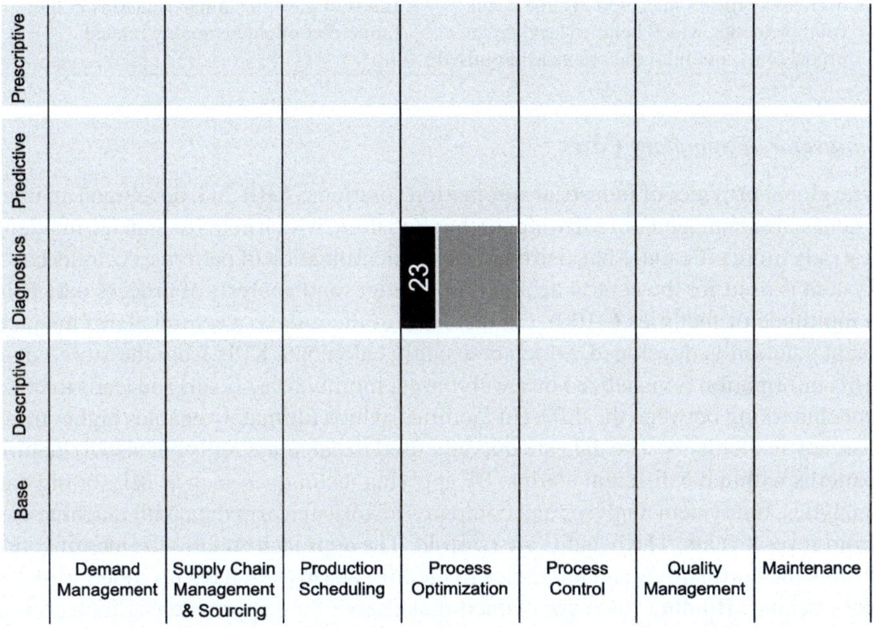

Fig. 3.48 Smart factory framework—use case 23

- Enterprise Resource Planning System
- Manufacturing Execution System
- SCADA System
- Cloud Computing
- Edge Computing
- Internet of Things
- Sensors and Actuators
- Big Data Analytics
- AI, ML & Simulation

Impact on Competitive Priorities

The impact of this use case on competitive priorities is shown in generic form in the Fig. 3.49. The specific impact depends on the context of the company.

Impact on flexibility • Helps to prevent similar scenarios in the future, which might be flexibility related	Impact on costs • Reduces costs due to fewer incidents • Improves safety and capacity utilization
Impact on delivery • decreases process interruptions that occur from incidents, which helps to keep up the productivity and fulfil the orders as promised	Impact on quality • Helps to prevent similar scenarios in the future that might be quality related

Innovators/Exemplary Cases

The global provider of industrial automation solutions, ABB AG, developed an intelligent alarm management software for the shop floor, which helps to manage incidents not only more efficiently but also enables the identification of patterns. A cloud-based system is used for the central acquisition, storage, and analysis of process data from a multitude of facilities (~100). On the basis of these data, a central alarm management solution is developed, which constantly calculates KPIs from the stored data. This information is visualized on a web-based, intuitive dashboard and leads to better benchmarking between the different facilities, which ultimately enables higher transparency in incidents. The data are not only used to calculate KPIs but also to identify patterns within the different alarms. By applying techniques such as ML for big data analytics, the system analyses and compares historical alarm data with machine data from several years. The benefits are twofold. The approach enables the identification of specific correlations and also helps to identify alarms that have no added value for the operator. Binding rules are defined that enable the suppression of unnecessary rules that would activate alarms to little purpose. This makes the management of the alarms more effective and less complex (ABB Research Center Germany, 2014; Treichel, 2022).

When and How to Apply This Use Case?

This use case is especially useful in the following situations:

- **Frequent incidents**: The more frequently alarms from incidents occur, the more a manufacturer might want to improve its incident management by integrating intelligence into its processes.
- **A multitude of different alarms**: Many different alarms may lead overtaxing employees, which could result in either a need for additional personnel or poor management of the said incidents.
- **High process complexity**: The more complex certain processes are, the more complicated it might be to trace back the different alarms to their root causes.

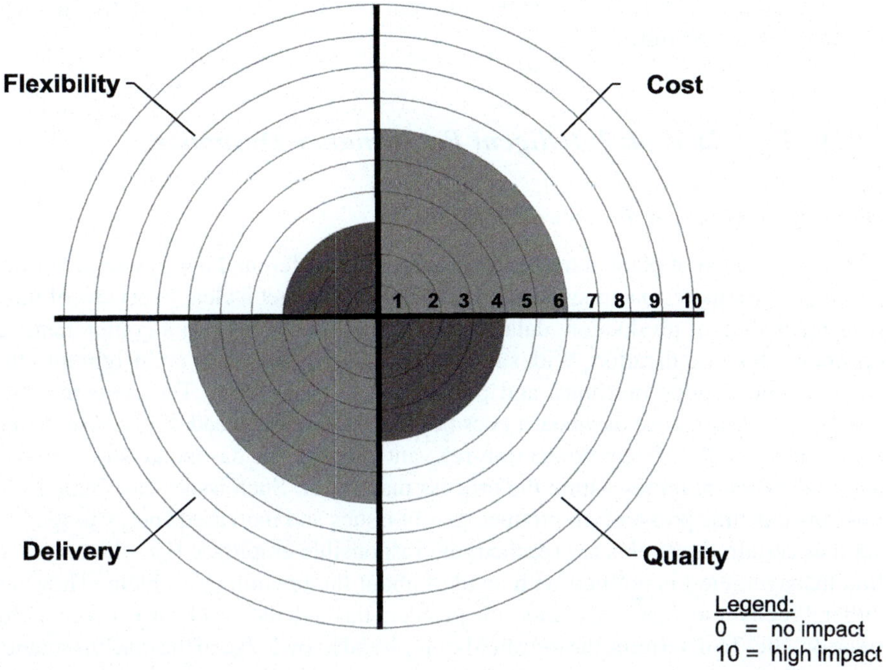

Fig. 3.49 Impact—use case 23

By enabling a pattern recognition software, correlations can be identified, which otherwise would have remained hidden.

Questions to Ask

- Where do we assume correlations between machine parameters and incidents exist?
- How can we analyze the above-mentioned correlations to identify patterns and help to mitigate those issues in the future?
- How can we securely aggregate the different data sources within a central incident management platform?

3.6.3 Use Case 24: Intelligent Performance Dashboard

Description Along the Framework Categories

Within a production plant, a tremendous amount of different data is generated from processes, products, machines, and other instances. To get a clear overview of their respective state of production, data are often brought together in a comprehensive dashboard for visualization. With such a tool, the employee gets an insight into the status of the various machines and products at any one time. The concept of an intelligent performance dashboard goes beyond this idea. Instead of just visualizing an aggregation of different production-relevant datasets, the dashboard shows where and how different issues within the process may occur. Such issues can range from possible machine problems to product quality issues and increased energy usage. As soon as certain thresholds are reached, the responsible employee (i.e., the operator, line manager, etc.) is notified with an alert about the upcoming problem. They can either directly intervene and resolve the problem through the dashboard or correct the issue on-site. To determine the thresholds, empirical knowledge of the machine manufacturer, organizational and empirical experience, and also expert knowledge can be integrated into the system. Intelligent performance dashboards build a foundation for faster and better decision-making, which leads to the optimization of processes, either in a specific factory or across multiple manufacturing sites. The relevant data are acquired by a multitude of different sensors, which are all connected to a central system.

Picture of the Model/Framework Explanation

The relevant data are recorded, stored, and processed centrally. They are evaluated in such a way that any trends and upcoming issues can be identified at an early stage. Although this could be interpreted as predictive analytics, the algorithm only uses rigid, predefined rules without analyzing more complex relationships between different instances in a big data cluster. Therefore, the use case is classified as diagnostical (see Fig. 3.50).

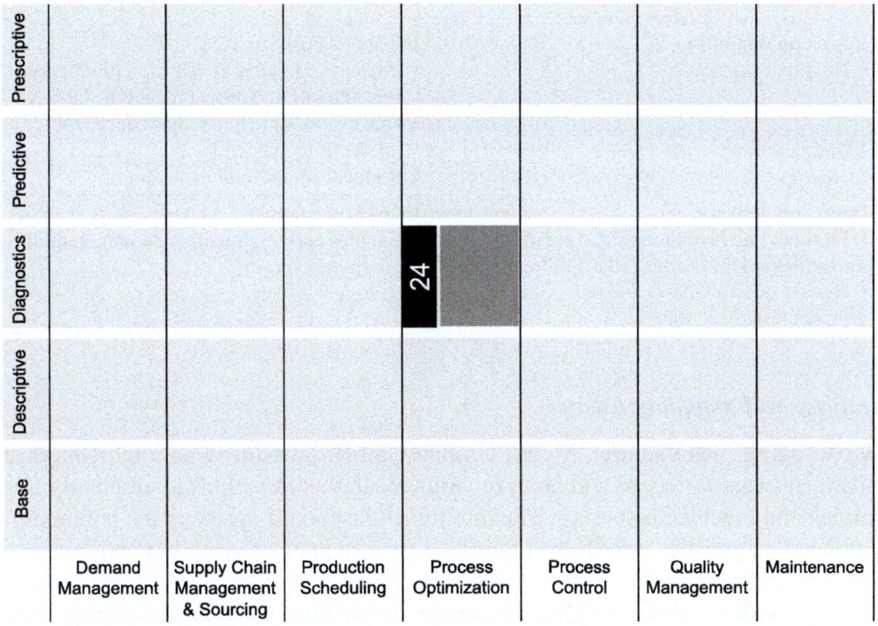

Fig. 3.50 Smart factory framework—use case 24

Relevant Technologies

The following technologies are considered relevant for this use case:

- Enterprise Resource Planning System
- Manufacturing Execution System
- SCADA System
- Cloud Computing
- Edge Computing
- Internet of Things
- Sensors and Actuators
- Track & Trace Technologies
- Cyber-Physical Systems
- Big Data Analytics

Impact on Competitive Priorities

The impact of this use case on competitive priorities is shown in generic form in the Fig. 3.51. The specific impact depends on the context of the company.

Impact on flexibility • No direct impact	Impact on costs • Prevents costs that result from production breakdowns and other process deviations • Detects possible flaws and reduces the production cycle time • Reduces monitoring expenditure
Impact on delivery • Prevents line breakdowns and scrap, leading to a higher and shorter order fulfilment (rate) • Detects possible flaws and reduces the production cycle time	Impact on quality • Enables early notification about potential quality issues

Innovators/Exemplary Cases

A packaging manufacturer, Mondi Gronau GmbH, introduced an intelligent dashboard software to record and analyze various datasets as well as to efficiently align human and machine resources. The intelligent dashboard opens up the potential for extended diagnostics—from the individual performance of machines and sensors to the broader process level. This enables the continuous improvement of processes. The approach is based on methods from the area of data analytics and pattern recognition. To initially detect anomalies, the available process data is fed into the model. Any anomaly that exceeds the predefined threshold is detected and observed in more detail. The patterns of these data are analyzed and respectively classified in a traffic light system, ready to be visualized by the operator/manager. Such a solution reduces scrap, downtime, energy consumption and enables high flexibility with large product variation. Additionally, the collected data can be further analyzed to find specific patterns which can lead to process performance prediction and thus enable more optimized processes. The continuous ongoing monitoring system and its early warning alerts help to predict possible process errors, which allows intervention in a proactive manner (Kohlert, 2022).

Yokohama Rubber, a manufacturer of tyres and other rubber-related products, set itself the goal to optimize the operational performance of its production sites across its global production network, as opposed to making disjointed individual adjustments to local sites. To this end, the company planned to improve its operational visibility, automate the data capturing, and enhance efficiency. While the company had the capabilities to analyze and monitor the processes of a single factory, the heterogeneity of its products across different sites made it challenging and time-consuming to make cross-sectional comparisons between the different manufacturing sites. As a solution, the company opted for the implementation of an intelligent dashboard software to aggregate and visualize the production statuses of the 15 biggest production sites. The introduced technology enabled real-time comparison of the production results, operating rates, and achievement levels between multiple plants at the same time. The solution was used to visualize processes and address bottlenecks in real-time while also improving inventory and throughput. The aggregated data was used to track specific issues within the processes and thus facilitated the performance of root

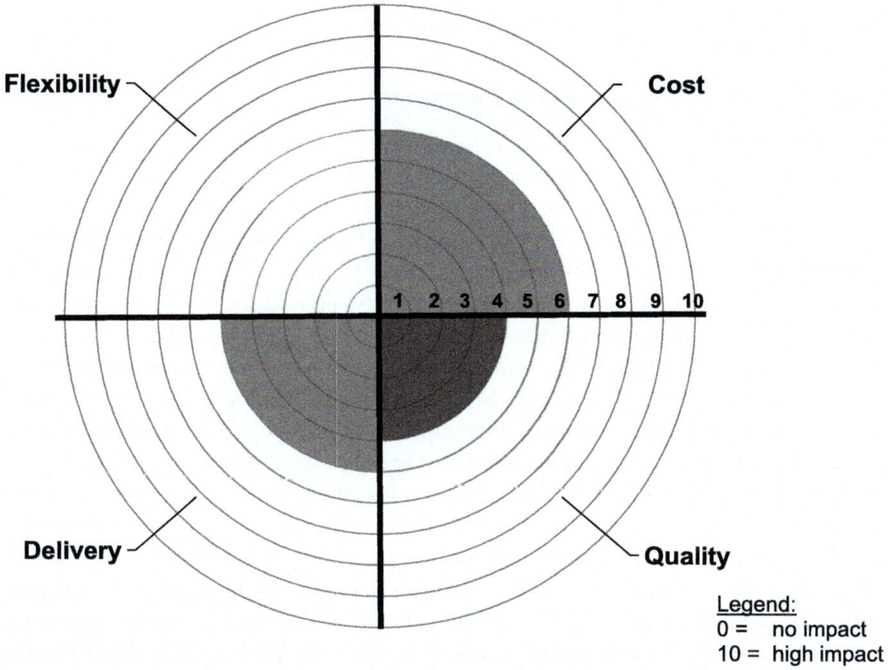

Fig. 3.51 Impact—use case 24

cause analysis. In summary, the advantages amounted to constant visibility, real-time information sharing, automated data retrieval, and aggregation of multiple plants that led to successful factory optimization on a global scale (Fujitsu, 2018).

When and How to Apply This Use Case?

An intelligent performance dashboard is recommended for every manufacturing plant. But there are scenarios where such software is especially useful.

- **Long process chains**: In longer, more complex process chains, possible issues can emerge in more places that need to be monitored continuously.
- **High monitoring expenditure**: Intelligent dashboards with an automated alert system can reduce the required effort to monitor a manufacturing plant.
- **Multiple different datasets of a similar source**: An intelligent performance dashboard can be used to make cross-plant comparisons to improve specific operational quality and efficiency.

Questions to Ask

- What do we want to learn from these data?
- How are thresholds and performance criteria determined?
- Are these valid across our production network?

3.6.4 Use Case 25: Process Mining

Description Along the Framework Categories

The continuous management and evaluation of processes within an enterprise is a fundamental component of maintaining time and cost-efficiency. Processes in the manufacturing environment, however, tend to be complex and span a variety of business functions that are rarely fully integrated end-to-end (Davenport & Spanyi, 2019). With the digitalization trends described throughout this book, operational processes "are becoming more and more intertwined with information systems", and vast amounts of process data are being generated (van der Aalst, 2016). At the same time, however, a disconnection remains between such data and process design and evaluation activities (van der Aalst, 2016). Process mining has recently emerged as a promising use case to address this gap. Especially processes in manufacturing and its support processes are predestined to apply process mining techniques.

The process mining approach is based on the analysis of "event data", which is usually collected by the information systems carrying out (or supervising) specific process operations. It is a set of techniques used for obtaining knowledge of and extracting insights from processes by the means of analysing the given data, generated during the execution of the process. This allows a continual comparison between what is happening in the physical processes to what should be happening according to a predefined process model (Burattin, 2015). This way, greater visibility into the performance of current processes is fostered, and data-driven improvements can be considered. The evaluation process enabled by process mining is described in Fig. 3.52, where process executions are iteratively compared to a theoretical model to measure performance (Burattin, 2015).

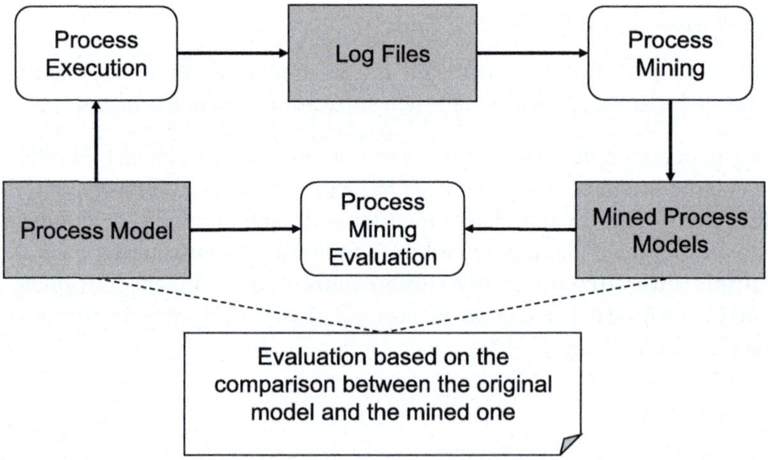

Fig. 3.52 Process mining for evaluation and optimization (Burattin, 2015)

Importantly, the event data (log files) are pivotal to this analysis as they relay the needed knowledge basis to "discover, monitor and improve processes" (van der Aalst, 2016). These data are acquired from traditional information systems. Alternatively, the evaluation and data-driven insights are produced by AI and machine learning—i.e., the application of AI algorithms to find causes for variation (Davenport & Spanyi, 2019).

Picture of the Model/Framework Explanation

Process mining approaches allow companies to link the vast process data at their disposal with process modelling and optimization activities. The analytical techniques are sophisticated but remain limited to the historical and actual event logs for calculations. For this reason, the use case is characterized as diagnostic (see Fig. 3.53).

Relevant Technologies

The following technologies are considered relevant for this use case:

- Enterprise Resource Planning System
- Manufacturing Execution System
- SCADA System
- Cloud Computing
- Edge Computing

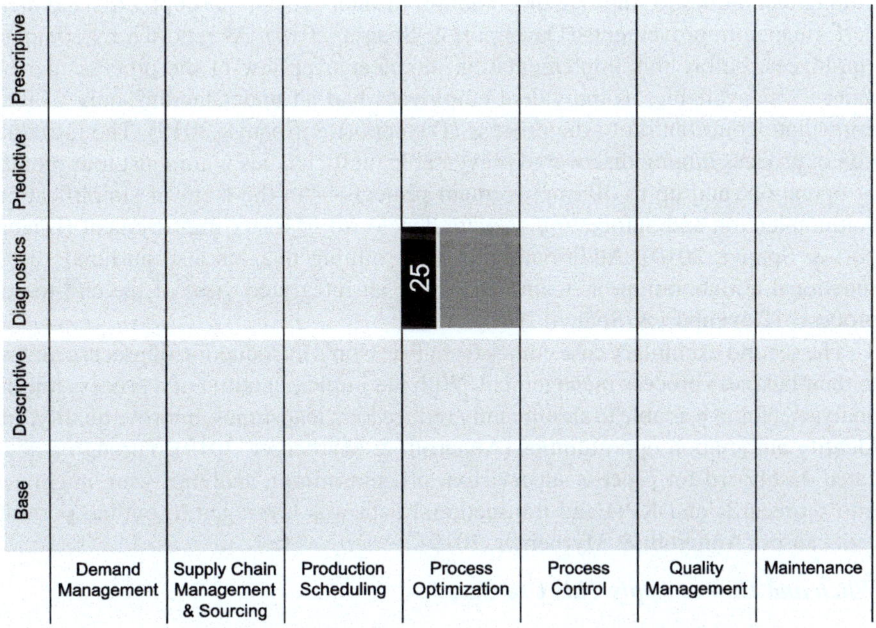

Fig. 3.53 Smart factory framework—use case 25

- Internet of Things
- Sensors and Actuators
- Cyber-Physical Systems
- Big Data Analytics
- AI, ML & Simulation

Impact on Competitive Priorities

The impact of this use case on competitive priorities is shown in generic form in the Fig. 3.54. The specific impact depends on the context of the company.

Impact on flexibility • Shortens the time needed to modify/adapt processes	Impact on costs • Prevents costs that result from production breakdowns and other process deviations • Reduces monitoring expenditure
Impact on delivery • Accelerates delivery of product/service due to process efficiency improvements • Detects bottlenecks and reduces process inefficiencies, decreasing lead time	Impact on quality • Improves quality due to data-driven process improvements—i.e., fewer defects due to fewer process deviations

Innovators/Exemplary Cases

The American chemical company Chemours turned to the implementation of process mining techniques to improve the visibility of their order-to-cash process and drive performance improvements (Davenport & Spanyi, 2019). As reported by company employees, before this implementation, no clear overview of the process' performance was available, as individual employees had an understanding only of their immediate contribution to the process (Davenport & Spanyi, 2019). The introduction of process mining discovered preventable inefficiencies within just four months of operation, and up to 40 improvement projects—"in the form of simplification, communication and automation"—were developed to foster improvement (Davenport & Spanyi, 2019). Additionally, process mining insights also nurtured "cross functional collaboration, as teams could see an integrated view of the end-to-end process" (Davenport & Spanyi, 2019).

The second exemplary case comes from Ericsson's introduction of process mining to their business process management. With the implementation of a process mining analyser, Nokia was able to significantly reduce long lead-times, improve quality, and identify automation opportunities (Anneroth & Myrberger, 2019). Through a dedicated dashboard for process supervision, process mining analytics were integrated into scorecards and KPIs and transactional data was leveraged to outline possible root causes (Anneroth & Myrberger, 2019).

When and How to Apply This Use Case?

This use case is especially useful in the following situations:

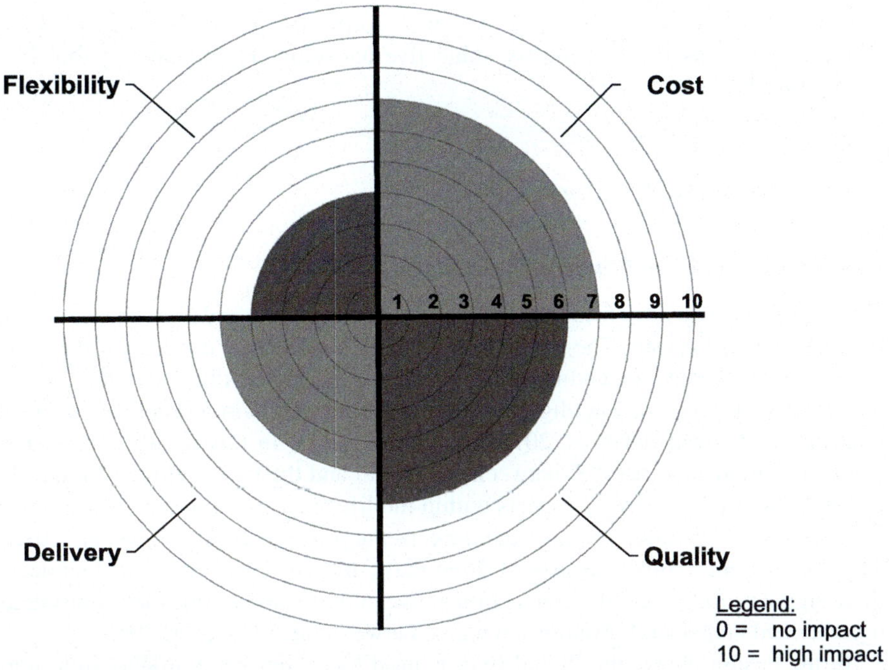

Fig. 3.54 Impact—use case 25

- **Complex products/processes**: Complex products usually require multiple process steps, which all must be done properly. Process mining not only can help to supervise and optimize the processes but also helps to document and validate the working steps done at each station via event logs.
- **High volatility of the processes**: The less predictable the outcomes of certain processes are, the more it is in the interest of the manufacturer to apply technology to increase visibility and improve current operations.
- **Cross-functional nature of process**: Processes that require the input of diverse functions within the organization can be harder to keep track of and evaluate accurately. Process mining increases visibility and integrates evaluation of cross-functional processes.

Questions to Ask

- Has a change in the environment led to the need for an update of a given process?
- Does the process that needs to be evaluated generate enough data to allow for process mining? Are all steps of the process logged?
- Is there end-to-end integration of all the relevant actors?

- Is there uncertainty with regards to the performance of a specific process? Are the conclusions reached mostly subjective and only interpretable by involved employees?

3.6.5 Use Case 26: Digital Twin

Description Along the Framework Categories

The digital twin can be considered one of the most prominent and promising concepts in smart manufacturing. The concept builds to a large extent upon the use case of the digital workpiece, extending the notion and potential of digitized pieces in the manufacturing process. The digital twin is a (connected) virtual replication of the real, physical world (Grieves, 2015; Qi & Tao, 2018). In this regard, the concept can be regarded as a "bridge" between the virtual and the real world, which has the potential to improve various aspects within the life cycle of a product—such as the product design, manufacturing, scheduling, or the product service (Tao et al., 2018). The digital twin mainly consists of three parts, the physical objects and subjects, their digital replicas, and the connection between those two worlds that brings them together and enables information exchange between them (Grieves, 2015).

In the design phase, the digital twin is used to engineer the product in a more efficient and precise way. Instead of iteratively calculating, building, testing, and improving the model of the product with its properties, all physical world interactions can be simulated virtually. Therefore, the product specifications, as well as its manufacturability, can easily be forecasted, tested, and validated (Qi & Tao, 2018; Schleich et al., 2017).

In the manufacturing process, the digital twin can perform a variety of different functions. On the one hand, it is used to ensure and improve one or multiple objectives, such as product quality, process stability, or overall efficiency (Qi & Tao, 2018). On the other hand, digital twins of production machines, tools, etc., can also help to proactively identify their degradation (Qi & Tao, 2018). To enable the digital twin to carry out complex measures, the whole shop floor needs to be digitized and all relevant elements of the system—operators, tools, machines, and material—as well as their behaviour, rules, and other non-physical aspects need to be integrated into the model (Qi & Tao, 2018; Tao et al., 2017).

Picture of the Model/Framework Explanation

The digital twin builds the general framework for further improvements in various process steps, such as process optimization, process control, quality management and maintenance (Qi & Tao, 2018). In this use case, the different objects and subjects, along with their properties, are mapped and mirrored. This builds the foundation for various high analytical smart manufacturing processes, and, therefore, it is classified as predictive (see Fig. 3.55).

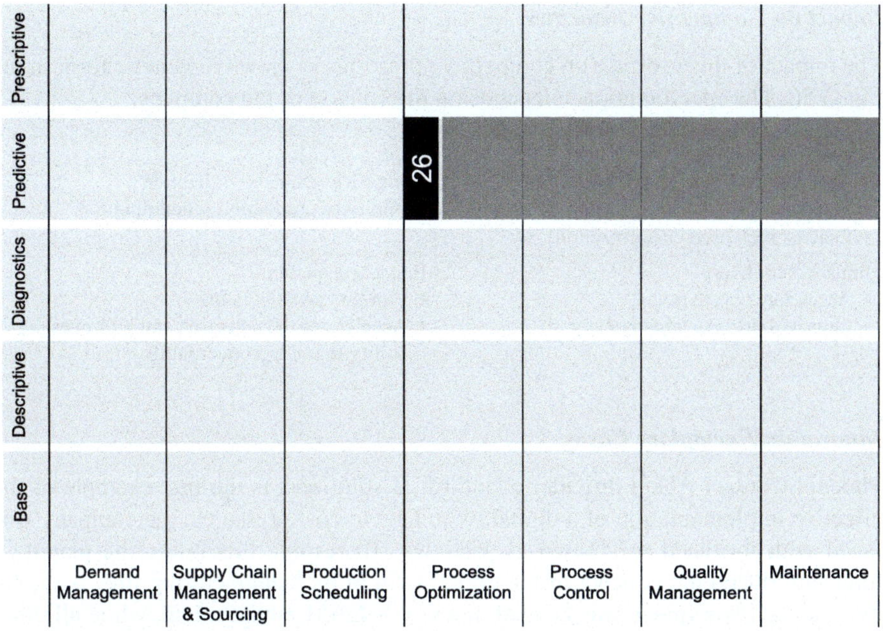

Fig. 3.55 Smart factory framework—use case 26

Relevant Technologies

The following technologies are considered relevant for this use case:

- Enterprise Resource Planning System
- Manufacturing Execution System
- Product Lifecycle Management System
- SCADA System
- Sensors and Actuators
- Track & Trace Technologies
- Cyber-Physical Systems
- Big Data Analytics
- Cloud Computing
- Edge Computing
- M2M Communication
- Internet of Things
- AI, ML & Simulation
- Wearables
- Virtual Reality

Impact on Competitive Priorities

The impact of this use case on competitive priorities is shown in generic form in the Fig. 3.56. The specific impact depends on the context of the company.

Impact on flexibility • Accelerates the development of new products • Enables increased customization	Impact on costs • Improves capacity utilization • Improves production scheduling
Impact on delivery • Accelerates cycle times • Improves process transparency	Impact on quality • Prevents product faults • Accelerates the detection and improves the ability to trace product faults

Innovators/Exemplary Cases

Phoenix Contact AG, a provider of industrial solutions, is the first example of the effective implementation of a digital twin. Due to cost pressures, the company was faced with the need to shorten cycle times. To resolve this issue, the manufacturer introduced the concept of the digital twin in its manufacturing line. A digital twin of each workpiece was created in a CAD-/CAM-environment, while all other required technological process data were automatically transferred to the process

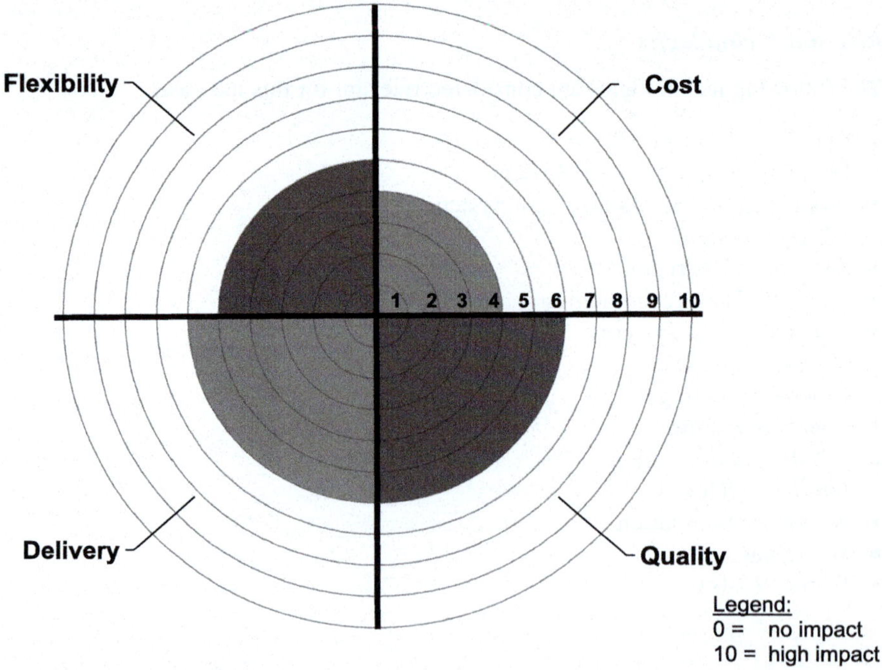

Fig. 3.56 Impact—use case 26

control system of the production line. In the manufacturing process, the physical assembly pieces were automatically identified with RFID and went through the single working stations autonomously. Their virtual counterparts were connected, via real-time monitoring, with the relevant machines and software. This approach replaced the locally automated process steps with a process-mixed automation approach. The benefits included 30% shorter cycle times, an increase in the capacity utilization by 20%, a semi-automated creation of CNC programs, and the possibility to constantly monitor the working steps in real-time. Furthermore, the new system enabled the remote control of machines, mechanized various manual working steps and increased the traceability of processes dramatically (Flachmann, 2022).

The second example is provided by MAN Truck & Bus AG, which implemented the concept of the digital twin together with an SAP module in order to increase transparency in its processes. In addition to the classic ERP functions, SAP maps all logistics processes for the provision of parts on the assembly line. This, together with the control loops within a simulation, enables a holistic mapping of the system behavior in a digital twin. Through the additional integration of RFID chips on material containers, the communication from the production process to the planning system became highly automated. The benefits included a significant reduction in errors in the commissioning process as well as the optimization of the logistic processes (Erk, 2022).

When and How to Apply This Use Case?

This use case is especially useful in the following situations:

- **High customization**: A high customization rate leads to higher complexity. Digital twins can help to guide the different ordered working pieces (semi-) autonomously through the processes.
- **Complex products/processes**: Complex products usually require multiple process steps, which all have to be done properly. Digital twins not only can help to control and guide the working pieces through the processes but also help to document and validate the working steps done at each station.
- **Decrease of documentation costs**: Autonomous recording and mapping of the objects and subjects within the manufacturing floor help to reduce documentation costs.
- **Need to increase process transparency**: Digital twins improve insight into the different processes

Questions to Ask

- Which assets are involved in the manufacturing process?
- Which subject matters are involved in the manufacturing process?
- Which properties, rules, etc., do the abovementioned subjects and objects have?
- What data are important?
- How will the data be used/analyzed in the future?
- What do we want to learn from such data?

3.6.6 Use Case 27: Process Performance Prediction

Description Along the Framework Categories

The following use case seeks to predict the future properties or states of ongoing manufacturing processes. Predictions regarding the final states and outcomes of a process are made at a predefined point in the process. This point is called a checkpoint (Marquez-Chamorro et al., 2017). Multiple variables have an influence on the process, such as the environmental temperature, the degradation of the different tools, the quality of the raw materials, and many other factors. To forecast the outcomes of the process, a calculation model based on past logs and other historical data is created. (Marquez-Chamorro et al., 2017; Verenich et al., 2019). Sensors record the relevant properties and send them to the calculation model, where likely outcomes are determined.

The predictions can either be used to alert the operators about upcoming issues or can be used as a foundation for further process improvements. The predicted values can either be numeric, Boolean, or categorical (Marquez-Chamorro et al., 2017). An overview of the different prediction types is provided in Fig. 3.57. Depending on the value to be predicted, the practicability of the models may vary. A model which is based on decision rules might be useful for the prediction of certain risks, whereas a regression tree is used to determine output values (Marquez-Chamorro et al., 2017). The more relevant variables are considered, the more exact and reliable the process predictions (see Fig. 3.57).

The general structure of an approach to predict process properties is depicted in Fig. 3.58. In the first phase, event logs—optionally a process model—as well as other

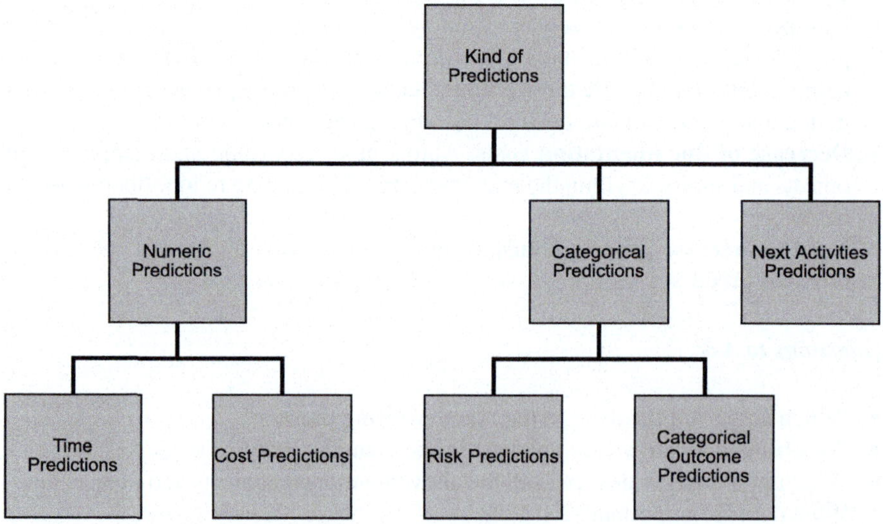

Fig. 3.57 Prediction types (Di Francescomarino et al., 2018)

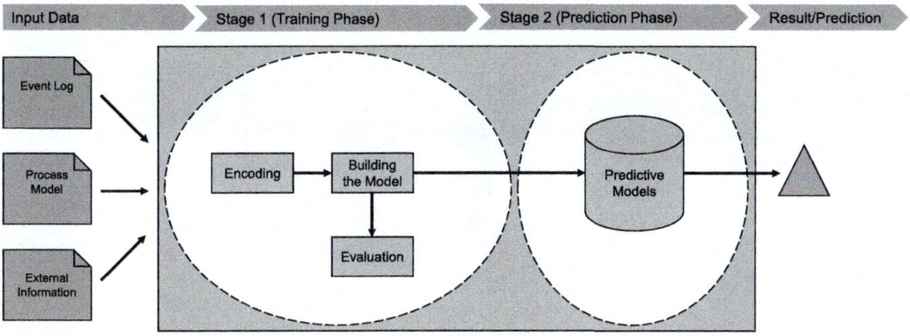

Fig. 3.58 Experimental procedure of a general prediction approach (Marquez-Chamorro et al., 2017)

external data are used as a foundation for the algorithm. Next, those data sources are encoded into feature vectors that can be read and interpreted by the algorithm, which leads to the creation of the predictive model. At the end of the first phase, the validity of the model is tested. If the model passes the validity check, it is applied to ongoing processes in order to forecast future metrics (Marquez-Chamorro et al., 2017). This use case is showing great potential, the complexity is significant due to an amount reliable data and processes complexity (e.g., machines, tools, equipment's, people, systems) (Wick et al., 2020).

Picture of the Model/Framework Explanation

In the context of this book, the application of process prediction within the manufacturing process is central. Because the predictions can be used to improve various aspects, such as product quality, cycle time, or the prevention of machine breakdowns, they can be classified within the process optimization and process control column. The use case is characterized as predictive (see Fig. 3.59).

Relevant Technologies

The following technologies are considered relevant for this use case:

- Enterprise Resource Planning System
- Manufacturing Execution System
- Product Lifecycle Management System
- SCADA System
- Cloud Computing
- Edge Computing
- Internet of Things
- Sensors and Actuators
- Track & Trace Technologies
- Vision Technologies
- Cyber-Physical Systems
- Big Data Analytics

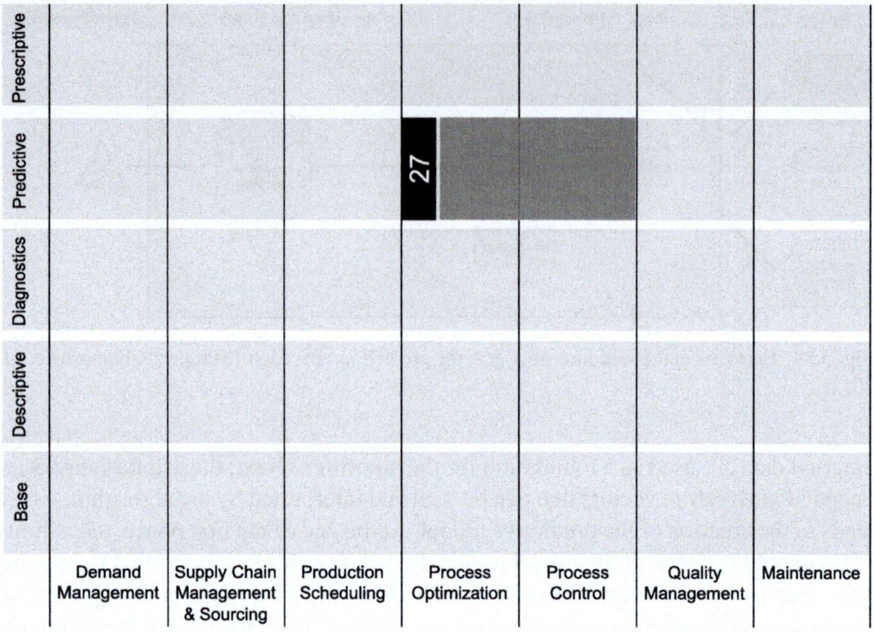

Fig. 3.59 Smart factory framework—use case 27

- AI, ML & Simulation

Impact on Competitive Priorities

The impact of this use case on competitive priorities is shown in generic form in the Fig. 3.60. The specific impact depends on the context of the company.

Impact on flexibility • Improves the planning of different product groups	Impact on costs • Improves resource allocation • Enables early intervention, leading to fewer breakdowns, shorter cycle times, and more efficient allocated processes • Improves process stability, reducing the need for excessive safety stock
Impact on delivery • Improves order fulfilment rate through early detection of issues and optimization potentials, which leads to better processes (shorter cycle time, less scrap, etc.)	Impact on quality • Prevents scrap and postprocessing due to the prediction of quality issues

Innovators/Exemplary Cases

A steel manufacturer was faced with different quality grades of the ores brought to its mill. Depending on the degree of purity or quality of the ores, the process parameters

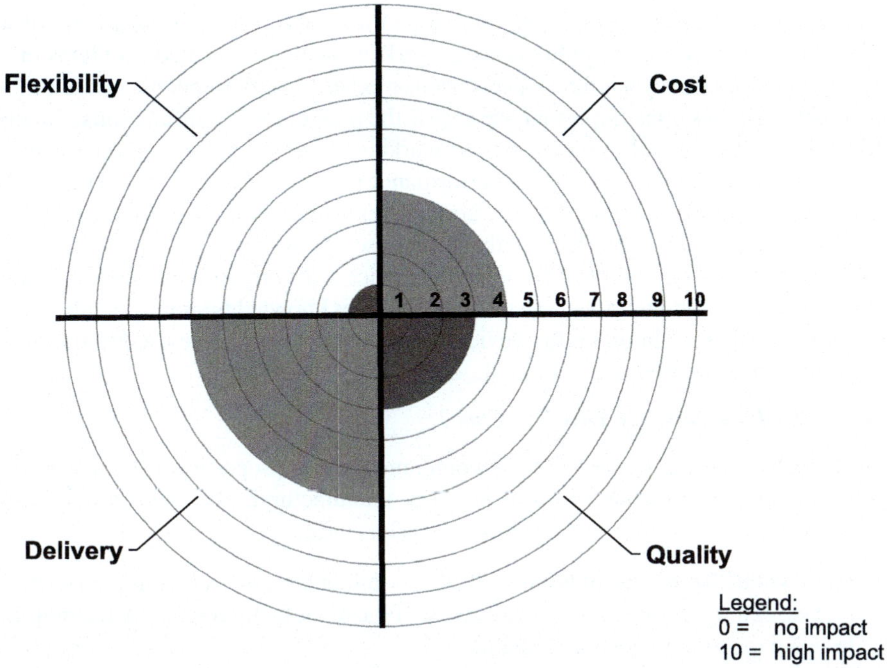

Fig. 3.60 Impact—use case 27

needed to be adjusted. The wrong parameters led to flawed processes, which eventually led to unexpected downtimes. These downtimes amounted to production losses of up to 30,000 USD an hour and high repair costs. Laboratory tests were not possible due to the long turnaround time, which led to an insufficient warning speed. As a solution, the company implemented an AI software that helped to identify patterns that indicated ore with inferior quality levels. This allowed the operators to mitigate blockages within the mill by adjusting the grinding speed. The introduced solution led to annual savings of 900,000 USD (Falkonry, 2022a).

A gas extraction company was challenged by high tank vapor pressures as well as volatile organic compounds requiring constant monitoring and adjustment to keep them under a certain level. Existing approaches could measure the emissions, but the company could not adequately predict regulation violations. Constant monitoring of the emissions was also not a viable solution due to cost considerations and margins for human error. If the emissions exceeded the permitted level, the company had to either shut the company down for a certain time—which could also have resulted in reservoir damage—or pay a high fine for keeping the operations going. Either way, a disturbance would have led to increased costs for the company. The introduced solution learned to recognize patterns that identified indicative conditions for increased emission values. This enabled the company to comply with regulations and reduce the monitoring expenditures and potential profit setbacks (Falkonry, 2022b).

A Swiss industrial company has developed a solution to use a smart watch to communicate performance predictions to workers, such as expected problems on a machine. The company has developed various models to predict specific incidents for different machines. One important use case is the prediction of malfunctions. Timely notification of potential malfunctions to workers is critical for appropriate action. Therefore, in addition to developing performance predictions, they focused primarily on how to communicate them to workers. Several options for communicating the results were evaluated. Worker feedback on ease of use led to the smart watch as the preferred option. A prototype of such a device led to important insights for the device, but also brought up a key issue of how worker knowledge should be integrated through a feedback option on the device. This feedback should be used to optimize performance prediction (Baldauf et al., 2021).

When and How to Apply This Use Case?

The benefits of the prediction of process outcomes or their properties are very versatile and thus are recommended to almost every manufacturer. But there are specific circumstances or situations where this use case is particularly beneficial.

- **High volatility of the processes**: The less predictable the outcomes of certain processes are, the more it is in the interest of the manufacturer to apply technology to make the outcomes more feasible.
- **Large consequences of process deviations**: The more serious the consequences of the experienced process deviations are, the more crucial the prediction of certain events/processes.
- **Frequent occurring deviation-based issues**: The more often certain (issue-triggering) deviations occur, the more important the prediction.
- **Available countermeasures**: The prediction of the processes is only useful if there are countermeasures available that help to optimize the processes.

Questions to Ask

- What outcomes do we want to predict?
- What internal and external factors are relevant to predict the specific outcomes?
- How does variable X influence outcome Y?
- How much does variable X influence outcome Y?
- What countermeasures are available to resolve the issue or optimize the process?
- How can we learn and experiment without a fully scaled IT-Infrastructure?

3.6.7 Use Case 28: Optimal Process Parameter Identification

Description Along the Framework Categories

Modern production lines often have a variety of different setting parameters allowing the setup of the different process configurations. Often, these parameters are only set once for different product types and variants. However, there are other non-machine-related dynamic aspects, such as the quality of the material or environmental conditions (e.g., humidity or temperature). These dynamics can also have an influence on the processes and their outcomes. This issue demands an intelligent system to determine the best machine parameters under the given circumstances for each product type. While the sibling use case, *Process Parameter Prediction*, forecasts the specific product- and process results, this use case pursues a more sophisticated approach, where algorithms not only predict the specific outcomes but also provide the process parameters needed to achieve the desired results. Csalódi et al. (2021) investigated the requirements and factors that contribute the most to optimization solutions with regard to Industry 4.0. Apart from the vertical and horizontal integration, other important determinants are concepts such as real-time capability, corporate social responsibility (CSR), sustainability, flexibility, virtualization as well as the concept of the smart factory itself. Depending on the number and nature (linear or nonlinear) of the constraint and objective functions, more or less sophisticated methods have to be applied in order to be able to optimize the said goals (Csalódi et al., 2021). These approaches can range from classical tools of mathematical analysis to highly complex computation methods (Csalódi et al., 2021). To achieve fast and reliable solutions, AI-based approaches (Csalódi et al., 2021) such as ML (Sai, 2022) are often applied.

Picture of the Model/Framework Explanation

As the name of this use case indicates, the focus lies on the optimization of processes, particularly manufacturing and assembly operations. The identification of the parameters is realized by advanced mathematical and statistical methods, which identify patterns within big data sets. A common method to analyze this multitude of data involves AI-based approaches. In this regard, this concept requires tremendous analytical capabilities to determine the right process parameters for a multitude of different configurations. Hence, the use case is classified as predictive (see Fig. 3.61).

Relevant Technologies

The following technologies are considered relevant for this use case:

- Enterprise Resource Planning System
- Manufacturing Execution System
- Product Lifecycle Management System
- SCADA System
- Cloud Computing
- Edge Computing
- Internet of Things
- Sensors and Actuators
- Track & Trace Technologies
- Vision Technologies
- Cyber-Physical Systems
- Big Data Analytics
- AI, ML & Simulation

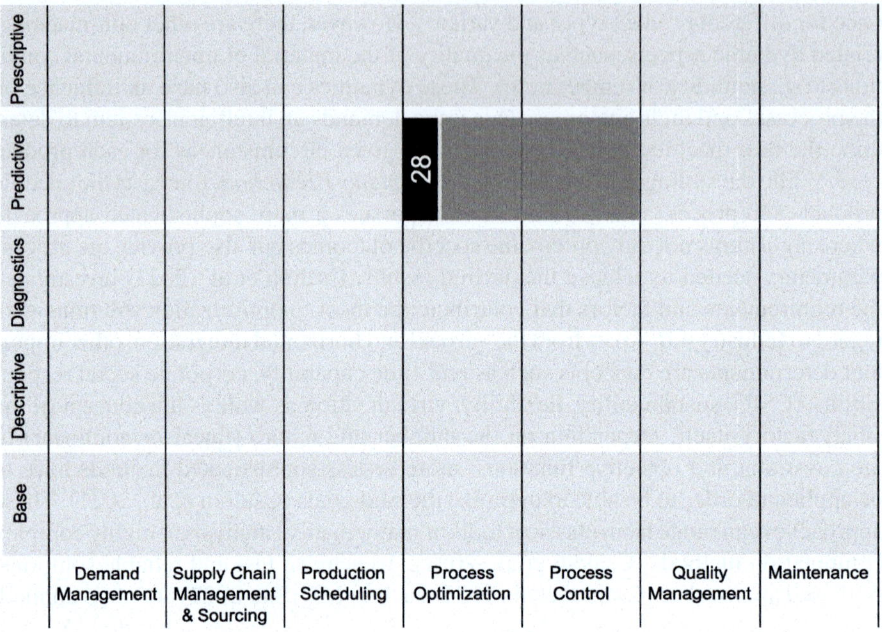

Fig. 3.61 Smart factory framework—use case 28

Impact on Competitive Priorities

The impact of this use case on competitive priorities is shown in generic form in the Fig. 3.62. The specific impact depends on the context of the company.

Impact on flexibility • Improves fitting, dynamically adjusting process parameters allows the processing of customized workpieces in low batches	Impact on costs • Improves process stability, leading to fewer problems and improved cost-efficiency • Improves yield
Impact on delivery • Avoids postprocessing and scrap, leading to faster order fulfilment • Stabilizes processes and compliance with production and delivery schedules	Impact on quality • Reduces scrap and postprocessing • Enables proactive measures to ensure the desired (product) quality

Innovators/Exemplary Cases

The WZL institute of the RWTH Aachen University created a novel value stream-oriented production control system called WoPS 4.0. The solution not only evaluates the quality of actual process configurations but also enables an automated simulation of alternative control setups, allowing the identification and evaluation of the best-fitting parameters for a given process. The simulation is built online in the plant simulation environment of Siemens and uses ERP and MES data. KPIs such as the

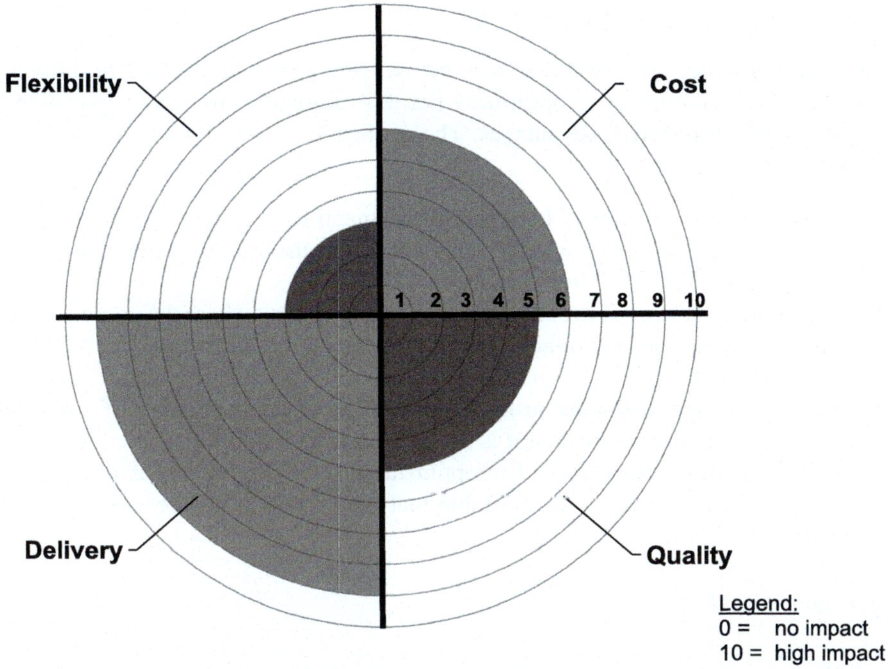

Fig. 3.62 Impact—use case 28

inventory level, lead time, capacity utilization, and meeting of the delivery dates can be optimized, which leads to overall higher competitiveness (Luckert, 2022).

A more scientific approach was carried out by Straßer et al. (2018). The researcher opted to improve the yield quality of produced parts in a multistage manufacturing company. To achieve this, a five-step approach was suggested. First, all related process and product parameters of each produced part were collected. Some 200 produced parts were recorded, and 14 product variables and ten process variables were collected to enable a description of their life cycle during the manufacturing process. Second, the relationship between the quality measurement and the product and process values were modelled and inverted to create a function to determine the process parameters for a given quality measurement. Third, before concrete analyses, the data were pre-processed (i.e., binarized, checked on collinearity, etc.) to avoid any eventual biases. Fourth, to model the relationship between the yield quality and the other variables efficiently, the researchers applied artificial neural networks (ANNs) and regression models. A second approach would have required a further selection of the most significant variables. Finally, the different models with different prediction variables were compared to select the most precise one.

When and How to Apply This Use Case?

The more variables are integrated into the computation model, the more different parameters can potentially be optimized. For this reason, this use case can lead to a broad number of different advantages. The application is particularly beneficial in the following situations:

- **High accuracy required**: The higher the demand for a product, and the greater the importance of process quality, the more it is in the manufacturer's interest to optimize processes by identifying the right parameters.
- **Large consequences of process deviations**: In a context where the consequences of process deviations are rather serious, it is crucial to optimize and stabilize the initial parameters.
- **Frequent occurring deviation-based issues**: The higher the number of impactful deviations, the more important it is to set the correct parameters.
- **Available countermeasures**: The optimization of the processes is only useful if there are countermeasures available that help to optimize the processes and mitigate the issues.

Questions to Ask

- Which process parameters do we want to optimize?
- Which internal and external variables are relevant to optimize the specific outcome metrics?
- How does variable X influence outcome Y? (qualitatively)
- How much does variable X influence outcome Y? (quantitatively)
- Is it possible to achieve the targeted parameters with the proposed process parameters (regarding restrictions in the asset functionality, etc.)?
- In the case of the simultaneous optimization of multiple goals, are there any interdependencies between the different objects (e.g., target harmonies or target conflicts)?

3.7 Use Case 29: Autonomous Process Optimization

Description Along the Framework Categories

The following use case can be considered the pinnacle of the 44 use cases. While other use cases, such as autonomous controlled manufacturing/assembly, focus solely on the issue of automation, and use cases such as autonomous quality management focus only on the optimization of individual processes, this use case deals with the networking, automation, and optimization of (ideally) the complete operational process chain. Ideally, this begins with planning and scheduling and continues through production to quality inspection. Therefore, this use case is very close to the concept of the smart factory (Lee, 2015). The use case is heavily dependent on the vertical integration of all relevant entities. The smart factory, as a connected factory, employs autonomous process optimization and calls for the connection and networking of machines, production and inventory control systems, and all other relevant agents. It uses modern sensor and communication technologies. Once the vast pool of information is collected and ready for analysis, intelligent software carries out various process optimization calculations. By connecting this central intelligent system with the actuators of the respective operational assets, the system not only allows for the determination of the right process parameters but also enables the autonomous control and optimization of said assets and their parameters. This way, a self-regulating and self-optimizing end-to-end networked process system is established (Rüedy et al., 2021; Wick et al., 2021).

As presented by Pietri (2020), a self-optimizing plant incorporates three important characteristics. First, it is self-learning. The system uses data from various sources to become smarter and increase the accuracy of its analyses and predictions. Second, it is self-adapting, which enables parameter adjustment in case of changing conditions to meet the objectives. Third, a self-optimizing plant sustains itself. This means that it can autonomously detect anomalies and react early by triggering countermeasures to prevent possible issues.

One of the challenges when implementing this use case revolves around the vertical integration of the factory, as there is no industry-wide standard for data formats and interfaces (Hops, 2022).

Picture of the Model/Framework Explanation

This use case describes the autonomous control of a multitude of interconnected processes and enables the optimization of a single objective or a group of objectives. Because the system not only recommends actions for optimization but also executes them autonomously, this use case is considered prescriptive (Fig. 3.63).

Relevant Technologies

The following technologies are considered relevant for this use case:

- Enterprise Resource Planning System
- Manufacturing Execution System
- Cloud Computing
- Edge Computing
- M2M Communication
- Internet of Things
- Sensors and Actuators
- Product Lifecycle Management System
- SCADA System
- Track & Trace Technologies
- Vision Technologies
- Cyber-Physical Systems
- Big Data Analytics
- AI, ML & Simulation

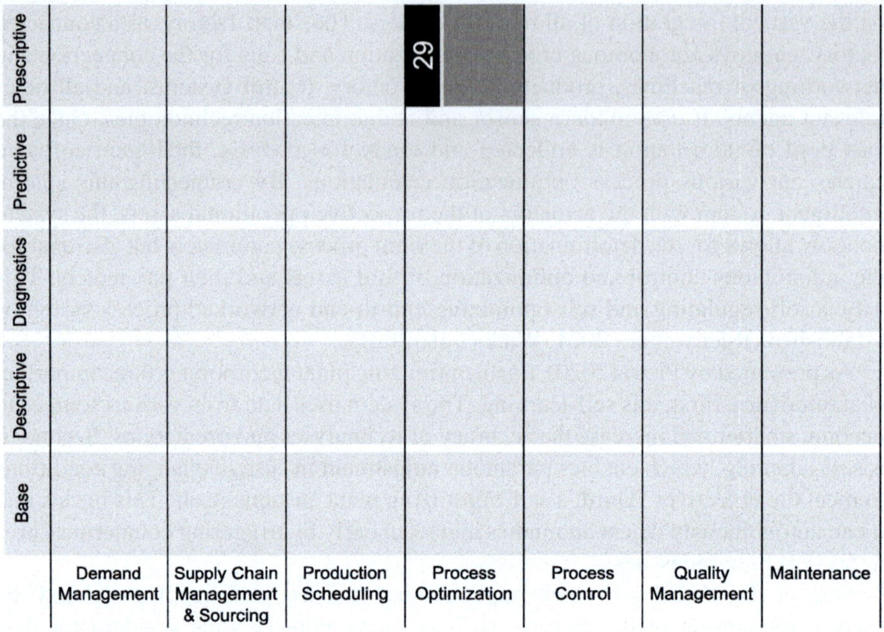

Fig. 3.63 Smart factory framework—use case 29

Impact on Competitive Priorities

The impact of this use case on competitive priorities is shown in generic form in the Fig. 3.64. The specific impact depends on the context of the company.

Impact on flexibility • Accelerates set-up times between different product categories • Enables fast and easy adjustment of the production schedule, leading to increasingly flexible handling of the manufacturing orders • Improves customizability	Impact on costs • Improves capacity utilization of the assets and the plant as a whole • Increases overall yield due to early interventions that increase efficiency • Reduces human-related errors due to the omission of the human workforce
Impact on delivery • Enables the quick adjusting of production schedules, enhancing compliance with delivery deadlines • Reduces set-up and cycle times due to optimized manufacturing processes and machine configuration	Impact on quality • Enables constant monitoring and early intervention, allowing a continuous control and optimization of the product and process quality • Reduces scrap due to optimized processes • Enables informed and traceable decisions

Innovators/Exemplary Cases

The Smart Electronic Factory, a German information and demonstration platform for industry 4.0 solutions that is integrated into a factory of electronic parts, implemented

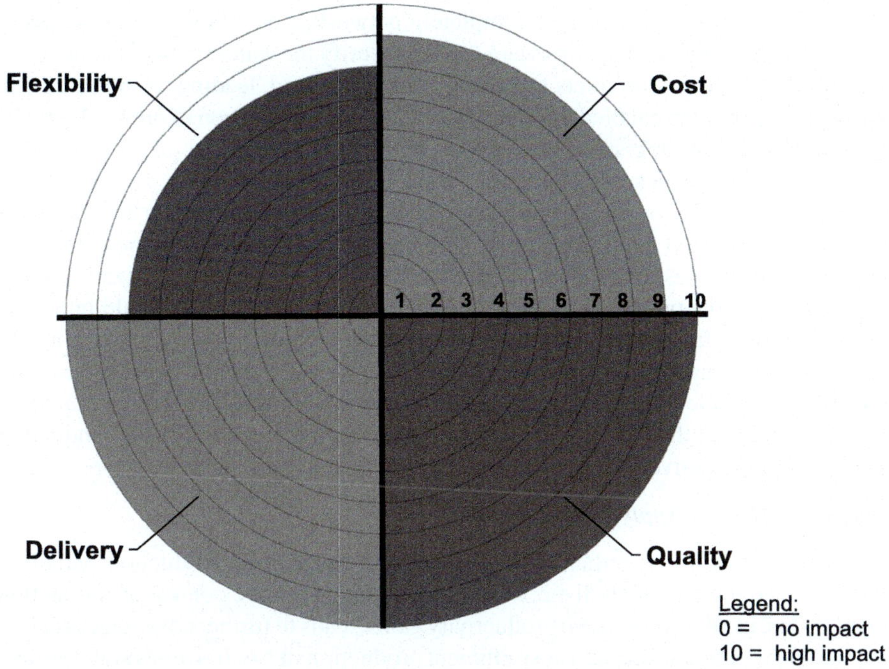

Fig. 3.64 Impact—use case 29

a holistic smart factory solution. The objective was to investigate the requirements for industry 4.0 in the electronics sector. By collaborating with software, manufacturing, control technology, and propulsion solutions companies, the association was able to develop a holistic smart factory solution. Concerning the production processes, the system could autonomously record process errors, document them, and analyze the causes via independent learning systems in order to be able to intervene preventively in the production process. The data were analyzed by correlation analyses that looked for patterns within the relevant data sets to identify outcomes and process errors. To effectively generate the information, older machines were equipped with sensors and actuators. Furthermore, in addition to the individual machine and process data, other data relevant to production, such as humidity and temperature, were also included. An APS system also handled the detailed planning of production orders, taking into account machine and material availability among other factors, and responded to new requests in real-time. This solution made it possible to produce a batch size of one as economically and error-free as possible (Bienek, 2022).

Infineon, a producer of semiconductors, wanted to optimize its yield by increasing capacity utilization on the one hand and its first time-through yield on the other. For these performance specifications, it was necessary to identify faults as quickly as possible and to address them before they happened. One of the main requirements of this approach was to ensure a completely networked data flow within the plant. Among other things, RFID chips were used to identify the water transport containers. This not only allowed products and processes to be traced, leading to an increase in transparency but also enabled the error-free, automatic production of orders. By intelligently linking intralogistics, production control, and system operation, the highly complex production of semiconductors, which in some cases consisted of up to 1,000 process steps, could not only be automated but also highly optimized. In the new smart factory, optimization algorithms controlled the various robots, machines, transport systems, and batches throughout the entire production process. Visualization functions displayed the production data to the relevant personnel so they could intervene if the algorithms failed to self-regulate. In addition, the system communicated in real-time with other manufacturing sites globally to optimize and control production. As a result, this semiconductor factory is one of the most highly automated manufacturing facilities in the world. This helped to increase resource utilization and led to a 100% on-time delivery rate, with increased quality and productivity (Hops, 2022).

When and How to Apply This Use Case?

This use case brings together all the advantages of the new production paradigm Industry 4.0, such as the holistic, autonomous, and proactive control of production plants. This, in addition to large productivity gains, leads to further advantages such as increased process stability and cost-efficient production of batch sizes as small as one. For this reason, the use case is extremely profitable for any company. However, there are various requirements that must be met to implement this solution successfully. Colli et al. (2019) defined the following requirements:

- **Governance**: The first point describes the organizational level necessary to determine factors such as the allocation of the resources, the hierarchical structures within the organization, and the strategy.
- **Technology**: To realize the generation, processing, and utilization of the data, the required assets must first be installed. These technologies include conventional ones such as ERP systems and rather novel approaches and technologies such as IoT, or AI.
- **Connectivity**: The infrastructure has to allow the transmission (and processing) of the various data within and outside the organization. Parts of connectivity are, for example, the data sharing capabilities and architectures or the structuring of the data.
- **Value creation**: To capture the value generated from the recorded and processed data, effective models have to be applied, such as utilization-based revenue models, predictive maintenance, etc.).
- **Competencies**: The last point concerns the relevant internal and external stakeholders of the organization, whose mindset and skills are needed to conduct the digital transformation process. This characteristic contains aspects such as digital competence, the learning culture, and more.

Questions to Ask

- What benefits would occur from this use case?
- From which assets and processes do we already acquire data? Which ones do we still have to exploit?
- What is the investment necessary to realize an autonomous and self-optimizing production?
- What is our digital maturity level?
- What is the focus area for process optimization?

3.8 Process Control

3.8.1 Use Case 30: Digital Shopfloor

Description Along the Framework Categories

The shopfloor can be defined as the area in the factory in which assembly and production take place (Gruber, 2020). The fundamental aspects of process control and optimization occur at shopfloor level. Shopfloor management approaches include the direct observation and analysis of production and process data to enable "process improvement, identification of deviations, and structured problem solving directly on the production shopfloor" (Tauber et al., 2019). Traditionally, shopfloor management activities are assisted by visual cues such as whiteboards or paper-based information visualization tools. This use case aims to digitalize the presentation of the production and process data being generated in real-time during factory operations. This use case is being discussed and implemented widely.

The digital shopfloor ought to be understood as a fundamental component of the smart factory transformation. Once data is generated via the integration of cyber-physical systems and analyzed with sophisticated analytics, the information must necessarily be relayed back to the key decision makers. Often, these key decisions need to be made and implemented on the shopfloor level. The digital shopfloor provides the lens through which the combination of the physical with the virtual world can be observed, as key metrics and actionable insights are presented for workers to implement in real-time (Tao & Zhang, 2017). The digitalization of the traditional knowledge visualization tools is achieved with a variety of digital technologies: digital panels, mobile computing, and augmented reality. To enable the real-time presentation of insights and data, these digital objects are connected to a cloud IoT platform from which the insights are analyzed and prepared for presentation.

Notably, the ultimate utility of the digital shopfloor depends on the quality of the information it is presenting. If the information prepared is not exhaustive or not representative, the digital shopfloor loses its productivity and process optimization potential.

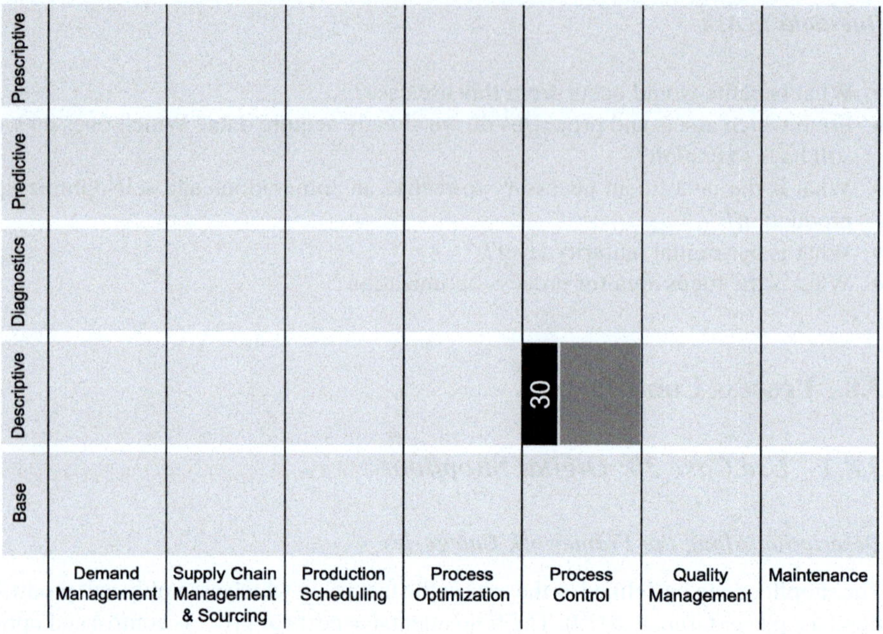

Fig. 3.65 Smart factory framework—use case 30

Picture of the Model/Framework Explanation

The digital shopfloor is an enabling technology to visualize exhaustive production and process data, together with analytics backed actionable insights presented by the intelligent factory. The digital visualization tools do not carry out any analytical functions themselves. For this reason, the use case is categorized as descriptive (see Fig. 3.65).

Relevant Technologies

- Enterprise Resource Planning System
- Manufacturing Execution System
- SCADA System
- Cloud Computing
- Track & Trace Technologies
- Big Data Analytics
- Automated Guided Vehicles
- Edge Computing
- Mobile Computing
- Internet of Things
- Sensors and Actuators
- Wearables

- Augmented Reality

Impact on Competitive Priorities

The impact of this use case on competitive priorities is shown in generic form in the Fig. 3.66. The specific impact depends on the context of the company.

Impact on flexibility • Facilitates quick and responsive decision-making by providing a real-time overview of production data	Impact on costs • Reduces downtime/time to repair • Reduces documentation costs • Increases labour productivity
Impact on delivery • Improves delivery-in-full on-time score due to better quality assurance and process control	Impact on quality • Improves quality testing and improvement procedures, as digital visualization tools enable real-time presentation of descriptive, diagnostic, and prescriptive insights • Enables the effective implementation of quality improvements due to facilitated knowledge transfer

Innovators/Exemplary Cases

Lorenz et al. (2019), examined the positive impacts of digitalizing shopfloor management by looking at the case of the Norwegian company Kongsberg Maritime AS. The maritime technology company recently made the switch from shopfloor meetings relying on analogue information tools such as whiteboards and printed data to digital shopfloor management based on data connectivity and digital information visualization. Following the switch from analogue to digital tools, the authors observed an overwhelming increase in improvement suggestions in three out of four sampled teams (Lorenz et al., 2019). Additionally, results demonstrated more qualitative improvements in the effectiveness of shopfloor management. According to company employees, digitalization allowed for greater horizontal collaboration, easier vertical escalation with the collaboration and sharing of knowledge with other employees, more optimized workflows due to reduced administrative costs, and better insights due to the possibility of seamlessly integrating historical machine data in comparison studies. (Lorenz et al., 2019).

The potential of the digital shopfloor is also aptly demonstrated by NavVis's digital solution NavVis IVION Go. With regards to shopfloor management, lean manufacturing philosophy suggests that many of the process improvements come from physically visiting and closely monitoring the shopfloor. This procedure—often referred to as the Gemba walk—is best complemented with the 5S from a second lean manufacturing strategy (Yusuf, 2021). With NavVis IVION Go solution, shopfloor workers can carry out the abovementioned routines with the additional help of information visualization technologies. Operators are accompanied by mobile computing technologies as they walk through the machine floor and review data at predetermined or spontaneously defined points of interest (Yusuf, 2021). The digital visual aid can provide key information to the operator as they carry out their

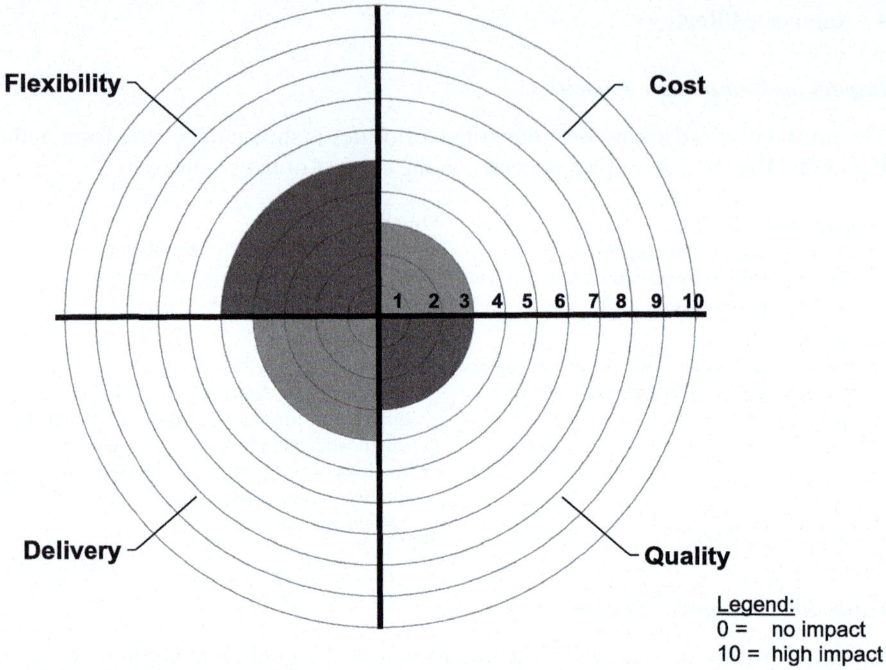

Fig. 3.66 Impact—use case 30

inspection and optimization activities. Concurrently, the operator is also able to record and share data for review by other employees through an enterprise-wide platform (Yusuf, 2021). Such a solution demonstrates the full potential of the digital shopfloor in terms of supporting shopfloor activities and facilitating knowledge transfer.

When and How to Apply This Use Case?

- **Availability of data-driven insights**: Digital shopfloor elements are only as useful as the information they are presenting. Once analytics-backed insights are produced, the information visualization tools can help turn these insights into action.
- **Long process chains**: In longer, more complex process chains, possible issues can emerge in more places, which need to be monitored continuously.
- **Complex products/processes**: Complex products usually require multiple process steps, which all have to be done properly. Digital shopfloor can help control and validate the working steps done at each station.
- **Decrease of documentation costs**: Autonomous presentation and mapping of the activities of the manufacturing floor can help to reduce documentation costs.

Questions to Ask

- Are useful data-driven insights being produced and not implemented effectively?
- Does the company possess the right level of digital maturity to switch to a digital shopfloor in a cost-effective way?
- Will all employees be able to interact with and utilize the new digitalized information visualization systems?
- How can we integrate the digital shopfloor into the management analogue shopfloor process?

3.8.2 Use Case 31: Low Batch Size Manufacturing

Description Along the Framework Categories

As the name suggests, this use case is about manufacturing small batches of products in an economically viable way. With increased demands for customization from consumers and the proliferation of lean manufacturing methods, manufacturing firms are seeking new ways to reduce the batch size of their product lines lines (HOMAG, 2020; Wuthrich & Hof, 2019). The key to an economically viable production of low batches lies in the optimization of batch sizes. To this end, a variety of analytical methods can be applied. For instance, simulations are often used to gain insights into key process performance metrics with different batch sizes, relying on both historical and present data for their analysis. The impact of batch size optimization on operational performance is significant. Simulation quantifies this impact.

Once optimized theoretically, the low batch size can requires specific technology in order to be manufactured efficiently, focusing on fast and flexible delivery (Wuthrich & Hof, 2019). In this regard, the most promising technology is additive manufacturing. Although a uniform assignment to one of the six DIN 8580 manufacturing processes is not possible, additive manufacturing is often categorized as primary forming (Kumke, 2018). While additive manufacturing can be utilized for several purposes, such as the production of spare parts and tools or the rapid prototyping of new products—it has recently been utilized for low batch production as well (Gibson et al., 2021; Khoshnevis, 2004; Liu et al., 2014; Schilling, 2022).

Picture of the Model/Framework Explanation

Simulation methods to determine the optimized batch size do not rely on overly complex analytical tools. Often discrete event simulation can be used while the additive manufacturing process appears very complex, the principle behind it is relatively simple: by inputting a CAD or CAM model, the 3D printer is instructed to manufacture the object according to the specified dimensions. Therefore, this use case can be classified as descriptive (see Fig. 3.67).

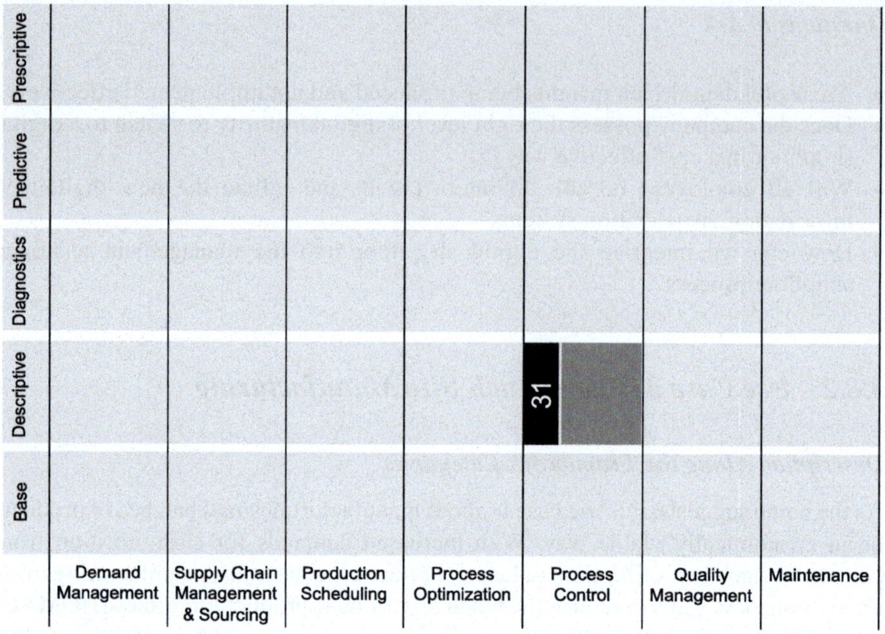

Fig. 3.67 Smart factory framework—use case 31

Relevant Technologies

The following technologies are considered relevant for this use case:

- Enterprise Resource Planning System
- Manufacturing Execution System
- Cloud Computing
- Internet of Things
- Sensors and Actuators
- Track & Trace Technologies
- Big Data Analytics
- AI, ML & Simulation
- Additive Manufacturing
- Wearables

Impact on Competitive Priorities

The impact of this use case on competitive priorities is shown in generic form in the Fig. 3.68. The specific impact depends on the context of the company.

Impact on flexibility	Impact on costs
• Increases flexibility regarding small batch sizes • Increases flexibility regarding the development of new products (rapid prototyping) • Increases customization potential	• Increases the cost-efficiency of producing small batch sizes • Increases scalability • Increases the automation of manufacturing processes • Reduces downtimes of machines due to fast printing of spare parts
Impact on delivery • Enables the fast development of new/altered products	Impact on quality • Increases first-time-right quote due to standardized processes • Increases process reliability due to standardized processes

Innovators/Exemplary Cases

A first exemplary case is provided by Siemens. The company revolutionized the development and production of several machine parts in one of its divisions by introducing additive manufacturing as precise laser melting. Additive manufacturing was used to integrate the testing of components such as turbine blades for gas turbines in the product development process. This led to an overall faster evaluation of novel

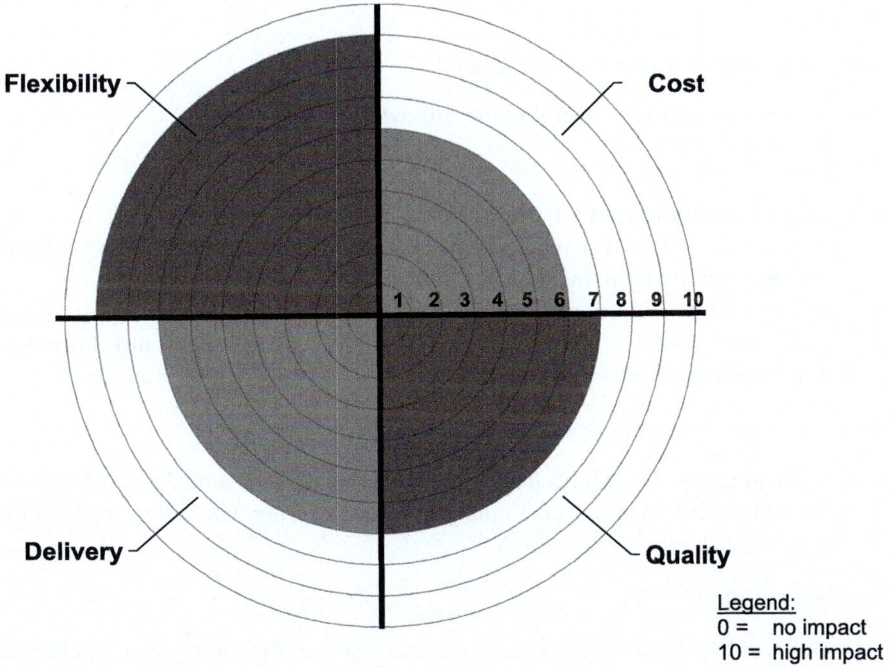

Fig. 3.68 Impact—use case 31

product designs, resulting in the overall faster and cheaper production of small batches. Furthermore, the selective laser melting process sometimes delivered better material properties than the conventional casting process (Piegert, 2022).

A second example of the benefits of additive manufacturing is Robot Bike co., an innovative manufacturer of bicycles. Because bicycle frames are moulded, most bicycles have only three to four different sizes. To change this, the company introduced an additive manufacturing approach. However, it was challenged by quality requirements. For example, the printed parts had to be as robust and lightweight as possible. By partnering up with multiple engineering companies, Robot Bike co. was able to optimize the design of the bicycle. The solution now allows customers to enter their body measurements, which are automatically transferred to the advanced metal additive manufacturing system. As a result, the company produces one of the first customizable AM-based bicycle frames, which are also strong, lightweight, and reliable (Renishaw, 2016).

A medium-sized Swiss manufacturer of fine parts used discrete event simulation to quantify the impact of different lot sizes on operational performance of on-time delivery and production throughput. The simulation showed a significant improvement in the company's profitability by reducing lot size by a factor of two while improving on-time delivery and production throughput. This helps management prioritize flexibility and optimization of the production process over the focus area of setup costs (Budde et al., 2022). The implementation further improved the logistics process with a key supplier by optimizing the ordering and forecasting process.

When and How to Apply This Use Case?

This use case is especially useful in the following situations:

For R&D

- **Complex products/parts**: If physical properties of the product/part to be developed are rather complex, it results in the challenge that they cannot be individually calculated and therefore must be tested in simulations.
- **Fast product development**: Specific orders for products sometimes require fast development. Additive manufacturing for product development, also known as rapid prototyping, supports the iterative building and testing of prototypes.

For spare parts and tools

- **Long lead times**: Some tools have an enormous lead time, sometimes over a year. If an unforeseen outage occurs, additive manufacturing can help to replace the defective part/tool expeditiously.

For product(-parts)

- **Produced lot sizes**: Smaller lot sizes increase the flexibility of a production process.

- **Small batch sizes**: Additive manufacturing is useful for the production of smaller batch sizes.
- **High customization**: Additive manufacturing can support the customization of products regarding their shape, colour, and other properties.
- **Small warehouse**: A small warehouse for semifinished and finished goods requires a shift towards a make-to-order system. Additive manufacturing can help here by enabling the fast production of parts.

Questions to Ask

- What available data di I have that can help simulate my production process
- Does the 3D-printed product have the same required properties as the conventionally manufactured model?
- Has the CAD system transferred the design correctly to the 3D printer?
- How many parts are needed? Is it possible to produce such a quantity within the required time?
- What are the benefits of 3D printing, compared to producing parts conventionally or ordering them? (e.g., lead time, costs, energy consumption, etc.)
- Is it possible to produce the workpiece with a 3D printer?

3.9 Use Case 32: Human-Centred Flexible Automation

Description Along the Framework Categories

Despite the increasing trend of process automation in the context of industry 4.0, about 90% of all new emerging tasks cannot properly be executed with conventional automation technologies such as robots (Bloss, 2016). This can be traced back to existing robot technologies not meeting the flexibility requirements to properly execute these tasks (Weckenborg et al., 2020). Therefore, human labour is indispensable in areas where an automated solution is either not possible or not profitable.

While the strengths of human workers lie in things such as flexibility, the recognition of visual scenarios, and creativity, robots are particularly useful in hazardous situations and for repetitive and exhausting tasks (Weckenborg et al., 2020). To resolve this issue and reap the benefits of both human and mechanical workforce, human-centred automation systems that allow for flexibility in a variety of different tasks have been introduced. The relatively novel paradigm of human–robot collaboration describes the abovementioned concept, where the strengths of both players are combined in order to create a symbiotic relationship (Wang & Zhang, 2017). The robots, which are designed to interact and collaborate directly with humans on the shopfloor, are referred to as collaborative robots, or cobots. A cobot can be defined as "a robot specifically designed for direct interaction with a human within a defined collaborative workspace" (Beaupre, n.d., p. 6). Nowadays, collaborative robots are deployed in a multitude of different jobs. For example, for process automation, machine tending, or pick and place tasks (Association for Advancing Automation, n.d.).

To enable efficient and safe collaboration between cobots and human workforces, the design of the cobot is adjusted in order to prevent any possible hazards to humans. The collaborative robots are, for example, often equipped with sensors (and corresponding software, often employing vision technologies), which allow them to recognize humans (and other objects) to prevent collisions. Additionally, casings have no sharp edges and the lifting capacities are decreased (Pilat et al., 2019). Cobots can often either be appointed to work autonomously (e.g., for machine tending) or collaboratively (e.g., for supporting a welder by moving the workpiece). In collaborative functions, the assistance of the labour workforce has priority (Beaupre, n.d.).

Picture of the Model/Framework Explanation

Human-centred automation systems are often designed in a way that the automation parts (i.e., collaborative robots) can operate flexibly to fulfil a variety of different working tasks and/or different product variants. As mentioned above, in such systems there is a particular focus on the secure collaboration between the human and the robotic workforce. To realize such an adaptable system, a variety of analytics has to be applied, such as machine vision, robotics and more. In this regard, this use case implies an advanced level of analytics and is, therefore, classified as diagnostic (see Fig. 3.69). Furthermore, human-centred automation systems, collaborative robots in this context, are employed on the shop floor, which leads to the classification of this use case in the process control column.

Relevant Technologies

The following technologies are considered relevant for this use case:

- Manufacturing Execution System
- Cloud Computing
- M2M Communication
- Mobile Computing
- Internet of Things
- Sensors and Actuators
- Track & Trace Technologies
- Vision Technologies
- Cyber-Physical Systems
- Big Data Analytics
- AI, ML & Simulation
- Collaborative Robotics
- Wearables

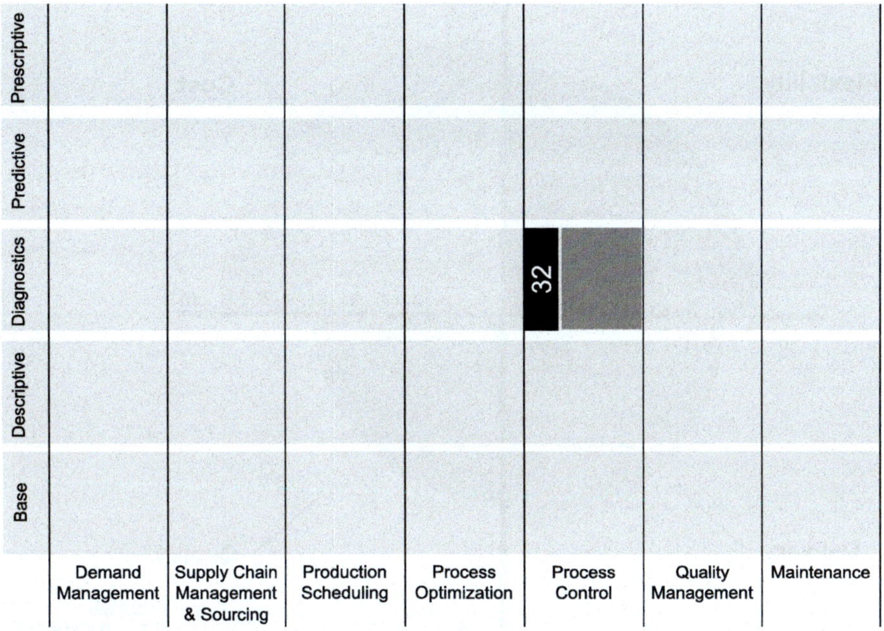

Fig. 3.69 Smart factory framework—use case 32

Impact on Competitive Priorities

The impact of this use case on competitive priorities is shown in generic form in the Fig. 3.70. The specific impact depends on the context of the company.

Impact on flexibility • Recognizes and processes different parts/product types autonomously • Reduces reprogramming tasks. Cobots can often be applied for a multitude of tasks without the need to be reprogrammed	Impact on costs • Increases productivity because no breaks etc. are required • Increases capacity utilization • Improves motivation and labour productivity. Cobots take over tedious and repetitive tasks, which allows appointing the operator to superior tasks
Impact on delivery • Accelerates order fulfilment through a reduced production lead time	Impact on quality • Achieves a more consistent process quality

Innovators/Exemplary Cases

Swiss Productions, a manufacturer of components for various industries, such as the aerospace, medical, and electronics sector, was looking for ways to grow its business. Already producing over two million parts a month, the company sought a way to further increase its productivity by increasing labour productivity as well as maximizing the capacity utilization of its machines. Amongst other things, the

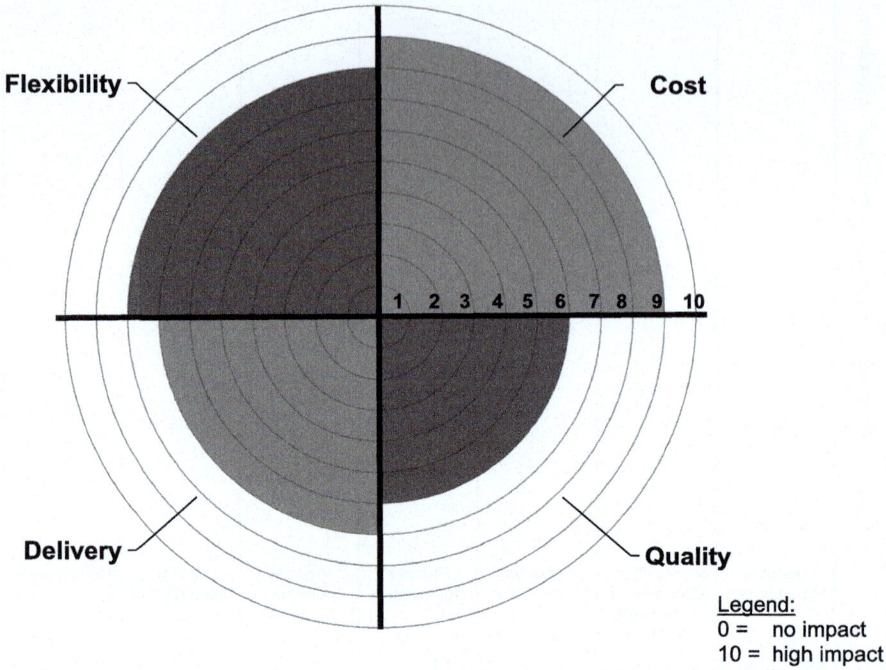

Fig. 3.70 Impact—use case 32

company opted to maximize the spindle time of its CNC machines. The issue was that, up until then, employees loaded and unloaded the machines manually, whereby they waited next to the machine until the processing was finished. This approach led to tremendous losses in time. To eliminate these inefficiencies, the company deployed a cobot with seven axes in front of the CNC machine. The introduced solution was able to assist a human by loading, unloading the CNC chamber, starting the process, and even cleaning the collets all by itself. This replaced most of the manual work by the employees and allowed them to simultaneously operate two additional machines. As a result, the company almost doubled the weekly machine running time from 40 to 76 h (Crowe, 2020).

Another promising application area for cobots is bin-picking. Allied Moulded, a manufacturer of non-metallic electrical boxes, faced labour shortages. As a solution, the company aimed to automate its bin-picking processes in order to relieve its employees of these repetitive and tedious tasks and appoint more sophisticated and complex duties to them. Due to severe space restrictions, the employment of a robot was impractical, because the additional safety measurements would prevent employees from working in this area simultaneously. Therefore, the company employed a cobot to fulfil the task. Due to the self-learning capabilities of the cobot, the solution was incredibly easy to implement. The cobot was, for example, able to recognize the relevant products autonomously just by loading the CAD data into its system. The introduced solution could pick up the right parts and place them in the correct direction on a conveyor while simultaneously sorting the incorrect ones out. The cobot was able to generate constant output while the human operator became tired and increasingly inattentive during the day. A second benefit was the freed-up labour capacity that was then used for higher-ranking tasks, which not only provided added value for the company but also for the employee (Crowe, 2021).

When and How to Apply This Use Case?

This use case is especially useful in the following situations:

- **Repetitive tasks that require human labour**: Repetitive and tedious tasks, which cannot solely be solved by automation systems, can be addressed by using collaborative robots
- **Hazardous tasks**: To protect one's employees, collaborative robots can be used to automate the dangerous working steps

Questions to Ask

- Where does our working force struggle? Which tasks could be facilitated by the deployment of collaborative robots?
- How can collaborative robots facilitate the abovementioned issues by supporting the employee?
- What would be the requirements for the cobot (regarding movement, strength, number of axes, product handling devices etc.)?
- Are fully automated processes or collaborative robots the best solution?
- Are there any disadvantages or hazards in the implementation of a cobot?

3.9.1 Use Case 33: Digitally Guided Operations

Description Along the Framework Categories

The shortage of skilled workers in industry is posing a serious problem for manufacturing firms. This trend can be traced back to various root causes, such as an ageing workforce and the lack of interest of younger generations to work in the industrial sector, especially in science, technology, engineering and math (STEM) areas (Göttlich, 2020; O'Rahilly, 2021). These trends result in manufacturing firms often employing underqualified workers to keep up with their workforce requirements. This skill shortage can lead to increased risks in various operational tasks due to human errors. To mitigate this risk and ensure high quality and efficiency, companies rely on paper-based documents to lead employees through the assembly process. Recently, this paper-based approach has been replaced by a variety of novel technologies achieving the same purpose. Digitally guided operations is an umbrella term for these technologies.

Probably the most promising technology in this use case is the application of augmented reality to support assembly tasks (Seeliger et al., 2022). The same paper-based information traditionally utilized can be overlaid onto the visual field of the employee in an integrated and seamless manner. To visualize that information, devices with a display such as tablets, mobile phones, or HMDs are required (Radkowski, 2015). This enables the active guidance of the worker through the assembly process or set-up process of a machine. Information about the single working steps (e.g., identification, handling, alignment, joining, and adjustment) is displayed on the assembler's device either via text (descriptive) or via 2D or 3D illustrations (depictive) (Radkowski, 2015). Depending on the design, such devices often include sensors (e.g. cameras, etc.) to control either the movements of the employees (Klein, 2022) or the assemblies themselves (Zheng et al., 2020).

Other prominent solutions may include technologies such as smart tools, which support the employee in the execution of the assembly tasks by fulfilling tasks such as the autonomous adjustment of their settings. Examples include a clamping tool, which autonomously adjusts its rotary moment, or a drilling machine that monitors and documents the drilling depth and recalibrates itself for every new material (Boria, 2015).

Finally, pick-by-x solutions are also rapidly developing as worker assistance tools. For example, pick-by-light consists of displays and lights, which are installed on the shelves in warehouses, and/or on the shop floor and support the operator by giving information about how many parts of which type are needed (Baechler et al., 2016). Due to the field of application being predominantly in warehouses, these solutions are classified as a separate use case—digital picking assistance.

Picture of the Model/Framework Explanation

As explained above, this use case addresses the assistance and guidance of the employees through the assembly process. Often, a quality control element is incorporated in the solution. The context-based guidance of the worker as well as the quality control approaches require an advanced form of analytics, which is why this use case is classified as diagnostic (see Fig. 3.71).

Relevant Technologies

The following technologies are considered relevant for this use case:

- Manufacturing Execution System
- Cloud Computing
- Edge Computing
- Mobile Computing
- Internet of Things
- Sensors and Actuators
- Track & Trace Technologies
- Vision Technologies
- Cyber-Physical Systems
- Big Data Analytics

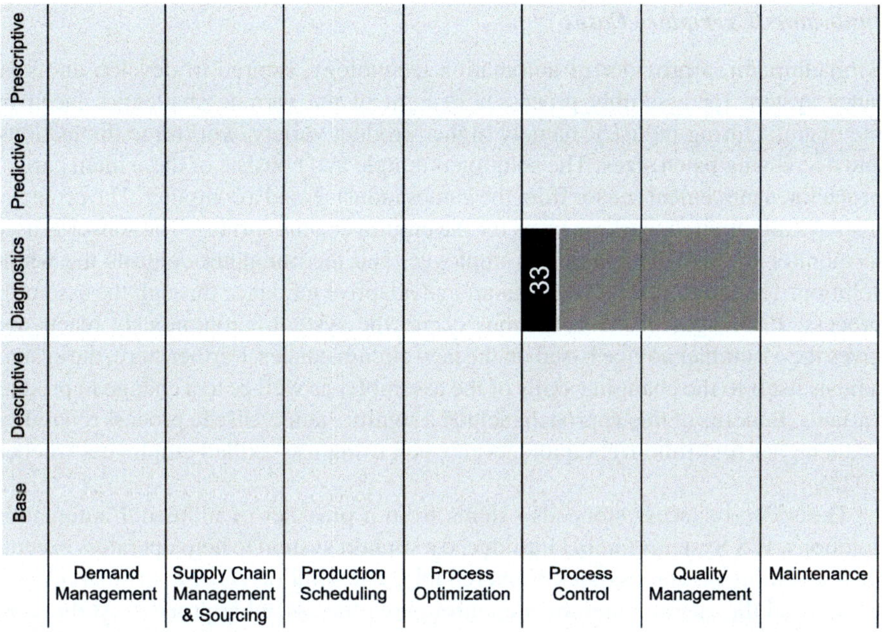

Fig. 3.71 Smart factory framework—use case 33

- AI, ML & Simulation
- Augmented Reality
- Virtual Reality

Impact on Competitive Priorities

The impact of this use case on competitive priorities is shown in generic form in the Fig. 3.72. The specific impact depends on the context of the company.

Impact on flexibility • Improves flexibility regarding product variants • Improves flexibility regarding the variety of products	Impact on costs • Lowers the wage cost due to the employment of less-skilled workers • Improves process stability leading to fewer errors and thus enhances the yield
Impact on delivery • Helps to increase the efficiency of the assembly processes while simultaneously avoiding any errors regarding the product specifications and the quality due to interactive guidance • Improves delivered-in-full on-time score	Impact on quality • Improves process stability, leading to fewer errors and thus enhances the overall quality • Avoids further processing by detecting faults quickly due to in-line monitoring

Innovators/Exemplary Cases

Schnaithmann, a provider of automation technology, aspired to develop an assistance system for assembly processes in light of the various challenges faced by the manufacturing industry, namely higher product variety, workforce fluctuations, and decreasing batch sizes. The solution is simple and consists of three main parts, a projector, a movement sensor from the games industry, and a computer. The projector displays all required working steps for the employee on a surface. The sensor is used to monitor the movements of the employee, and the computer controls the whole solution. The system can provide smart and adaptive guidance through the assembly process. If changes in the conditions occur, the system autonomously reacts and gives the assembler advice based on the new circumstances. Furthermore, the system adjusts itself to the changing skills of the assembler as well as to a change in product variants. Benefits of this approach include a significant increase in process reliability, and a higher first-time-right quote, even when using less-skilled employees (Sieber, 2022).

The second success story also stems from a provider of industrial automation solutions. WS Systems GmbH introduced a support system to help operators execute their assembly processes quickly and reliably. The solution consists of smart glasses which lead the operator step-by-step throughout the assembly process. At the same time, the system simultaneously detects and analyzes occurring faults, which are

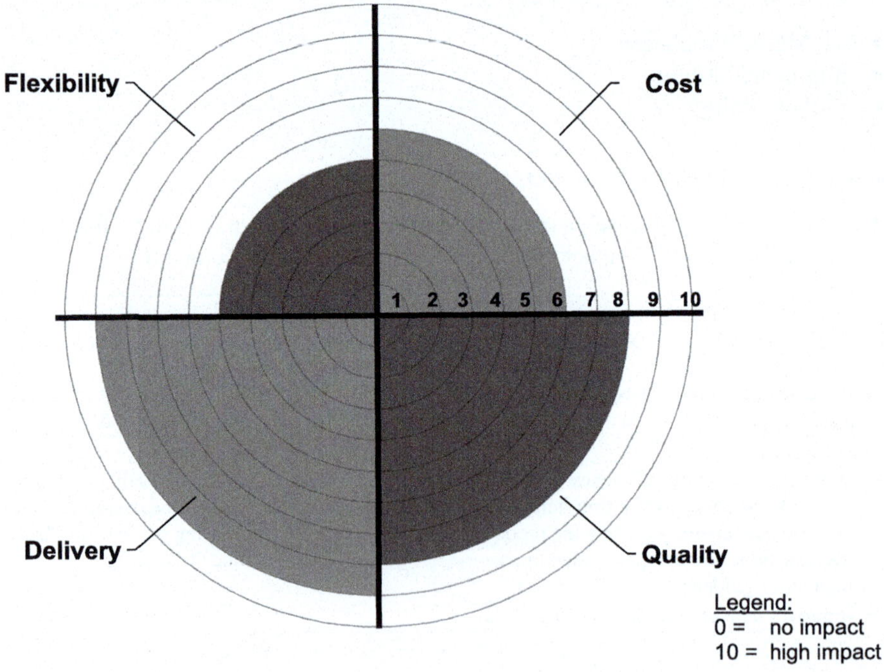

Fig. 3.72 Impact—use case 33

automatically relayed to the operator in real-time. In addition, the solution contains a multi-sensor system able to detect whether the employee has picked the right component out of a container or not. This is achieved with RFID-wearables, whereas other sensors are built-in for other purposes, such as cameras, proximity and heat sensors. With this solution, the benefits amounted to more reliable processes, the prevention of errors, faster training of new workers, and increased motivation of the employees (Saeidi, 2022).

When and How to Apply This Use Case?

This use case is especially useful in the following situations:

- **Many different products**: The wider the product portfolio, the more know-how is required to put the different assemblies together correctly. Digital tools can help to guide the assembler throughout the process by providing instructions.
- **Many different product variants**: The wider the product portfolio, the more know-how is required to assemble the different product variants correctly. Digital tools can not only help to guide the operator throughout the process but also ensure that the right product configurations have been built.
- **Complex products and assembly processes**: The more complex an assembly process (regarding the number of parts, the number of required process steps, etc.), the harder it is for the operator to maintain an overview and conduct every process step correctly.
- **Lack of qualified personnel**: A shortage of skilled personnel can be counter-acted with the employment of underqualified personnel equipped with digital technologies to compensate for their lack of skills.

Questions to Ask

- Where do our employees struggle the most?
- In which working steps occur the most faults?
- Do we have the needed amount of skilled working personnel?
- Where should we increase process quality?

3.9.2 Use Case 34: Autonomously Controlled Material Flow

Description Along the Framework Categories

Apart from the actual assembly of the prefabricated, auxiliary, and raw materials into finished products, the unobstructed transport of the needed resources within the facility is also one of the most important processes of a manufacturing company. The relevant intralogistics processes include the transport between the warehouse and the production facility, the guidance of the subassemblies through the different stations on the shop floor, and, at the end of processing, transport back to the warehouse or the shipping company. Depending on individual flexibility requirements, the automation of these processes poses a minor or major challenge. Successfully mastering it can lead to significant increases in efficiency. The following use case, autonomously controlled material flow, addresses the issue of effort intensive transportations of the workpieces through and between the production facility and the warehouse. The principle itself is older than manufacturing. In the early fifteenth century, the Venetians used waterpower to guide ships through the construction process and with the introduction of conveyor belts in vehicle manufacturing, Henry Ford revolutionized not only vehicle manufacturing but also revolutionized the paradigm of industrial manufacturing itself.

In addition to the classic conveyor belt, novel technologies have been introduced in the last few decades with the objective of making the processes more flexible and efficient. In this regard, automatic guided vehicles (AGVs) have been introduced as a solution.

According to the Association of German Engineers, AGVs can be defined as follows: "automated guided vehicles are fixed area means of conveyance with their own drive, which are automatically steered and driven without direct contact. They serve to transport goods, namely, to pull or carry freight with active or passive means of load transfer. This guideline will consider only wheeled vehicles, excluding track-guided vehicles, air-cushion vehicles or walking machines" (VDI, 2005 cited in Ullrich, 2015, p. 98).

The AGV-system consists of three parts: the vehicle itself, the guidance system, which allows the vehicle to maneuvers itself throughout the facilities, and floor-mounted equipment (VDI, 2005 cited in Ullrich, 2015). For the sake of simplicity, we will generally refer to AGVs that use both fixed guidance tracks and digital methods for navigation. The technologies used to navigate throughout the area are very versatile and can consist of cameras, laser-based techniques, optical sensors, Bluetooth receivers, magnetic sensors, and more (dos Reis & Morandin Junior, 2021). The main parts of AGVs are the chassis, load handling devices, warning and safety devices, energy supply supplies, vehicle control, navigation elements, data transmission, and other control elements (VDI, 2005).

Picture of the Model/Framework Explanation

The autonomous guidance of finished and unfinished products, as well as other materials, has two areas of application. It can be used to transport the material between the production facility and the warehouse and lead the goods to be manufactured throughout the manufacturing process. Since the process is automatically planned using complex algorithms and technologies, it is considered prescriptive and part of the intralogistics and process control (see Fig. 3.73).

Relevant Technologies

The following technologies are considered relevant for this use case:

- Enterprise Resource Planning System
- Manufacturing Execution System

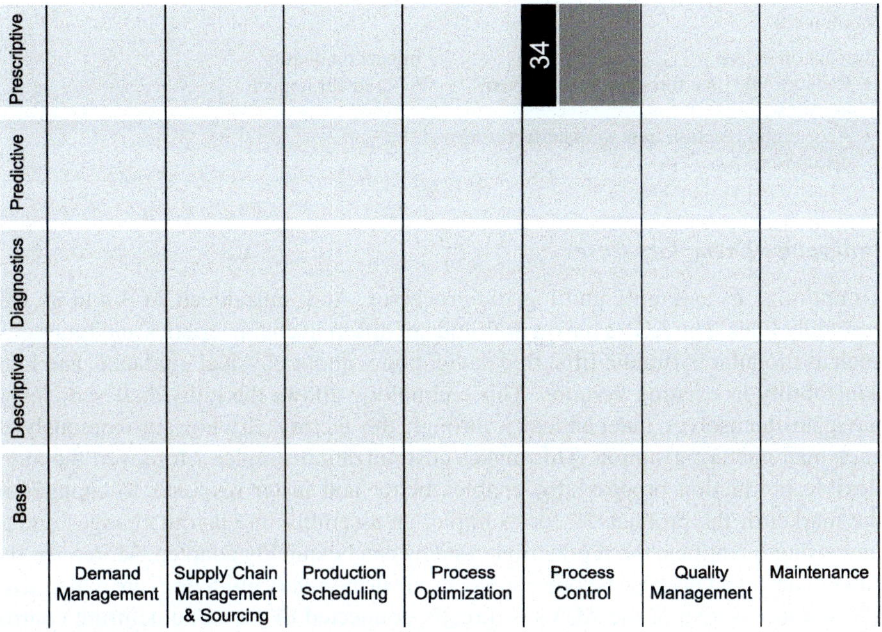

Fig. 3.73 Smart factory framework—use case 34

- Product Lifecycle Management System
- Cloud Computing
- M2M Communication
- Internet of Things
- Sensors and Actuators
- Track & Trace Technologies
- Vision Technologies
- Cyber-Physical Systems
- Big Data Analytics
- AI, ML & Simulation
- Automated Guided Vehicles

Impact on Competitive Priorities

The impact of this use case on competitive priorities is shown in generic form in the Fig. 3.74. The specific impact depends on the context of the company.

Impact on flexibility	Impact on costs
• Changes the conveying sequence • Adapts and reconfigures the conveyor system • Adapts to fluctuating transport performance requirements	• Relieves employees, freeing up personnel capacity due to the automated material flow • Increases capacity utilization due to the autonomous material flow

(continued)

(continued)

Impact on delivery	Impact on quality
• Reduces idle time through more efficient logistics • Accelerates reaction time to manufacturing disruptions	• No direct impact

Innovators/Exemplary Cases

To optimize its assembly and logistic processes, Audi introduced AGVs in its R8-assembly lines. The AGVs have a multitude of different characteristics and functions, such as modular hydraulic lifts, free navigation without physical guidance, and high adaptability to existing systems. This technology allows the individual vehicles to navigate themselves independently through the factory, driving autonomously to each manufacturing station. This makes customization simpler. Moreover, the more flexible production process also enables better and faster response to changes in the market or the product. If, for example, an assembly line layout changes due to increasing quantities, the production systems can be quickly adapted. Moreover, the AGVs can detect people and other subjects, ensuring safety guidelines are adhered to. The control system of the AGVs is directly connected to the manufacturing control system. This way, the automated guided vehicles are directly updated as soon as changes in production occur (Kirmse, 2022).

One of the most well-known and exemplary cases of the use of AGVs is their application in the warehouses of e-commerce giants such as Amazon or Alibaba. China's biggest robot warehouse in Wuxi, owned by a subsidiary of Alibaba, utilizes 700 AGVs that can autonomously pick up and deliver parcels throughout the warehouse. Although humans still work together with the robots, the robots already have brought vast time savings. In 2019, Alibaba processed over 1.3 billion orders on "singles day", the world's biggest shopping holiday. AGVs played a key role in processing and fulfilling the various orders (Kharpal, 2018; Meisenzahl, 2020).

The AutoStore company has developed an autonomous intralogistics and storage system. The system consists of a small number of basic components that can be easily implemented in existing facilities. Simulation software continuously optimizes the warehouse and logistics processes. The growing demand for e-business has led to more than 900 installations of this autonomously controlled warehouse system around the world (Gapper, 2022).

When and How to Apply This Use Case?

Modern technologies for transporting products and materials within and between the production plant and the warehouse can have decisive advantages over the conventional conveyor belt. The following points lead to an increased added value through the application of modern transportation technologies:

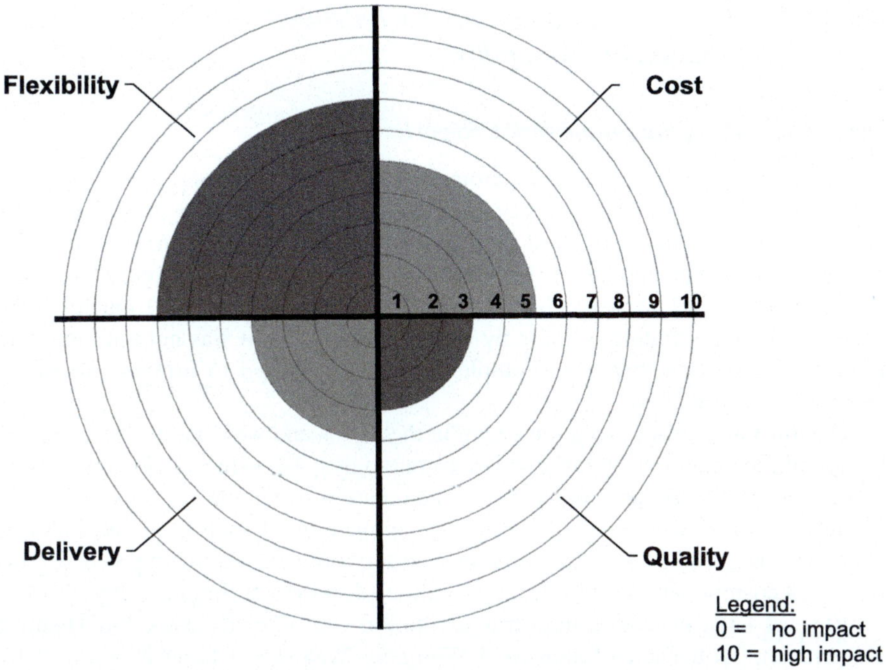

Fig. 3.74 Impact—use case 34

- **High throughput**: A higher throughput may lead to an increased effort in intralogistics, especially if there are many customers with rather small orders. AGVs help to process these orders faster.
- **Existence of a cyber-physical system**: The combination of AGVs and the production control within a cyber-physical system leads to synergies and helps to achieve added value.
- **High required flexibility**: A high need for flexibility within the manufacturing process (i.e. due to high customization of the product) can be made more feasible by deploying technologies that can transport single workpieces/batches independently.
- **Regular changeover of the production layout**: AGVs are more flexible when it comes to changes in the manufacturing layout than conveyor belts.

Questions to Ask

- How many orders do we have to process on a daily basis?
- Are there any changes during the manufacturing/assembly process?
- Where do we need individual transportation of our products?

3.9.3 Use Case 35: Autonomously Controlled Manufacturing/Assembly

Description Along the Framework Categories

Mechanization and, in particular, automation are possibly among the most fundamental paradigms of industrialization or of the industrial revolutions. The Cambridge Dictionary defines the concept of automation as follows: "the use of machines and computers that can operate without needing human control" (Cambridge, n.d.). Although processes were mechanized (and debatably automatized) much earlier, for example, the grinding of grain by wind or watermills, it was not until the third industrial revolution that more complex operations started to become automated (Sharma & Singh, 2020).

The third industrial revolution began in the 1970s and was mainly characterized by the implementation of information technologies. By using computers as well as memory-programmable controls, the automation of simple manufacturing and production processes became possible. The application of such novel technologies led to various benefits such as mass production, more standardized processes and increased customization of product possibilities. (Sharma & Singh, 2020).

40 years later, the fourth industrial revolution was officially coined by Henning Kagermann, Wolf-Dieter Lukas, and Wolfgang Wahlster (Chung & Kim, 2016; Kagermann et al., 2011) at the Hannover Messe. The main concept behind Industry 4.0 is the connection of all relevant subjects and objects of the different processes within cyber-physical systems, which is enabled by a multitude of novel information and communication technologies (Chung & Kim, 2016). This leads to more intelligent automation of the processes, which, in turn, allows the mass production of personalized products in small batches or even single pieces (Chung & Kim, 2016; Cohen et al., 2019). To this end, a variety of different technologies are applied, such as AI, IoT, or vision systems. Digital and analogue labels, such as RFID, matrix codes, or bar codes, which give the specific batches/products a unique identity within the shop floor, are also key enablers of this revolution and enable the automatic and smart production of the respective goods (Cohen et al., 2019).

Picture of the Model/Framework Explanation

In this use case, the manufacturing and/or assembly processes are automatized by intelligent working machines. For this, a variety of rather novel digital technologies and methods are used, which enable mass customization. Due to the high level of analytical capabilities and autonomy of the technologies involved, on top of its predictive and self-regulating elements, this use case is classified as prescriptive (see Fig. 3.75).

Relevant Technologies

The following technologies are considered relevant for this use case:

- Enterprise Resource Planning System

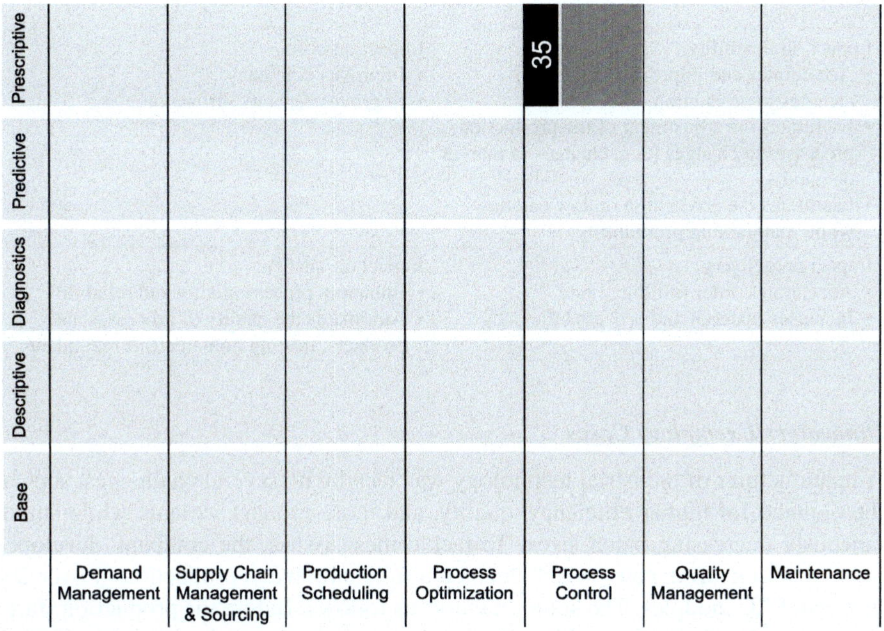

Fig. 3.75 Smart factory framework—use case 35

- Manufacturing Execution System
- Product Lifecycle Management System
- Long Range Wide Area Network
- 5G
- M2M Communication
- Mobile Computing
- Internet of Things
- Sensors and Actuators
- SCADA System
- Cloud Computing
- Edge Computing
- Track & Trace Technologies
- Vision Technologies
- Cyber-Physical Systems
- Big Data Analytics
- AI, ML & Simulation
- Collaborative Robotics

Impact on Competitive Priorities

The impact of this use case on competitive priorities is shown in generic form in the Fig. 3.76. The specific impact depends on the context of the company.

Impact on flexibility • Accelerates and improves reaction to changes in production • Facilitates the adjustment of the production processes to changes (e. g. changes in market demand) • Facilitates the production on low batches while maintaining profitability	Impact on costs • Improves efficiency • Improves capacity utilization
Impact on delivery • Accelerates order fulfilment rate • Increases material and process efficiency	Impact on quality • Enhances process quality and reliability • Automates the quality of processes and products, making human effort redundant

Innovators/Exemplary Cases

A manufacturer of industrial technology was faced with several challenges, such as the demand for higher efficiency, quality, and more product variants while simultaneously decreasing batch sizes. To tackle these issues, the company developed novel manufacturing cells, which enabled a fully automated production of over 200 different PLC modules. The solution works as follows. Before the production starts, a serial number for each specific order/batch is generated, which is then added to the workpiece. While the piece goes through several assemblies, testing, and labelling stations, sensors detect the serial numbers and match them with the data of the ERP. This way, the different stations are automatically instructed on how to process the specific workpiece. After each station, the data are stored within the ERP. Additional to the automated processes, the process data are also automatically gathered and stored and can further be used for tasks such as the creation of preventive maintenance plans and further process optimizations. Ultimately, this system of fully automated machines led to higher productivity, an increase in transparency, quality, and traceability, as well as a decrease in unplanned downtimes of the machinery (B&R Industrie-Elektronik GmbH, 2016).

The second use case concerns the autonomous assembly process of automotive parts on an ongoing conveyor belt. The introduced solution consists of two industrial robots on each side of the vehicle chassis. When a chassis is conveyed between the robots, a smart camera detects the centre of the rim, as well as all other relevant spatial parameters such as the type, size, and colour of the wheel, and records them. Intelligent software checks whether the wheel complies with the order data within the system and if the order data matches the wheel. When all checks have been successfully completed, the robotic arms place the wheel on the vehicle and fasten the bolts. All this happens in under a minute. Apart from a significant increase in efficiency, the benefits of the solution are enhanced flexibility, fewer errors, and higher process reliability. Furthermore, the developed system is highly flexible and, therefore, can be used for a multitude of different vehicle and wheel types (Knappe, 2015).

When and How to Apply This Use Case?

Manufacturing/assembly automation is a use case that can be favourably applied to (almost) every process. There are, however, a few exceptions where manual labour is preferred to automated processes:

- **Versatile processes**: For processes with very versatile circumstances, full automation might not be feasible or efficient due to the high requirements of the system.
- **The process cannot be done better by machines**: There are very few exceptions where labour skills might be better when it comes to quality-related aspects.
- **Hand-crafted as a unique selling position (USP)**: Some companies label their products as "handcrafted". As such, manufacturing/assembly automation may not be suitable.

Questions to Ask

- Which process do we want to automatize?
- Which machines are required to automatize this?
- Is it possible to automate (parts of) this process?
- Can we achieve a better outcome with automation?

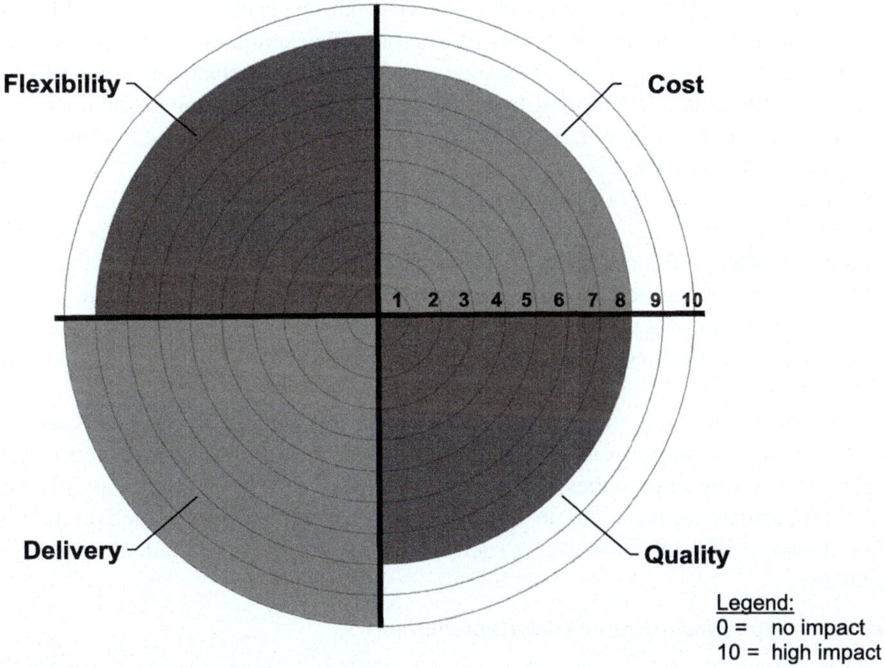

Fig. 3.76 Impact—use case 35

- Which track & trace technologies are suitable for your process and add minimum of cost to it?

3.10 Quality Management

3.10.1 Use Case 36: Fault Detection

Description Along the Framework Categories

The term quality refers to how well a product fulfils its intended functions as well as how well it matches its design characteristics. Two aspects mainly determine the quality of a product. Firstly, the design has a determining role in defining quality goals. Secondly, the process from which the product results influences its quality (Das & Miller, 1999).

To ensure the promised quality for the customer, increasingly complex techniques for quality management have been introduced. This use case focuses on the simplest approach for ensuring quality, namely the detection of faults in products. Although the roots of the quality inspection of products can be traced back to the Middle Ages, it was not until the early twentieth century that quality inspections were introduced (Quality Management System, n.d.; Weckenmann et al., 2015). Fault detection, the oldest quality management approach, concerns the inspection of final products in order to detect product faults to discard or repair (Weckenmann et al., 2015). Even though quality management methods have become more sophisticated and holistic over time, the direct detection of faulty pieces within or at the end of the manufacturing/assembly process remains one of the most important approaches to ensure the promised quality for the customer.

According to Mitra (2008), there are four quality metrics for products and services: structural, sensory, time-oriented, and ethical. Structural metrics concern the physical properties of the products, such as the elasticity of a steel beam or the weight of a bolt. The sensory metrics describe the perceived characteristics, such as the look of a car or the scent of a perfume. Time-oriented metrics include aspects such as the reliability and robustness of a shoe, and ethical metrics are concerned with ethical values such as sustainability.

Depending on the metrics to be tested, different methods for quality inspection exist. These can range from simple image processing to multi-stage functionality tests. The testing approaches can be either non-destructive or destructive. Novel methods, which are the focus in this use case, are mainly non-destructive and are based on novel digital technologies such as machine vision, laser, and other types of sensors.

Picture of the Model/Framework Explanation

Although some fault detection applications require advanced analytical algorithms (e.g., machine vision-based quality control methods), the approaches merely detect

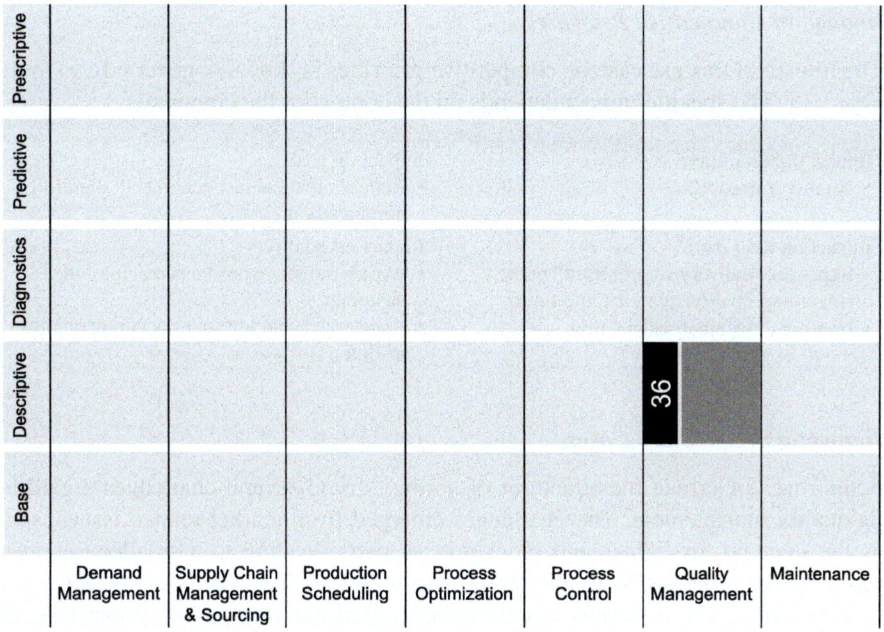

Fig. 3.77 Smart factory framework—use case 36

the flaws of the products without comparing them to other data for purposes such as the identification of the root causes. Therefore, the use case can be considered descriptive (see Fig. 3.77).

Relevant Technologies

The following technologies are considered relevant for this use case:

- Enterprise Resource Planning System
- Manufacturing Execution System
- Product Lifecycle Management System
- SCADA System
- Cloud Computing
- Internet of Things
- Sensors and Actuators
- Track & Trace Technologies
- Vision Technologies
- Cyber-Physical Systems
- Big Data Analytics
- AI, ML & Simulation
- Augmented Reality

Impact on Competitive Priorities

The impact of this use case on competitive priorities is shown in generic form in the Fig. 3.78. The specific impact depends on the context of the company.

Impact on flexibility • No direct impact	Impact on costs • Reduces manual inspection and, therefore, human labour
Impact on delivery • Improves "delivery as promised" quote (regarding quality, quantity, and time) • Improves customer satisfaction	Impact on quality • Avoids defective products reaching the customer • Accelerates and improves accurate quality testing

Innovators/Exemplary Cases

Schnöring, a German manufacturer of springs, faced several challenges regarding its quality management. The challenges emerged from market-related issues, such as the demand for lighter and more precise parts, leading to a smaller tolerance range for the products. With an output of up to 60,000 parts per hour and up to 100 metrics per part to be tested, reaching the desired quality was no simple task. At the same time, the company had employee-related difficulties. The people responsible for the quality assurance, which was still predominantly done analogously and manually, had differing levels of know-how and expertise. Training workers to adequately measure and document parts was not only time-intensive but also inefficient, leading to increased efforts in monitoring and reviewing. To mitigate these issues, while also increasing process reliability, the company implemented a novel machine vision-based quality testing method. To measure the metrics, the single parts were displayed on a profile projector, measuring all relevant characteristics within seconds and simultaneously comparing them with the associated CAD data. Such an approach enabled the company to conduct quality measurements faster, more reliably and with better precision while also becoming less dependent on the workers' individual skills (Spinnarke, 2015).

Manz AG, a producer of manufacturing solutions, encountered issues in the quality assurance process of its lithium batteries. To ensure stable and secure weld connections between the different battery blocks, destructive test methods, such as cross-section abrading, or force pull-off tests, were conducted on samples. This not only led to increased waste but was also an inadequate method of conducting a quality check for each block. To resolve this problem, a non-destructive machine vision-based approach was applied. The solution, based on AI, contains deep- learning mechanisms with many associated benefits, such as a decreased effort in coding, better detection of errors, and the autonomous detection of flaws. Due to the scarcity of valid images to train the algorithm, the company employed a module that enables the creation of a detection model of only one permillage of the normally required images. Overall, the integration of this technology-based solution allowed the introduction of an autonomous, non-destructive, 100% quality control approach, leading

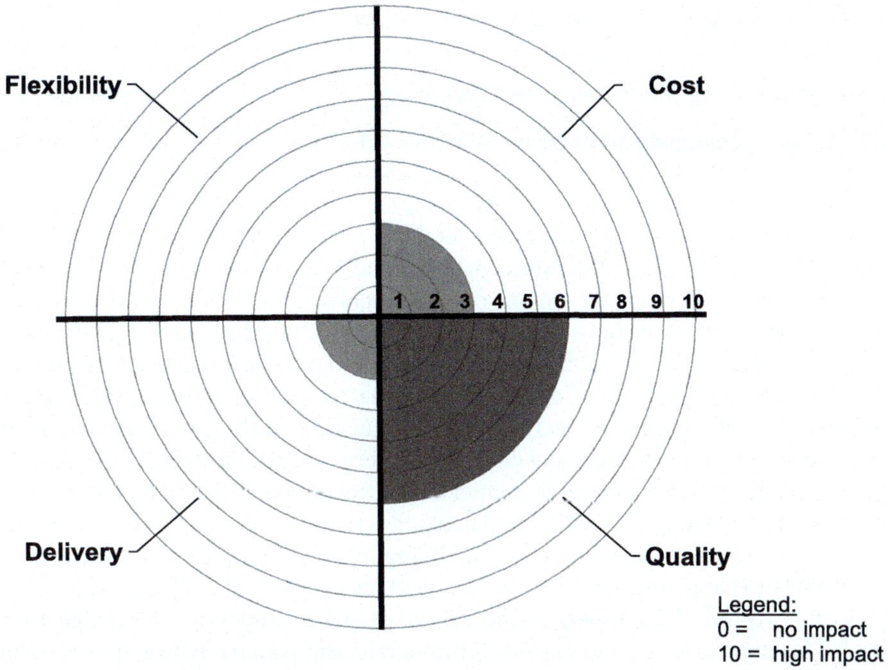

Fig. 3.78 Impact—use case 36

to a reduced scrap rate, faster processes, reduced costs, and overall increased quality and reliability (Manz AG, n.d.).

When and How to Apply This Use Case?

The inspection of products for possible faults is the most fundamental approach to ensure the quality of the produced goods for the customer. For this reason, quality inspection is fundamental for any production.

Questions to Ask

- Which quality metrics are important for our customers?
- Do these metrics differ in their importance?
- At what rate do our products not meet these quality standards?
- What are the consequences if a faulty piece reaches the customer?
- How can we measure these metrics faster and more reliably?

3.10.2 Use Case 37: Quality Diagnostics

Description Along the Framework Categories

During the operational part of the product life cycle, a multitude of different information is generated. This information can concern the quality of raw materials, process and machine data, and many other aspects. Each of these factors can influence the final quality of a product (Last & Kandel, 2001). Hence, it is in the interest of the manufacturer to record and use these data for quality measurements of products and processes. In recent years, the amount of operational data has been increasing steadily (Miller, 2017). Parts of this phenomenon can be traced back to the recent developments in information and communication technology, which made the recording, transmission, and storage of data more feasible for industrial use (Miller, 2017; Schuh et al., 2017). Unfortunately, the mere acquisition of the data is not enough to gain meaningful knowledge about the factors determining the quality of a product. To get further insights, techniques to extract and discover knowledge in databases must be applied. Knowledge discovery in databases consists of five steps: (1) data preparation, (2) data pre-processing, (3) data mining, (4) interpretation and evaluation of the results, (5) implementation of the obtained results (Köksal et al., 2011).

In this regard, data mining is the central step for extracting knowledge from data. This approach helps to discover correlations and patterns between exogenous variables and product quality, which can be used for quality assurance and further quality improvement. Data mining methods employ pattern recognition technologies and mathematical techniques to analyse large amounts of data in order to discover trends, patterns, and correlations within and between the various datasets (Gartner, n.d.). These insights can be used to identify the root causes of specific quality metrics of a product.

Picture of the Model/Framework Explanation

This use case is about the identification of the root causes of a specific qualitative outcome. To effectively identify the patterns that lead to the specific product quality, advanced analysis techniques must be applied to the various data sets. For this reason, this use case is categorized as diagnostic (see Fig. 3.79).

Relevant Technologies

The following technologies are considered relevant for this use case:

- Enterprise Resource Planning System
- Manufacturing Execution System
- Product Lifecycle Management System
- SCADA System
- Cloud Computing
- Internet of Things
- Sensors and Actuators
- Track & Trace Technologies

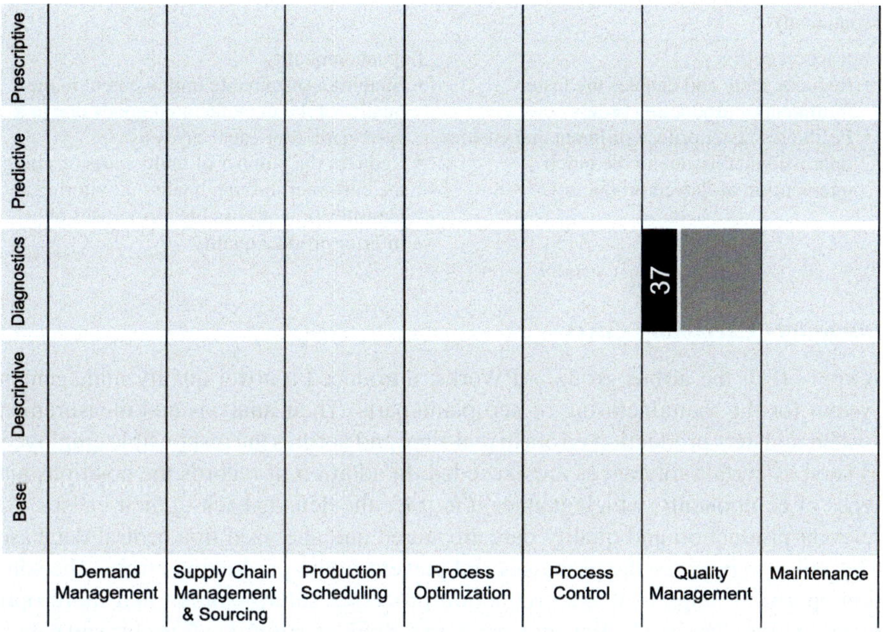

Fig. 3.79 Smart factory framework—use case 37

- Vision Technologies
- Cyber-Physical Systems
- Big Data Analytics
- AI, ML & Simulation

Impact on Competitive Priorities

The impact of this use case on competitive priorities is shown in generic form in the Fig. 3.80. The specific impact depends on the context of the company.

Impact on flexibility	Impact on costs
• Improves understanding of processes and facilitates the production of similar products	• Reduces waste, leading to an increased yield and thus helps to achieve a higher productivity • Improves productivity, reduces scrap and reduces the cost per item produced

(continued)

(continued)

Impact on delivery	Impact on quality
• Reduces scrap and enables the faster fulfilment of production orders • Facilitates faster order fulfilment and ensures improved quality due to the quick intervention of flawed products	• Enables a sustainable improvement in the quality of the manufactured goods due to a systematic root-cause analysis • Reduces the number of faulty parts reaching the customer through quality monitoring and identification of possible root causes of inferior product quality

Innovators/Exemplary Cases

A spin-off of the airbus group, APWorks, introduced a novel quality management system for the manufacturing of aeroplane parts. Their analysis and measurement system registers production errors in real-time and notifies the responsible employees as soon as certain tolerances are exceeded. In addition, it records the positions and types of components, which enables it to trace the defects back to their causes. All relevant production and quality data are stored and analyzed in a central database. This helps to improve the processes and prevent future production errors. The solution approach helps to make production processes safer, simpler, and more optimized. This makes production more resource-efficient, which leads to cost and weight savings. This approach is Airbus' first step towards centralized quality assurance, as the data and know-how gained can be shared with companies with similar production processes (Engel, 2022).

An international manufacturer of bread was facing quality issues in its toasted bread line, leading to an increased waste rate. Initially, the problems were investigated manually, and the root causes for the quality losses could not be identified. To address this failure, the manufacturer employed an analytics company to investigate the causes, optimize its processes and improve the quality. The applied AI platform enabled the identification of several causes, which led to the losses in its production. Subsequently, specific recommendations were extrapolated in order to prevent future inefficiencies. In a first step, all relevant process data, including raw material data, were consolidated and enriched to make the interdependencies visible. Secondly, the knowledge gained about the process characteristics was used to create predictive recommendations to optimize processes. To maximize utility, these recommendations defined proactive alerts highlighting the root causes of issues and alerting the production personnel as soon as quality-related process inefficiencies occurred. This way, troubleshooting was improved enormously. The introduced solutions led to a total of 138 new insights about the processes, which significantly increased the know-how of the production team. Furthermore, waste due to unsuitable size or weight was cut by at least two-thirds each. The reduced waste rate, as well as the process optimizations, led to an overall increase in production capacity which helped the company achieve annual savings of almost €1 million on a single production line (Seebo, 2018).

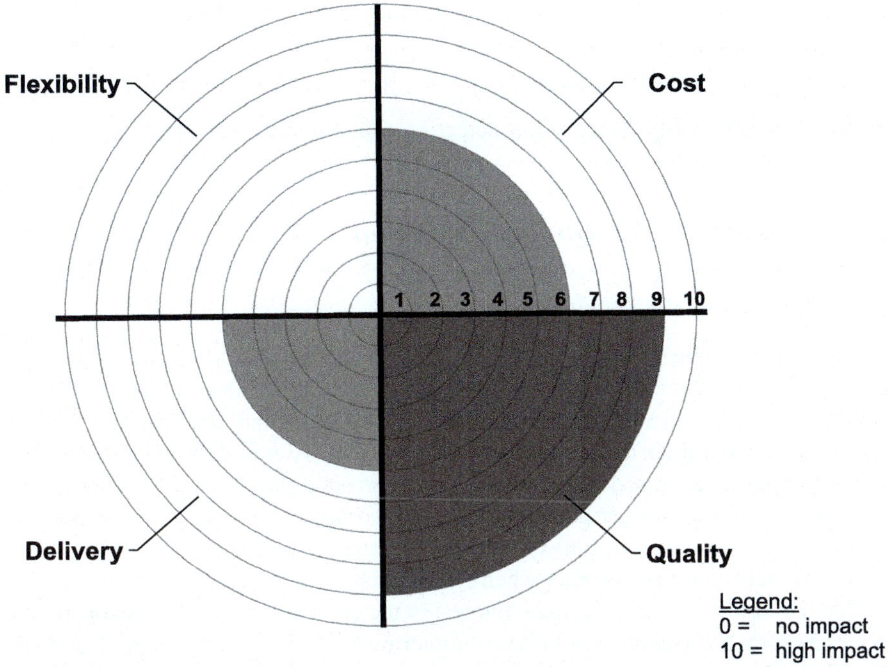

Fig. 3.80 Impact—use case 37

When and How to Apply This Use Case?

This use case is especially useful in the following situations:

- **High product fault rate**: A high product fault rate indicates flaws, either in the production process, the materials used, or the design itself. By applying quality diagnosis, the root causes of the flaws can be identified, and the knowledge gained used to prevent errors in the future.
- **Lack of knowledge about the processes**: Quality diagnostics also help to gain knowledge about the processes and, thus, also the parameters that influence the quality of the product. This might facilitate future process improvements.

Questions to Ask

- Which product faults occur on a regular basis?
- Do we know the root cause of these faults?
- If no:

  - Where do we assume there are correlations between the process parameters and the product quality?
  - How can we measure them?
- Do we store quality data or fault detection data comprehensively?

3.10.3 Use Case 38: Quality Prediction

Description Along the Framework Categories

In production, there is a multitude of determinants that can affect the subsequent quality of the product, such as tool wear (Jain & Lad, 2019), the quality of the materials used (Schmidt et al., 2018), and other process-related and -unrelated parameters. Often, however, hidden defects occur despite supposedly correct process parameters. For a company, late discoveries mean that the errors either have to be corrected or that correction is no longer profitable, and the workpiece ends up as scrap. Regardless which of these two options is chosen, the company experiences high additional costs, which become even higher when entire batches are affected by the same quality problem. For this reason, it is in the interest of the company to detect possible quality issues as soon as possible and initiate countermeasures at an early stage. Due to the interdependencies between various process-related and -unrelated parameters, the manual extrapolation of possible quality issues from plain sensor data is extremely difficult. (Wiegand et al., 2016). Therefore, novel sophisticated methods are required to detect possible quality issues and, thus, prevent them.

This is exactly where this use case comes in. Originally, statistical models were used to predict product quality. Here, an attempt was made to establish a statistical relationship between machining parameters and quality parameters. With the recent advancements in modern information technology, methods for the detection, and extraction of product and process features, which determine product quality, have become increasingly feasible. In particular, the application of AI is more commonly used for quality prediction (Bai et al., 2021).

Apart from the prevention of quality issues, the application of this use case has another major benefit. The prediction of product quality facilitates quality testing. This is especially useful with a high throughput, where additional machines are necessary to check the product quality of each piece. By utilizing quality prediction, however, only a selected number of pieces need to be tested.

Picture of the Model/Framework Explanation

In this use case, various advanced analytical techniques are applied to forecast product quality. Therefore, its classification is predictive (see Fig. 3.81).

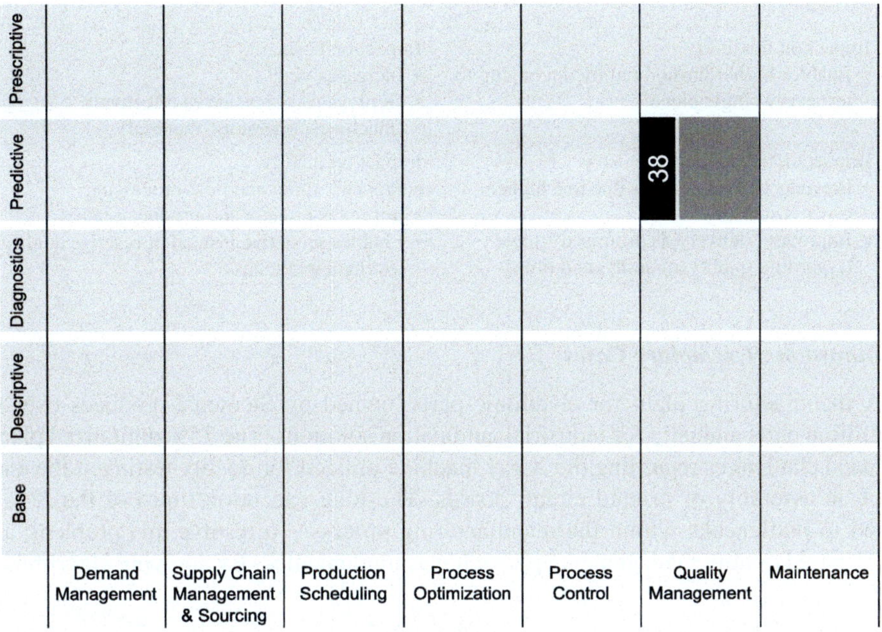

Fig. 3.81 Smart factory framework—use case 38

Relevant Technologies

The following technologies are considered relevant for this use case:

- Enterprise Resource Planning System
- Manufacturing Execution System
- Product Lifecycle Management System
- SCADA System
- Cloud Computing
- Big Data Analytics
- Internet of Things
- Sensors and Actuators
- Track & Trace Technologies
- Vision Technologies
- Cyber-Physical Systems
- AI, ML & Simulation

Impact on Competitive Priorities

The impact of this use case on competitive priorities is shown in generic form in the Fig. 3.82. The specific impact depends on the context of the company.

Impact on flexibility • Enables higher throughput by not having to test every single piece	Impact on costs • Increases yield • Deploys resources more effectively (machines, operators, material)
Impact on delivery • Prevents delivery delays due to a higher process stability • Improves "delivery as promised" quote (regarding quality, quantity, and time)	Impact on quality • Prevents scrap and post-processing • improves process and product quality • Enables proactive instead of reactive quality testing approaches

Innovators/Exemplary Cases

A manufacturing plant for electronic parts, owned by Siemens, produces over 6 million parts annually for industrial automation solutions. The 75% digitalized plant faced challenges regarding the X-ray machine utilized for quality testing at the end of an assembly of printed circuit boards. The long execution times of the X-ray led to bottlenecks within the manufacturing process. To resolve this problem, an AI-based quality management approach was implemented. By applying algorithms,

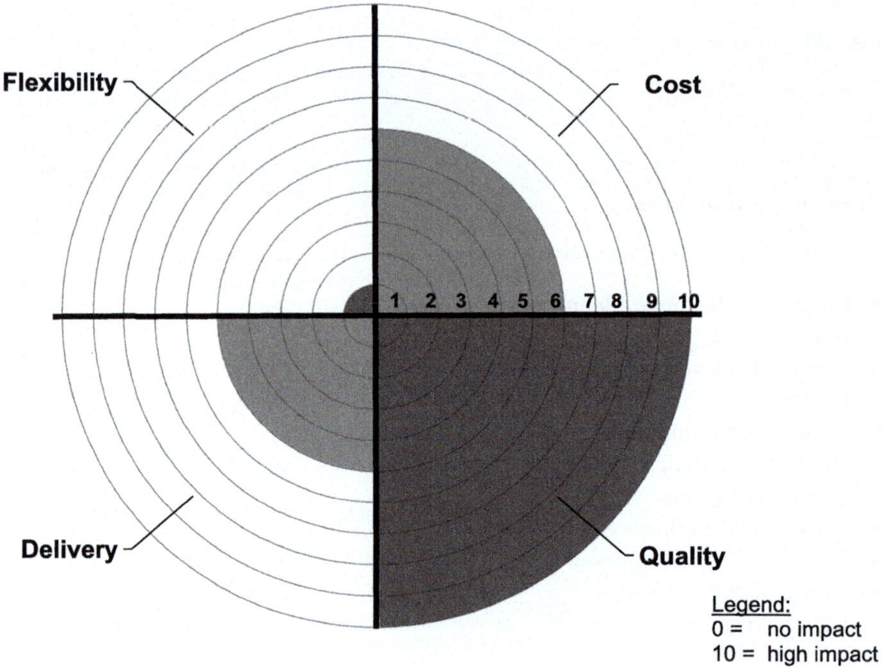

Fig. 3.82 Impact—use case 38

the probability of a production error was determined, enabling the decision as to whether a product/batch should go through the X-ray process or not. To establish the solution, the process data, as well as the quality data of the products, were collected and aggregated first. Subsequently, the multivariable dependencies between process data and X-ray results were modelled using an ML approach. Finally, the solution was implemented on the shop floor. The ML model was continuously "trained" to further increase the predictive power and reliability. This solution led to multiple advantages, such as the removal of the bottleneck without having to invest in another X-ray machine, a better understanding of the process, as well as a novel, continuously self-improving, quality management approach (Zemann, 2022).

It does not always take highly complex systems to proactively determine product quality. A manufacturer of copper wire was under strict quality requirements to satisfy its customer demand within the automotive industry. With a throughput of up to 1000 m of wire per minute, a reactive quality assurance approach was ineffective. By using the ERP to connect the data of the 300 production machines with the machine control system, information about the thickness of the wire, the external temperature, and the coil speed could be aggregated and analyzed. Such an approach enabled the company to make precise statements about the future quality of the wire and thus proactively intervene in the process (proALPHA Software GmbH, 2015).

When and How to Apply This Use Case?

This use case is especially useful in the following situations:

- **Post-processing is difficult/expensive or even impossible**: The postprocessing of certain parts might be either difficult or even impossible due to their highly complex structure (e.g., semiconductors). A proactive quality assurance approach can help to either timely correct or scrap the workpieces without further wasting resources.
- **Issues that affect whole batches**: If whole batches have been affected by the same issue(s), it might be advantageous to analyze the root causes to prevent the issues from reoccurring in the future.
- **High throughput**: If a machine/line achieves high throughput, individual parts cannot be tested without applying more monitoring machines. Quality prediction can be an economically favourable solution for this issue.
- **Quality cannot be measured within the process**: For some products, it might not be possible to monitor the quality during the manufacturing process. For reliable estimates, quality prediction approaches can help.

Questions to Ask

- Which process-related parameters influence the quality of our product?
- Which process-unrelated parameters influence the quality of our product?
- What are the benefits of a proactive quality management approach compared to a conventional, reactive approach?
- What is the scrap rate/ the expenditure of our existing quality assurance approach?

3.10.4 Use Case 39: Autonomous Quality Management

Description Along the Framework Categories

As mentioned in previous use cases, there are many different process-related and non-process-related parameters that can influence the quality of a product. The resulting complexity leads to several challenges for the manufacturers. On the one hand, the processes must be comprehensible to such an extent that the quality of the products (and the associated processes) can be understood, and proactive and targeted intervention can be accordingly taken. On the other hand, the adequate adjustment of the quality-relevant parameters can quickly become a stress test for employees, especially when multiple machines and adjustable variables exist. To circumvent this problem and to reduce errors caused by human input, approaches exist that allow a system to autonomously influence quality-relevant factors, thus ensuring and, ideally, improving product quality.

Lou and Huang (2001, cited in Lou & Huang, 2003) proposed an autonomous quality control approach to proactively control and optimize the yield quality (see Fig. 3.83). The model contains four main elements, which, significantly, overlap with the four specific quality use cases in the process step quality management of our framework:

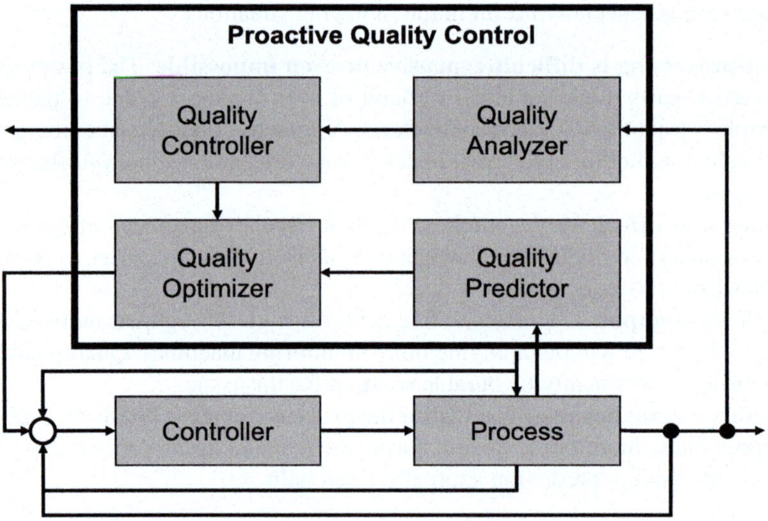

Fig. 3.83 Hierarchical process control for a proactive quality control approach (Lou & Huang, 2001, cited in Lou & Huang, 2003)

1. The **quality analyzer** module assesses the quality of the finished products. If the quality is insufficient, the component sends the metrics to the next module, the quality controller.
2. The **quality controller** module identifies the root causes of the flawed product. If the quality issue is caused by unsuitable process parameters, the relevant parameters are identified and transmitted to the fourth module, the process optimizer. Otherwise, if the quality problem is caused by other things, such as flawed equipment or human error, the diagnosis report is sent to off-line engineers for further measures.
3. The **quality predictor** module uses various operational data sources to forecast the quality of the product. Those data are then used for further quality assurances and improvements in module four.
4. The **process optimizer** module uses data of the quality predictor and the quality controller to calibrate the process parameters and ensure and optimize the quality of the manufactured products.

Picture of the Model/Framework Explanation

Due to the combination of the different quality management approaches, the high degree of self-regulation and its predictive applications, this use case is positioned within the prescriptive quality management field (see Fig. 3.84).

Relevant Technologies

The following technologies are considered relevant for this use case:

- Enterprise Resource Planning System
- Manufacturing Execution System
- Product Lifecycle Management System
- SCADA System
- Cloud Computing
- Edge Computing
- Internet of Things
- Sensors and Actuators
- Track & Trace Technologies
- Vision Technologies
- Cyber-Physical Systems
- Big Data Analytics
- AI, ML & Simulation

Impact on Competitive Priorities

The impact of this use case on competitive priorities is shown in generic form in the Fig. 3.85. The specific impact depends on the context of the company.

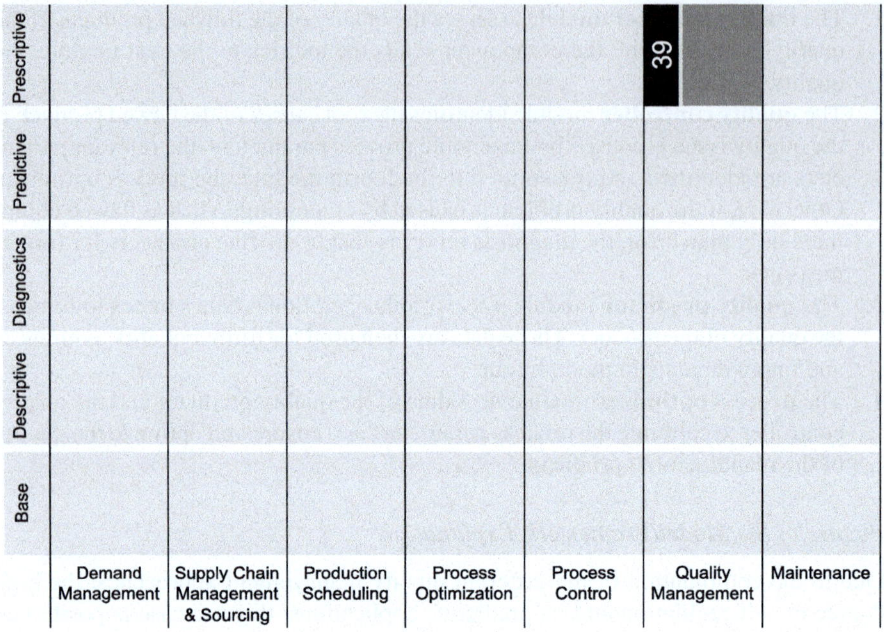

Fig. 3.84 Smart factory framework—use case 39

Impact on flexibility • Enables the production of more product variants due to autonomous process parameter settings • Facilitates easier switch between multiple product variants	Impact on costs • Increases yields • Accelerates the detection of process issues proactively, which otherwise might lead to additional costs
Impact on delivery • Prevents delivery delays due to a higher process stability • Improves "delivery as promised" quote (regarding quality, quantity, and time)	Impact on quality • Prevents scrap and post-processing • Improves process and product quality • Tests quality proactively instead of reactively • Automatizes quality assurance and optimization

Innovators/Exemplary Cases

Limtronik introduced an Industry 4.0 evaluation environment, focusing on automatic control and preventive optimization of the qualitative dimension of its products and processes. To this end, the data of various components, for example, plant modules and sensors, were coupled with each other. The process and machine data of the various plant modules were supplemented and merged with other measured values, such as the ambient temperature. This approach allowed the system to independently record and document product and process faults, and to process this information via

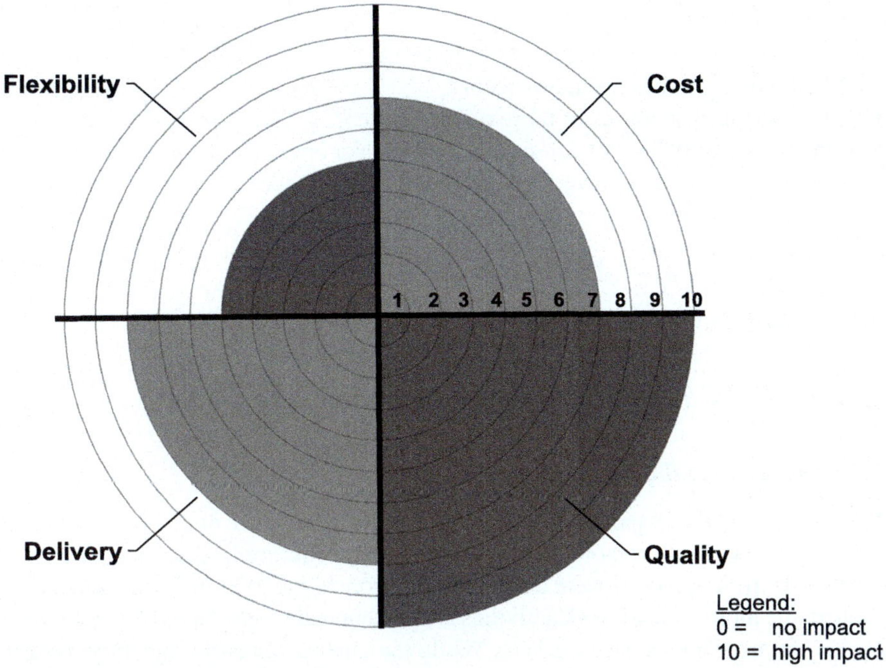

Fig. 3.85 Impact—use case 39

self-learning algorithms to determine the root causes of the faults. This enabled the system to intervene early in the processes to prevent possible errors. Such an approach facilitated the efficient and flawless production of goods in batches as small as one while also gaining process knowledge for future improvements (Bienek, 2022).

When and How to Apply This Use Case?

The scenarios where the application of autonomous quality management systems are useful are numerous. On top of the reasons mentioned in the previous use case, further cases exist whereby a company should think about implementing this use case within its manufacturing plant:

- **A multitude of process parameters to monitor and control**: The greater the heterogeneity of machines and influenceable parameters, the harder and more stressful it becomes for the responsible employees to monitor and adjust all parameters in time, especially if no central control system is available. Automated regulation of these parameters can help to resolve this issue.
- **Shortage of skilled personnel**: If the demanded personnel for quality-related tasks is not available, process automation addresses these shortages by circumventing the need for personnel.

Questions to Ask

- Which quality issues normally occur?
- What are the root causes of these quality issues?
- Can we measure the root causes adequately (aka operational parameters)?
- What is the impact of flawed products (e. g., lower prices, post-processing, scrap, etc.)?

3.11 Maintenance

3.11.1 Use Case 40: Remote Maintenance Assistance

Description Along the Framework Categories

The global COVID-19 pandemic demonstrated the importance of on-site presence in certain sectors. While remote working was rapidly embraced by service industries, workers from other sectors were unable to work from home. Within the manufacturing context, the more manual work activities are present in the operations of a factory, the greater the need for on-site workers. While the current industry trajectory towards smart manufacturing is increasingly reducing the need for physical labour, certain functions are still heavily dependent on human employees. One of these activities is the maintenance and repair of complex production equipment. Such tasks not only require manual skills but also require expert knowledge about the specific machines and their functionality. The shortage of highly skilled manufacturing engineers and the geographical distance between manufacturing sites means that knowledge about specific machines needs to be transferred remotely to the local maintenance personnel across the production network (Vorraber et al., 2020). A common way to consult and guide maintenance staff remotely is via (smart-)phones (Vorraber et al., 2020). Due to the complexity of the mechanics, verbal communication is sometimes insufficient to solve the issues of some machines, and visual aid is necessary to give the expert a greater insight into the issue and therefore enhance the troubleshooting process (Bottecchia et al., 2009).

This use case, remote maintenance assistance, is about the visual and verbal exchange of those issues so that maintenance can be carried out regardless of a lack of on-site experts. Novel industry 4.0 technologies such as AR and VR can support this function. By connecting an expert with a local maintenance worker, the latter can be guided through the maintenance processes without the expert having to be on-site (Wuest et al., 2020). The maintenance technician transmits what they see to the remote expert, who, in return, gives instructions on what to do (Walters et al., 2020).

Picture of the Model/Framework Explanation

Remote maintenance assistance relies on the transmission of real-time data, such as the visual properties/condition of the machine or the vocal communication between the responsible maintenance team and the expert. Therefore, no analytical applications are used with this use case, and its classification is descriptive (see Fig. 3.86).

Relevant Technologies

The following technologies are considered relevant for this use case:

- Enterprise Resource Planning System
- Manufacturing Execution System
- SCADA System
- Internet of Things
- Sensors and Actuators
- Vision Technologies
- Cyber-Physical Systems
- Cloud Computing
- Edge Computing
- Mobile Computing
- Wearables

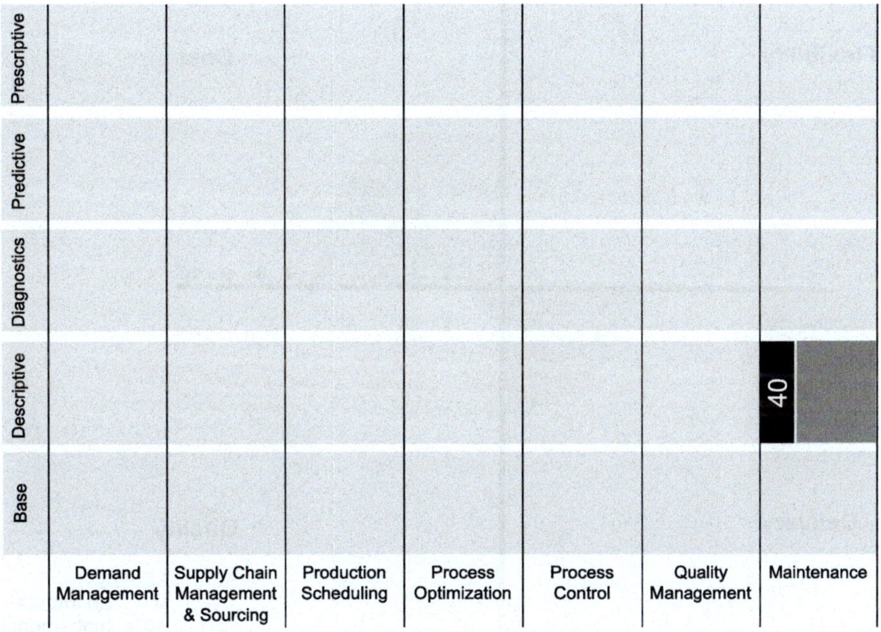

Fig. 3.86 Smart factory framework—use case 40

- Augmented Reality
- Virtual Reality

Impact on Competitive Priorities

The impact of this use case on competitive priorities is shown in generic form in the Fig. 3.87. The specific impact depends on the context of the company.

Impact on flexibility • No direct impact	Impact on costs • Shortens downtimes due to the prevention of depending on the arrival time of the expert • Avoids travelling costs
Impact on delivery • Improves reliability to deliver the products on due date due to shorter downtimes • Reduces downtimes of machines	Impact on quality • Accelerates and improves the detection and communication of error sources, which can lead to increased product quality

Innovators/Exemplary Cases

TECUMA Systems GmbH, a German machine manufacturer, developed a novel digital service concept for its machines for a customer with multiple international

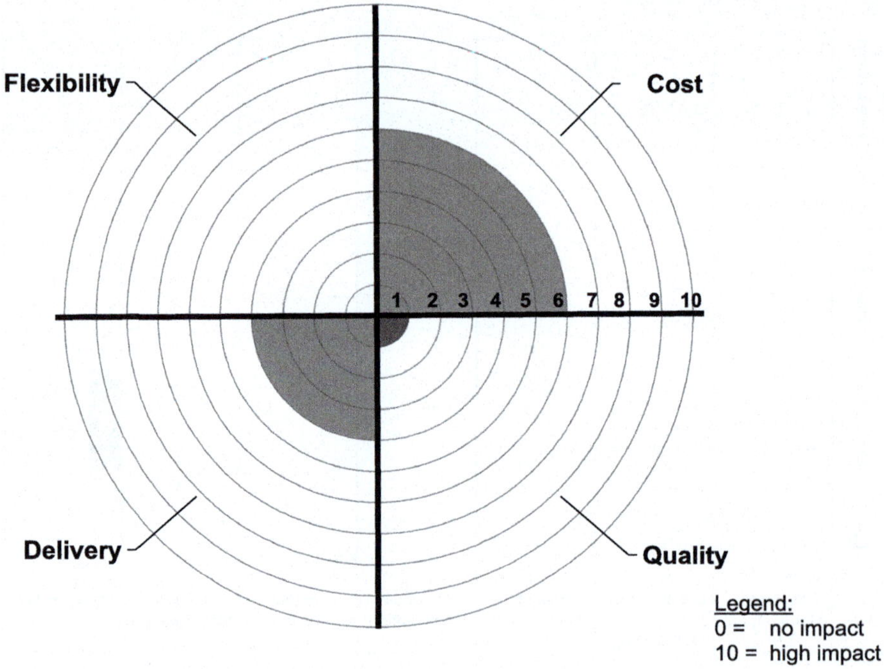

Fig. 3.87 Impact—use case 40

sites. For this purpose, the company used the smart service solution of the software producer Bitnamic. The solution contains a collaborative platform that facilitates the collaboration between an on-site machine technician and a remote expert through real-time multimedia interaction. During the repair/maintenance process, the technician wears industrial data glasses, which have several functions. The primary purpose is to visually transmit a real-time recording of what the technician sees to the remote expert, who instructs them about what to do. The smart glasses not only contain a camera but also have a small display so that the technician can see which part of the machinery gets visually transmitted to the expert. It is also possible to visualize the computer screen of the expert on the display so that the expert can visualize the next steps to solve the issue. To further enhance the instruction and the overall communication, the solution also contains an AR application. This allows additional informative content. To better instruct the on-site technician, the expert has a laser pointer to highlight specific points in the image. Furthermore, either a real-time or a recorded video, or a fixed image can be displayed. The software further records every video call together with its metadata with the purpose of eventually establishing a knowledge database (Deska et al., 2018).

When and How to Apply This Use Case?

There are several reasons for conducting remote maintenance assistance, such as not having sufficient qualified maintenance and repair personnel on-site and receiving specialized instructions remotely from experts. Furthermore, support could be outsourced as an alternative to employing internal technicians with the required knowhow:

- **Too few required maintenance activities**: It might not be worth hiring/training technicians or specific maintenance duties if the machine issues do not occur frequently enough.
- **Too few technicians with the required knowledge**: For some machines or technologies, there are not enough experts with the required knowledge. If a company cannot employ an expert, they could outsource this service.
- **Closed systems**: Not every operating system is disclosed to everyone. For reasons of confidentiality, some manufacturers might only utilize "closed systems", which makes them reluctant to employ "external" workers to maintain and investigate the hardware and/or software systems of their machines.
- **Different machines of different manufacturers**: Depending on the number of different machines a plant has; more or less specific knowledge is required to be able to handle them. A company with many different machines might want to deploy remote personnel because it cannot train its technicians for every machine.

Receiving external support instead of getting it directly remotely done by an expert.

- **Long travel time**: Longer travel time means also longer breakdowns of machines.
- **Travel restrictions**: travel restrictions, due to disruptions and/or regulations, for example, because of the COVID-19 pandemic, may hinder service personnel from leaving their country or region.

- **Additional costs**: Ordering an external maintenance technician usually leads to additional costs such as travel expenses, travel time, and accommodation.

Questions to Ask

- Do our in-house technicians possess the required knowledge about the maintenance of the relevant machine(-system)?
- Where are the experts who know how to repair said assets?
- Is it possible to train our technicians for said maintenance and repair duties? What are the costs?
- How long is the duration between the triggering of a service order and the arrival of the maintenance expert?
- How often does a breakdown in one of our machines occur? How often is expert knowledge required?

3.11.2 Use Case 41: Condition Monitoring

Description Along the Framework Categories

The regular monitoring of machinery and other manufacturing relevant equipment is fundamental for establishing proactive maintenance processes. This enables the shift from a preventive maintenance approach to a condition-based approach. Depending on the examined object, its condition can be evaluated in different ways. Possible methods to monitor the condition of assets include vibration analysis, thermography, tribology (the dynamics of the bearing-lubrication-rotor support structure of machines), visual inspections, or ultrasonics (Keith Mobley, 2002). To address machine and production issues as soon as possible and prevent accidents—both of which negatively affect productivity—machine monitoring is applied. Two aspects are important when it comes to the acquisition of machine data. On the one hand, the data must be displayed in real-time. This allows the quick reaction to defects and other anomalies before the damage increases. On the other hand, the data must be collected and visualized at a sufficiently high frequency. In addition to the improved reaction time, it also supports any subsequent analyses when it comes to determining the root causes of the defects. Ideally, the data acquisition is performed automatically by sensors that are non-invasive and non-destructive. In this way, the machine is not affected in terms of functionality and interruptions due to measurement. Additionally, the workers can focus on other, more important tasks.

Picture of the Model/Framework Explanation

In this use case sensors, such as cameras or thermometers, are used to measure specific properties of the manufacturing equipment. These measurements give insight into the condition of the said assets. Further analytics are required to investigate possible correlations of machine(-parts), products, and other process parameters. Therefore, condition monitoring is solely responsible for the acquisition of machine and tool data and, thus, falls within the descriptive category (see Fig. 3.88).

Relevant Technologies

The following technologies are considered relevant for this use case:

- Enterprise Resource Planning System
- Manufacturing Execution System
- SCADA System
- Cloud Computing
- Edge Computing
- Internet of Things
- Sensors and Actuators
- Vision Technologies
- Cyber-Physical Systems
- Big Data Analytics
- AI, ML & Simulation
- Augmented Reality

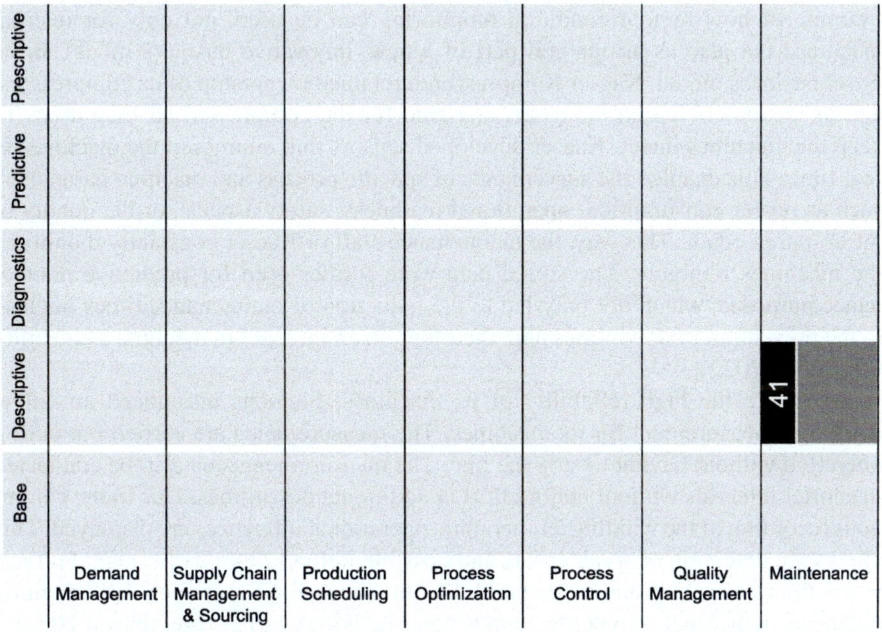

Fig. 3.88 Smart factory framework—use case 41

Impact on Competitive Priorities

The impact of this use case on competitive priorities is shown in generic form in the Fig. 3.89. The specific impact depends on the context of the company.

Impact on flexibility • Provides data for the evaluation of the remaining useful life (RUL) of a tool for later production changes	Impact on costs • Improves and facilitates the planning of maintenance, thanks to the possibility of monitoring and assessing the condition of machine parts and tools • Reduces downtimes
Impact on delivery • Improves and facilitates the planning of maintenance, thanks to the possibility of monitoring and assessing the condition of machine parts and tools • Reduces downtimes	Impact on quality • Enables the possibility of reducing scrap due to the monitoring of the tool wear

Innovators/Exemplary Cases

The German manufacturer of compressors, Kaeser Kompressoren, provides an example of how remote condition monitoring can be used, not only for internal machines but also as an integral part of a new, innovative business model. In its novel business model, Kaeser Kompressoren retained ownership of its compressors, and its customers leased the assets and paid for the compressed air they used. To keep the machines intact, Kaeser developed sensors that monitored the machines in real-time. This enabled the surveillance of specific process and machine parameters such as power consumption, operational readiness, safety aspects, or the quality of the compressed air. This way, the maintenance staff of Kaeser constantly monitored the machines remotely. The stored data were further used for predictive maintenance purposes, which not only led to the reduction of maintenance times but also to the prevention of downtimes and, therefore, to an increase in machine availability (Herzberg, 2022).

To ensure the high reliability of its machines, Siemens introduced an online discharge measurement for its machines. The measurements are carried out during operation without having to stop the line. The measurements can also be conducted in shorter intervals without major effort or additional downtimes. Due to the continuous recording of the winding temperature, operational influences are displayed. This allows the detection of weak points and possible failures at an early stage. Furthermore, the various measurements build the foundation for a more detailed condition diagnosis, which helps to define maintenance activities earlier (Steinbauer, 2022).

When and How to Apply This Use Case?

Condition monitoring should be applied when a preventive or corrective approach has negative consequences:

- **Possible negative consequences of premature maintenance**:

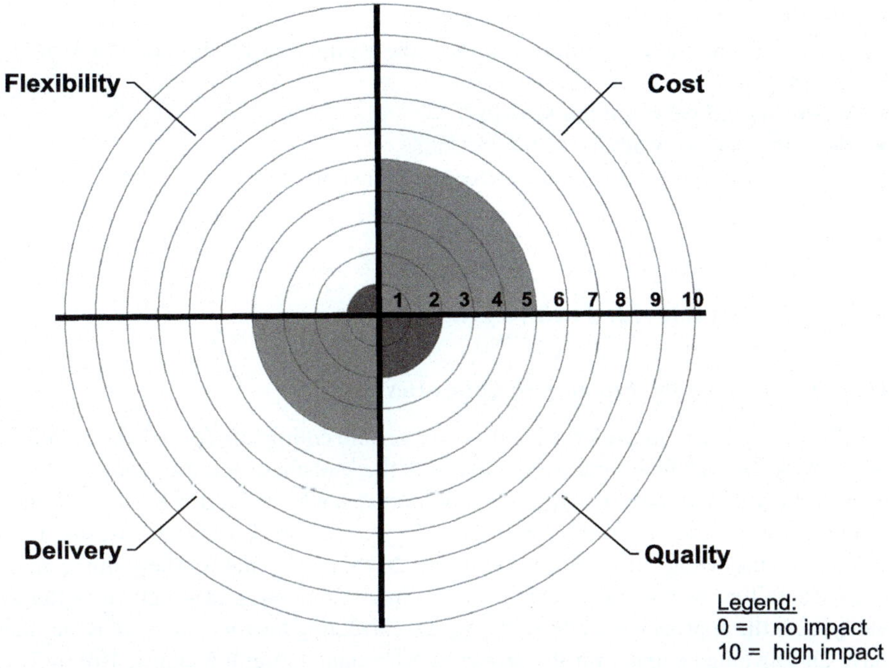

Fig. 3.89 Impact—use case 41

- Higher expenses due to unnecessary spare parts
- More downtimes due to maintenance
- Unnecessary deployment of workforce

- **Possible negative consequences of late maintenance**:
  - Product quality deterioration
  - Loss in productivity
  - Endangerment of the staff
  - Higher downtimes
  - Falling behind the schedule

The abovementioned issues are negative consequences if the maintenance activities are not timed correctly. In this regard, condition monitoring of assets is especially effective when equipment failures occur irregularly and when the consequences of badly timed maintenance are particularly impactful (regarding time, financial expenses for spare parts, lost productivity, etc.).

Questions to Ask

- What kind of data do we want to measure?
- Are these parameters useful regarding the evaluation of the condition of the specific asset?
- Where should we place our sensors?
- How often do we want to record the data?
- Do we have the infrastructure to record, transmit, store (and analyze) the said data?

3.11.3 Use Case 42: Failure Diagnostics

Description Along the Framework Categories

While sensors can help track the condition of certain equipment (parts), they often do not provide enough insight into the impacts of the said characteristics and/or the root causes of equipment failure. Failure in this context can be defined as follows: "failure is a limit state of the material behaviour and/or loss of carrying capacity of structural element or the whole structure" (Altenbach, 2015, p. 3). This is where this generic use case, failure diagnostics, is effective. Failure diagnostics, also known as failure analysis, is the approach to understanding and correcting the root causes of issues that have an unwanted impact on the process performance (Keith Mobley, 1999). This approach can be applied in real-time, periodically, or after certain events such as equipment failures or the exceeding of certain thresholds. The most commonly used approaches for failure diagnosis are pattern recognition-based, knowledge-based and model-based approaches (Tung & Yang, 2009).

The process to determine the root cause of an issue is referred to as root cause failure analysis. Common methods to investigate the potential root cause of a failure involve failure mode and effects analysis, fault tree analysis, cause-and-effect analysis, and sequence-of-events analysis (Keith Mobley, 1999). Modern information technology can help to reveal the root causes of certain defects and other asset-related issues by comparing and analysing different datasets. The identified patterns contribute to the understanding, how, why and/or where a particular error within a machine or manufacturing system occurs or has occurred. This knowledge is also further used for predictive maintenance (see predictive maintenance).

Picture of the Model/Framework Explanation

In failure diagnostics, advanced information technologies such as big data analytics are applied to analyse the root cause of (potential) machine and tool failures. Because of this analytical character, this use case is categorized as diagnostic (see Fig. 3.90).

Relevant Technologies

The following technologies are considered relevant for this use case:

- Enterprise Resource Planning System
- Manufacturing Execution System

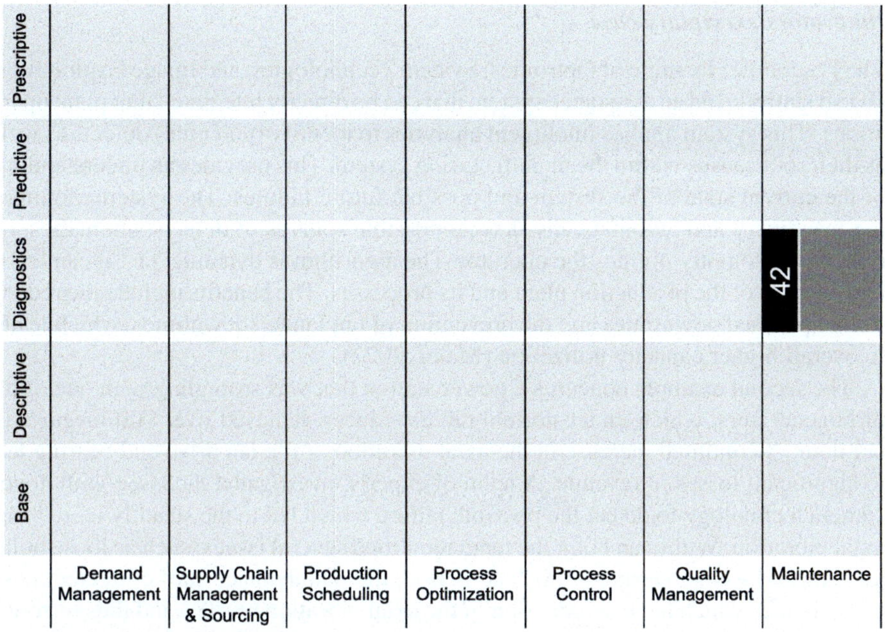

Fig. 3.90 Smart factory framework—use case 42

- Cloud Computing
- Internet of Things
- Sensors and Actuators
- Vision Technologies
- Cyber-Physical Systems
- Big Data Analytics
- AI, ML & Simulation

Impact on Competitive Priorities

The impact of this use case on competitive priorities is shown in generic form in the Fig. 3.91. The specific impact depends on the context of the company.

Impact on flexibility • Improves assessment of the RUL of a tool for future production changes	Impact on costs • Reduces maintenance and accelerates troubleshooting • Prevents downtimes through the early identification of potential failures
Impact on delivery • Reduces maintenance and accelerates troubleshooting • Prevents downtimes through the early identification of potential failures	Impact on quality • Enables the possible reduction of scrap and the prevention of postprocessing due to the monitoring of the tool wear

Innovators/exemplary case

The Fraunhofer Institute of Optronics, System Technologies, and Image Exploitation (IOSB) introduced an assistance system that can be directly integrated into plant operations. This system applies intelligent analytics to identify (potential) defects as well as their root causes within the manufacturing system. This provides an understanding of the current state of the system and possible future failures. The system monitors the machinery and automatically detects possible sources of defects, anomalies, or risks, and promptly notifies the operator. The algorithm is dynamic, i.e., it can learn new models of the production plant and its processes. The benefits include the reduction of planned downtimes and the prevention of unplanned downtimes, which leads to overall higher capacity utilization (Maier, 2022).

The second example concerns a power station that was struggling with one of its turbogenerators, which under normal circumstances achieved over 700 megawatts per hour. A continual increase in the axial vibration of the fan drive end bearing led to significant losses in revenue. A team of experts investigated the issue with novel camera technology to detect the possible failure which led to the steadily increasing axial vibration. Within an hour, the team identified several issues, such as loose bolts and cracked welds. During a low load period, the maintenance staff corrected most of the issues, which led to a reduction of the axial vibration by 70% and thus restored the performance of the power station's operations (RDI Technologies, n.d.).

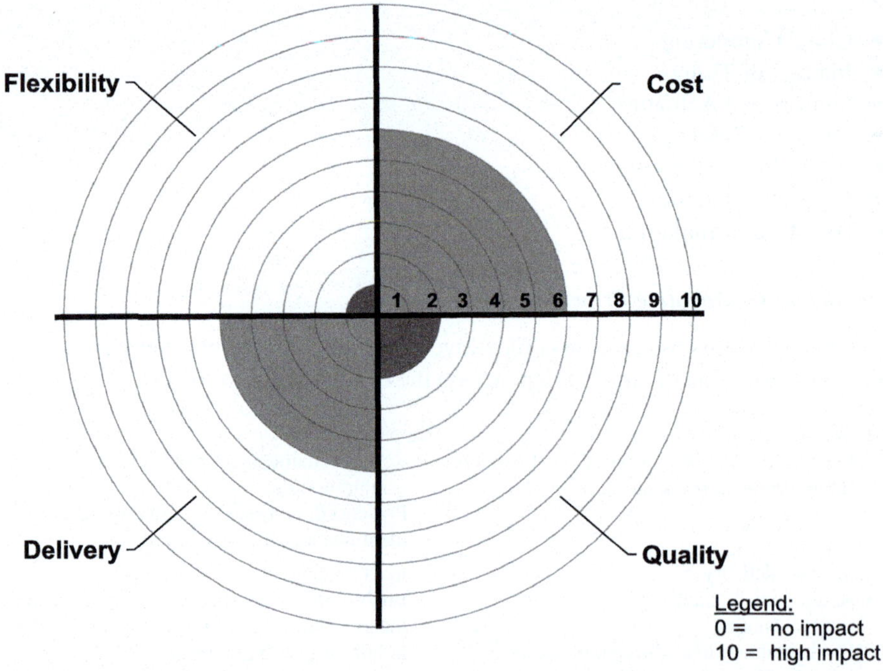

Fig. 3.91 Impact—use case 42

When and How to Apply This Use Case?

Failure diagnostics of manufacturing equipment can be applied to uncover correlations between different variables within the factory (machines, processes, etc.) to optimize maintenance activities. This leads to various benefits in comparison to a preventive or corrective approach. Depending on whether the timing of the maintenance is too early or too late can have various consequences on the operations, as well as on other variables.

- **Possible consequences of premature maintenance**:
  - Higher expenses due to unnecessary spare parts
  - More downtimes due to maintenance
  - Unnecessary deployment of workforce
- **Possible consequences of late maintenance**:
  - Product quality deterioration
  - Loss in productivity
  - Endangerment of the staff
  - Higher downtimes
  - Falling behind the schedule

Questions to Ask

- What kind of data do we want to measure?
- Are these parameters useful regarding the evaluation of the condition of the specific asset?
- What do we want to learn from these data?
- Where do we assume possible correlations?
- Which machine parts are problematic?

3.11.4 Use Case 43: Predictive Maintenance

Description Along the Framework Categories

Maintenance costs are a large part of the total costs of manufacturing operations. These costs can amount to up to 60% of the costs of the goods produced (Keith Mobley, 2002). This leads to the logical assumption that a reduction in the said expenses constitutes a major objective in operations management. In this regard, predictive maintenance approaches offer an effective way to reduce costs. Predictive maintenance can predict correlations, behaviour patterns, and trends in general. Machine learning or statistical models are used to anticipate upcoming failures and disruptions within the production environment (Zonta et al., 2020). This helps, among different things, to avoid downtime by improving the decision-making process for maintenance activities (Zonta et al., 2020). A crucial part of this is the assessment and prediction of the RUL of the asset (Bousdekis et al., 2020; Zonta et al., 2020). Various different methods exist to measure the RUL of equipment, such as the monitoring of vibrations, thermography, tribology (the dynamics of the bearing-lubrication-rotor support structure of machines), visual inspections, and ultrasonics (Keith Mobley, 2002). Predictive maintenance approaches can be classified into the following three categories: physical model-based, knowledge-based, and data-driven based approaches (Zonta et al., 2020). Independent of these methods, one of the main requirements for an optimized predictive maintenance approach is the acquisition and analysis of large amounts of real-time and historical data (Bousdekis et al., 2020; Kiangala & Wang, 2018).

Picture of the Model/Framework Explanation

As the use case name implies, predictive maintenance is all about the prediction of asset degradation, which is subsequently used to optimally schedule maintenance activities. Therefore, the case is clearly classified as predictive (see Fig. 3.92).

Relevant Technologies

The following technologies are considered relevant for this use case:

- Enterprise Resource Planning System
- Manufacturing Execution System
- Cloud Computing

Fig. 3.92 Smart factory framework—use case 43

- Internet of Things
- Sensors and Actuators
- Vision Technologies
- Cyber-Physical Systems
- Big Data Analytics
- AI, ML & Simulation

Impact on Competitive Priorities

The impact of this use case on competitive priorities is shown in generic form in the Fig. 3.93. The specific impact depends on the context of the company.

Impact on flexibility • Enables flexible process planning	Impact on costs • Extends equipment lifetime • Increases capacity utilization • Reduces maintenance costs • Optimizes process planning
Impact on delivery • Improves (process) reliability • Accelerates delivery and with the promised quality • Reduces and shortens downtimes	Impact on quality • Reduces machine-/tool-related product defects

Innovators/Exemplary Cases

The first success story concerns a case study of the Renault Factory, Cacia. With the introduction of a new gearbox project, the factory aimed to increase its annual production of crankcases from 500,000 to 650,000 pieces. To realize this objective, Cacia wanted to change the existing maintenance paradigm to increase productivity and reduce costs. In the first phase, scientific research was carried out to better understand the circumstances. In a second step, the experimentation phase, pilot lines were used to test the said concepts. It became quickly clear that a novel operating procedure regarding the different maintenance activities was needed. To resolve this issue, a more intuitive combination of preventive and predictive maintenance was introduced while also improving the communication of the maintenance processes to all relevant stakeholders to establish shared knowledge. The predictive maintenance approach worked as follows. Data from various sensors were constantly collected and sent to a central server for analysis. To create an efficient notification system, the engineers defined two upper limits for each part/machine. If a certain threshold was reached, an alert notification was sent to the maintenance team. If nothing was done to resolve the issue and the anomaly developed further, a second, more severe notification was triggered, which was sent in the form of an alarm to the maintenance staff. In one case, the team installed vibration and temperature sensors in the production line for differential boxes. Shortly afterwards, an anomaly was detected, caused by unwanted collisions of the boxes, and the maintenance team investigated the issue and analyzed the vibration values. Based on their expertise, they concluded that the machine could operate for a further six months, after which the process quality would eventually deteriorate. Although the predictive maintenance approach did not lead directly to specific measures, it informed the company of the necessity to buy the replacement material early while also optimally integrating the maintenance duties into the schedule. The benefits included the prevention of productivity losses as well as the maximization of the use time of the machine. The investment into the predictive maintenance system led to a return of about 200% of the initial investment (Fernandes et al., 2021).

The second success story concerns the application of smart sensors for predictive maintenance in the petrochemical industry. The biggest producer of chemicals in Japan, Mitsubishi Chemical Corporation, had issues with its pH sensors, which were used to monitor the neutralization of strong acids. Due to strongly fluctuating process conditions regarding temperature and the pH value, sensors could break within one to three months. Such failures had an impact on the process quality, product quality as well as productivity. To circumvent a preventive maintenance approach, which not only led to additional expenses but also to more process interruptions, the executives introduced a condition-based predictive maintenance approach to determine when the sensors needed to be recalibrated, cleaned, or replaced. With the introduction of intelligent sensor management, statements about the said issues could be made,

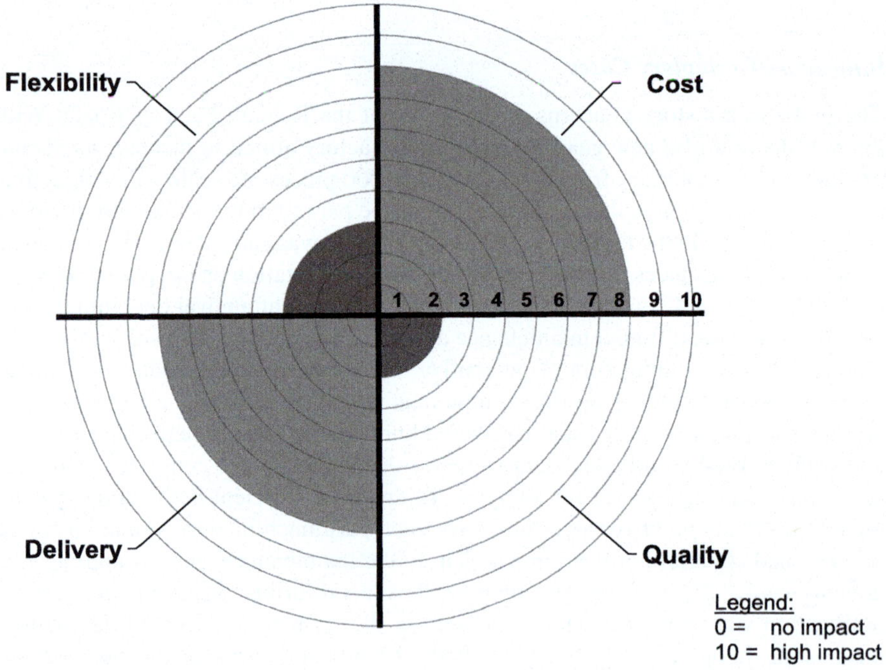

Fig. 3.93 Impact—use case 43

and it became possible to predict the RUL of each specific sensor (Mettler Toledo Gruppe, 2018).

When and How to Apply This Use Case?

Predictive maintenance can be used to determine the optimal time to schedule maintenance actions. A preventive approach, which implies the execution of maintenance tasks in strictly defined time intervals, can have several negative consequences because the tool/part is usually replaced too early or too late. In this regard, predictive maintenance helps to execute maintenance activities at the optimal time.

- **The possible negative consequences of premature maintenance**:
  - Higher expenses due to unnecessary spare parts
  - More downtimes due to maintenance
  - Unnecessary deployment of workforce
- **The possible negative consequences of late maintenance**:
  - Product quality deterioration
  - Loss in productivity
  - Endangerment of the staff
  - Higher downtimes
  - Falling behind the schedule

The abovementioned issues are the negative consequences of not timing maintenance activities correctly. In this regard, predictive maintenance is especially effective when equipment failures occur irregularly and when the consequences of wrong timed maintenance are particularly impactful (regarding time, financial expenses for spare parts, lost productivity, etc.).

Questions to Ask

- Which machines and/or parts break the most? What is their significance within the manufacturing process?
- Where do we assume the roots of equipment failures to be?
- What are the consequences of premature/ overdue maintenance approaches?
- What is the benefit of a predictive maintenance approach compared to the existing maintenance approach?

3.11.5 Use Case 44: Autonomous Maintenance

Description Along the Framework Categories

As can be seen in the previous use cases, the intelligent automation of processes constitutes one of the major pillars of the fourth industrial revolution. In addition to classic operational processes such as the scheduling, production, and quality control of industrial goods, the maintenance, repair, and overhaul (MRO) of machines is a core task in a production plant. The previous maintenance use cases illustrated that the paradigm of maintenance ranges from rather reactive to proactive approaches. Despite the increasing involvement of intelligent technologies, as exemplified by the concept of predictive maintenance, automation of MRO activities is still relatively rarely applied. This is mostly due to the fact that individual components of the machines often have to be physically accessed and that these components have to be either replaced or repaired. Nevertheless, there exist areas where the application of automated MRO processes might not only be achievable but also economically advantageous. The so-called self-maintenance methodology is defined by Lee et al., (2011, p. 115) as follows: "Self-maintenance refers to the ability to carry out regular quality and safety checking by the machine itself, to detect anomalies, and to make immediate repairs when needed by using stocked spare parts to avoid potential catastrophic loss". This implies that self-maintaining manufacturing assets not only trigger countermeasures in case of degradations and breakdowns but are also able to monitor themselves.

The idea of self-maintenance is (partially) included in superior, more holistic maintenance concepts, such as self-engineering or digitalized maintenance (Algabroun et al., 2020; Brooks & Roy, 2021). Contrary to concepts such as the engineering immune systems (Lee et al., 2011), self-maintenance addresses the actual maintenance paradigm in which the machine conducts actual services instead of just keeping up/recovering the functions. There are a multitude of possible methods and techniques allowing mechanical systems to conduct MRO services autonomously. These include approaches such as self-sealing (e.g. self-sealing pressure vessel panels), self-repairing (e.g. self-sharpening tool blades), or self-cleaning (e.g. self-cleaning filter systems) (Brooks & Roy, 2021). Importantly, the type of MRO approaches that can be applied depends heavily on the issues to be solved.

Picture of the Model/Framework Explanation

This use case deals with the autonomous monitoring of machine states and the execution of maintenance and repair work. Such approaches usually require an increased degree of automation intelligence. Hence, this use case is classified as prescriptive (see Fig. 3.94).

Relevant Technologies

The following technologies are considered relevant for this use case:

- Enterprise Resource Planning System
- Manufacturing Execution System
- Cloud Computing
- M2M Communication
- Internet of Things
- Sensors and Actuators
- Vision Technologies
- Cyber-Physical Systems

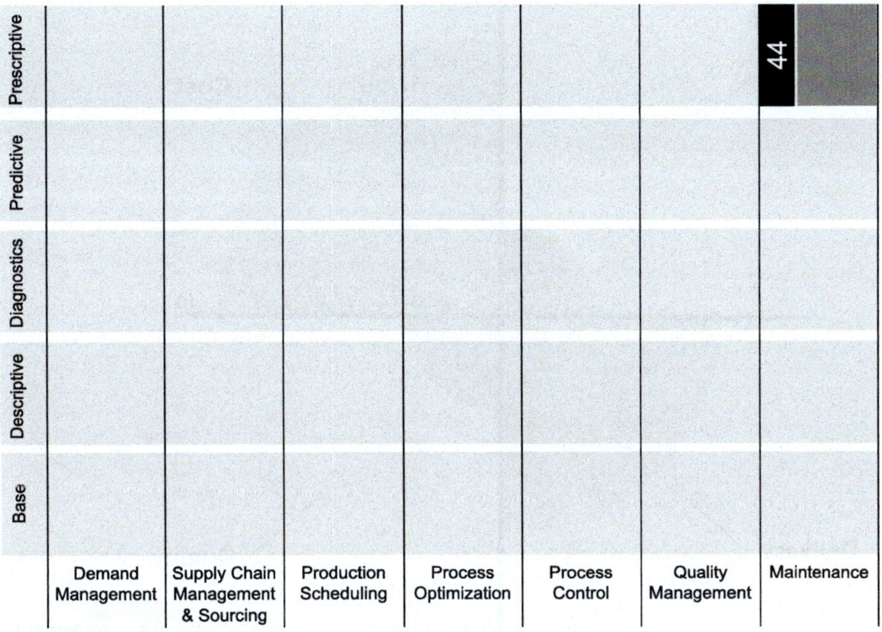

Fig. 3.94 Smart factory framework—use case 44

- Big Data Analytics
- AI, ML & Simulation

Impact on Competitive Priorities

The impact of this use case on competitive priorities is shown in generic form in the Fig. 3.95. The specific impact depends on the context of the company.

Impact on flexibility • Improves flexibility because the machine is not dependent anymore on MRO workers	Impact on costs • Reduces labour expense due to decreased manual maintenance efforts • Prevents downtimes and thus potentially increases the life of assets • Increases productivity due to fewer and shorter downtimes
Impact on delivery • Accelerates order fulfilment time • Improves delivered- in-full, on-time score • Improves (process) reliability	Impact on quality • Prevents product quality degradation by early intervention

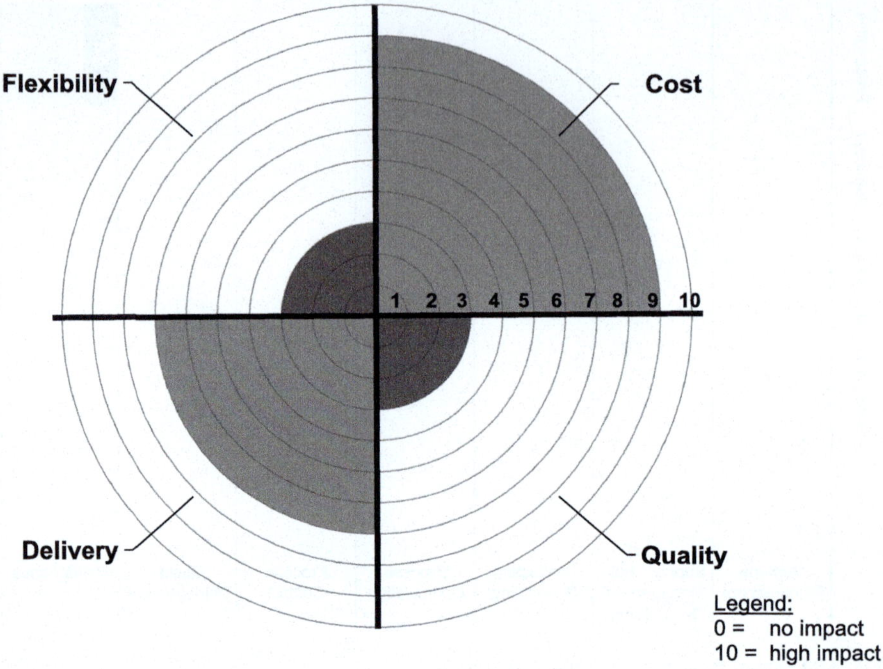

Fig. 3.95 Impact—use case 44

Innovators/Exemplary Cases

A leading semiconductor manufacturer collaborated with Lam Research Corporation, a provider of industrial equipment for semiconductor equipment, to implement self-maintaining edge-cutting manufacturing equipment in its plant. Etch process modules contain degradable parts that must be cleaned several times a month and must be regularly replaced to ensure a stable process performance. The related MRO processes were often labour and time-intensive due to the various working steps-opening, part replacing, cleaning, and requalification of the chamber. This resulted in more complicated scheduling and impacted productivity. The introduced solution consists of a process module, long-lasting chamber parts, an edge ring, and autonomous cleaning technology. The machine not only knows when the relevant parts need to be replaced but also conducts this process autonomously without the need to open the chamber. This approach has allowed the company to produce a whole year without the need for any manual maintenance-related cleanings, leading to increased uptime and enhanced productivity (Lam Research Organisation, 2019).

Schaeffler, a leading global provider of automotive components and industrial solutions, implemented a system for the smart maintenance of its intralogistics assets in its new warehouse and distribution centre. To realize this proposition, solutions for condition monitoring, as well as (partially) autonomous maintenance systems, were introduced. The system constantly monitors the engines within the warehouse and reliably detects possible defects in real-time. Centrally controlled lubrication tools continuously conduct the relubricating tasks. These tasks are performed quantitatively and spatially, with enough precision to avoid too much lubrication. In a further step, the company plans to integrate tools to calculate the RUL and for an overall increased analysis of the respective conditions. This enables the maintenance processes to be more predictive and offers more efficient maintenance approaches. The goal of the project was to avoid unplanned downtimes, increase plant availability, and significantly reduce maintenance costs. Ultimately, the targeted savings of 250,000€ was reached (Schaeffler, n.d.).

When and How to Apply This Use Case?

- **High costs caused by manual maintenance**: Manual maintenance causes direct costs (salaries etc.) and indirect costs (downtimes etc.). Embedded self-repairing mechanisms help to reduce these expenses.
- **Simple maintenance tasks**: Simpler maintenance and repair tasks are far more feasible for automated systems and, therefore, can more easily be achieved.

Questions to Ask

- What kind of MRO-related issues are we experiencing in our facility?
- What are the outcomes of these impacts?
- Which steps are required to resolve these issues?
- What are the technical requirements of a self-maintaining asset? How could the asset achieve its own maintenance?
- How feasible is this idea?
- What kind of applications can benefit from its successful realization?

References

ABB Research Center Germany. (2014). *ABB Research Center Germany annual report 2014* (Annual Report). ABB Research Center Germany. https://new.abb.com/docs/librariesprovider30/ueber-uns/jahresbericht-2014.pdf?sfvrsn=b026081e_4

Alexander Thamm GmbH. (n.d.). Nachfrageprognose zur Lageroptimierung | [at] Use Cases. *Alexander Thamm GmbH*. https://www.alexanderthamm.com/de/usecases/nachfrageprognose-zur-lageroptimierung/

Algabroun, H., Bokrantz, J., Al-Najjar, B., & Skoogh, A. (2020). Development of digitalised maintenance—A concept. *Journal of Quality in Maintenance Engineering* (ahead-of-print). https://doi.org/10.1108/JQME-04-2019-0039

Alicke, K., Rachor, J., & Seyfert, A. (2016, October 27). *Supply chain 4.0—The next-generation digital supply chain | McKinsey*. McKinsey. https://www.mckinsey.com/business-functions/operations/our-insights/supply-chain-40--the-next-generation-digital-supply-chain

Altenbach, H. (Ed.). (2015). *Failure and damage analysis of advanced materials*. Springer.

American Institute of Chemical Engineers. (1998). CCPS process safety incident database completed. *Chemical Engineering Progress, 8*.

Anguiano, D. (2021, September 23). A record number of cargo ships are stuck outside LA. What's happening? *The Guardian*. https://www.theguardian.com/us-news/2021/sep/22/cargo-ships-traffic-jam-los-angeles-california

Anneroth, C., & Myrberger, J. (2019). *Ericsson's journey of building and deploying process mining capability*. QPR Conference 2019.

Ardjmand, E., Ghalehkhondabi, I., Weckman, G., & Young, W. (2016). Application of decision support systems in scheduling/planning of manufacturing/service systems: A critical review. *International Journal of Management and Decision Making, 15*, 248. https://doi.org/10.1504/IJMDM.2016.080703

Arrow. (2018, July 31). The pros and cons of cloud vs in house servers. *Arrow Voice & Data*. https://arrowvoice.com.au/blog/cloud-vs-inhouse-servers/

Arviem Cargo Monitoring. (2017, June 6). Carbon footprint monitoring in the logistics chain to support sustainability. *Arviem Cargo Monitoring*. https://arviem.com/carbon-footprint-monitoring-logistics-chain-next-step-towards-sustainable-supply-chains/IPCCC. Report: 'Code red' for human driven global heating, warns UN chief. (2021, August 9). *UN News*. https://news.un.org/en/story/2021/08/1097362

Asprova AG. (2016). *Case Study bei GEBHARDT Fördertechnik—Asprova AG*. https://www.asprova.eu/de/branchen-und-referenzen/planungssoftware-case-studies/anwenderberichte/feinplanung-gebhardt-foerdertechnik/

Association for Advancing Automation. (n.d.). Emerging markets: Collaborative robots. *Automate*. https://www.automate.org/blogs/emerging-markets-collaborative-robots

AWS and BMW Group Team Up to Accelerate Data-Driven Innovation. (2020, December 8). BMWGROUP. https://www.press.bmwgroup.com/global/article/detail/T0322118EN/aws-and-bmw-group-team-up-to-accelerate-data-driven-innovation?language=en

Axxiv. (2019, January 29). *Cloud server versus inhouse server*. https://www.axxiv.ch/cloud-server-versus-inhouse-server/

B&R Industrie-Elektronik GmbH. (2016). *Lösungen für mehr Produktivität in der Fertigung*. https://industrie40.vdma.org/documents/4214230/5356229/Industrie%204.0%20konkret%202016/9912b1a7-be6b-4f32-a132-79aef5b8a11c

Bächler, A., Bächler, L., Autenrieth, S., Kurtz, P., Hörz, T., Heidenreich, T., & Krüll, G. (2016). *A comparative study of an assistance system for manual order picking—Called pick-by-projection—With the guiding systems pick-by-paper, pick-by-light and pick-by-display*. 2016 49th Hawaii International Conference on System Sciences (HICSS), 523–531. https://doi.org/10.1109/HICSS.2016.72

Bai, Y., Xie, J., Wang, D., Zhang, W., & Li, C. (2021). A manufacturing quality prediction model based on AdaBoost-LSTM with rough knowledge. *Computers & Industrial Engineering, 155*, 107227. https://doi.org/10.1016/j.cie.2021.107227

BAIRLE Druck & Medien GmbH. (2016). *BAIRLE Druck & Medien GmbH—ALLIANZ Industrie 4.0 Baden Württemberg | Industrie 4.0*. https://www.i40-bw.de/de/100orte/druckerei-bairle-gmbh/

Baldauf, M., Müller, S., Seeliger, A., Küng, T., Michel, A., & Züllig, W. (2021). *Human interventions in the smart factory—A case study on co-designing mobile and wearable monitoring systems with manufacturing staff*. Extended Abstracts of the 2021 CHI Conference on Human Factors in Computing Systems, 1–6. https://doi.org/10.1145/3411763.3451774

Banerjee, A., Bandyopadhyay, T., & Acharya, P. (2013). Data analytics: Hyped up aspirations or true potential? *Vikalpa: The Journal for Decision Makers, 38*(4), 1–12. https://doi.org/10.1177/0256090920130401

Baryannis, G., Validi, S., Dani, S., & Antoniou, G. (2019). Supply chain risk management and artificial intelligence: State of the art and future research directions. *International Journal of Production Research, 57*(7), 2179–2202. https://doi.org/10.1080/00207543.2018.1530476

Beaupre, M. (n.d.). *Collaborative robot technology and applications*. https://www.robotics.org/userAssets/riaUploads/file/4-KUKA_Beaupre.pdf

Benevides, C. (2016, November 14). 7 cost-saving benefits of automated guided vehicles. *Conveyco*. https://www.conveyco.com/7-cost-saving-benefits-automated-guided-vehicles-agvs/

Benfer, M., Peukert, S., & Lanza, G. (2021). Operations research in international manufacturing networks. In T. Friedli, G. Lanza, & D. Remling (Eds.), *Global manufacturing management: From excellent plants toward network optimization* (pp. 219–231). Springer International Publishing. https://doi.org/10.1007/978-3-030-72740-6_16

Bienek, M. C. (2022). *Smart electronic factory e.V.* platform industrie 4.0. https://www.plattform-i40.de/IP/Redaktion/DE/Anwendungsbeispiele/065-die-smart-electronic-factory-im-hause-limtronik/beitrag-die-smart-electronic-factory-im-hause-limtronik.html

Bloss, R. (2016). Collaborative robots are rapidly providing major improvements in productivity, safety, programing ease, portability and cost while addressing many new applications. *Industrial Robot: An International Journal, 43*(5), 463–468. https://doi.org/10.1108/IR-05-2016-0148

Bodendorf, F., & Zimmermann, R. (2005). Proactive supply-chain event management with agent technology. *International Journal of Electronic Commerce, 9*(4), 58–89. https://doi.org/10.1080/10864415.2003.11044340

Boria, S. (2015, April 21). *Intelligente Werkzeuge für das Airbus-Werk der Zukunft*. NI. https://www.ni.com/de-ch/innovations/case-studies/19/developing-smart-tools-for-the-airbus-factory-of-the-future.html

Bottecchia, S., Cieutat, J.-M., Merlo, C., & Jessel, J.-P. (2009). A new AR interaction paradigm for collaborative teleassistance system: The POA. *International Journal on Interactive Design and Manufacturing (IJIDeM), 3*(1), 35–40. https://doi.org/10.1007/s12008-008-0051-7

Bousdekis, A., Apostolou, D., & Mentzas, G. (2020). Predictive maintenance in the 4th industrial revolution: Benefits, business opportunities, and managerial implications. *IEEE Engineering Management Review, 48*(1), 57–62. https://doi.org/10.1109/EMR.2019.2958037

Brooks, S., & Roy, R. (2021). An overview of self-engineering systems. *Journal of Engineering Design*, 1–51. https://doi.org/10.1080/09544828.2021.1914323

Budde, L., Liao, S., Haenggi, R., & Friedli, T. (2022). Use of DES to develop a decision support system for lot size decision-making in manufacturing companies. *Production & Manufacturing Research, 10*(1), 494–518. https://doi.org/10.1080/21693277.2022.2092564

Burattin, A. (2015). Process mining. In A. Burattin, *Process mining techniques in business environments* (Vol. 207, pp. 33–47). Springer International Publishing. https://doi.org/10.1007/978-3-319-17482-2_5

Cambridge. (n.d.). *Automation*. Cambridge Dictionary. https://dictionary.cambridge.org/dictionary/english/automation

Campos Pinto, L., Lobo, J., & Tavares, A. (2013). GestInc—The incidents data base. *Renewable Energy and Power Quality Journal*, 449–454. https://doi.org/10.24084/repqj11.336

CDNetworks. (2021, January 29). Data transmission: What is it? Everything you need to know. *CDNetworks*. https://www.cdnetworks.com/enterprise-applications-blog/everything-you-need-to-know-about-data-transmission/

Chae, B. (Kevin), Olson, D., & Sheu, C. (2014). The impact of supply chain analytics on operational performance: A resource-based view. *International Journal of Production Research, 52*(16), 4695–4710. https://doi.org/10.1080/00207543.2013.861616

Chan, F. T. S., & Prakash, A. (2012). Inventory management in a lateral collaborative manufacturing supply chain: A simulation study. *International Journal of Production Research, 50*(16), 4670–4685. https://doi.org/10.1080/00207543.2011.628709

Chase, C. W. (2009). *Demand-driven forecasting. A structured approach to forecasting.*

Chung, M., & Kim, J. (2016). The internet information and technology research directions based on the fourth industrial revolution. *KSII Transactions on Internet and Information Systems, 10*(3), 1311–1320.

Climate Disclosure Project. (2021). *Transparency to transformation: A chain reaction* (CDP Global Supply Chain Report 2020). Climate Disclosure Project.

Cohen, I. (2022). *Top use cases for demand forecasting using autonomous forecast*. Anadot.Com. https://www.anodot.com/blog/demand-autonomous-forecast/

Cohen, Y., Faccio, M., Pilati, F., & Yao, X. (2019). Design and management of digital manufacturing and assembly systems in the Industry 4.0 era. *The International Journal of Advanced Manufacturing Technology, 105*. https://doi.org/10.1007/s00170-019-04595-0

Colli, M., Berger, U., Bockholt, M., Madsen, O., Møller, C., & Wæhrens, B. V. (2019). A maturity assessment approach for conceiving context-specific roadmaps in the Industry 4.0 era. *Annual Reviews in Control, 48*, 165–177. https://doi.org/10.1016/j.arcontrol.2019.06.001

Council, N. R. (1998). *Manufacturing process controls for the industries of the future*. The National Academies Press. https://doi.org/10.17226/6258

Crowe, S. (2020, June 4). How a CNC machine shop maximizes productivity with cobots. *Collaborative Robotics Trends*. https://www.cobottrends.com/cnc-machine-shop-productive-robotics/

Crowe, S. (2021, March 1). How ActiNav automated a challenging bin-picking task. *Collaborative Robotics Trends*. https://www.therobotreport.com/universal-robots-actinav-handles-challenging-picking-task/

Croxton, K. L., Lambert, D. M., García-Dastugue, S. J., & Rogers, D. S. (2002). The demand management process. *The International Journal of Logistics Management, 13*(2), 51–66. https://doi.org/10.1108/09574090210806423

Csalódi, R., Süle, Z., Jaskó, S., Holczinger, T., & Abonyi, J. (2021). Industry 4.0-driven development of optimization algorithms: A systematic overview. *Complexity*, e6621235. https://doi.org/10.1155/2021/6621235

Das, T. K., & Miller, W. A. (1999). Quality control in cellular manufacturing. In *Handbook of cellular manufacturing systems* (pp. 275–317). Wiley. https://doi.org/10.1002/9780470172476.ch11

Davenport, T. H., & Harris, J. G. (2017). *Competing on analytics: The new science of winning* (Updated, with a new introduction). Harvard Business Review Press.

Davenport, T. H., & Spanyi, A. (2019, April 23). What process mining is, and why companies should do it. *Harvard Business Review*. https://hbr.org/2019/04/what-process-mining-is-and-why-companies-should-do-it

De Vries, J., De Koster, R., & Stam, D. (2015). Exploring the role of picker personality in predicting picking performance with pick by voice, pick to light and RF-terminal picking. *International Journal of Production Research, 54*, 1–15. https://doi.org/10.1080/00207543.2015.1064184

Del Castillo, E. (2011). *Process optimization: A statistical approach*. Springer.

Deska, B., Schneider, H., & Wesel, M. (2018). *Innovationen in Wartung und Produktion – Assistenzsysteme, Wartungshelfer und erweiterte Realität*. Mittelstand 4.0-Agentur Prozesse. https://www.mittelstand-digital.de/MD/Redaktion/DE/Publikationen/Importiert/1538398724787-61-Publication.pdf?__blob=publicationFile&v=7

DHL Trend Research. (2018). *Logistics trend Radar. Delivering insight today, creating value tomorrow* (Logistics Trend Radar). DHL Customer Solutions & Innovation.

Di Francescomarino, C., Ghidini, C., Maggi, F. M., & Milani, F. (2018). Predictive process monitoring methods: Which one suits me best? In M. Weske, M. Montali, I. Weber, & J. vom Brocke (Eds.), *Business process management* (pp. 462–479). Springer International Publishing.

Dittmann, J. P. (2010). How to successfully implement projects to improve the demand prediction process. *The Business & Management Collection*. https://hstalks.com/t/1793/how-to-successfully-implement-projects-to-improve-/

dos Reis, W. P. N., & Morandin Junior, O. (2021). Sensors applied to automated guided vehicle position control: A systematic literature review. *The International Journal of Advanced Manufacturing Technology, 113*(1), 21–34. https://doi.org/10.1007/s00170-020-06577-z

Duong, L. N. K., & Chong, J. (2020). Supply chain collaboration in the presence of disruptions: A literature review. *International Journal of Production Research, 58*(11), 3488–3507. https://doi.org/10.1080/00207543.2020.1712491

Engel, F. (2022). *Digitalisierung in der Fertigung*. Platform Industrie 4.0. https://www.plattform-i40.de/PI40/Redaktion/DE/Anwendungsbeispiele/268-digitalisierung-in-der-fertigung-apworks/digitalisierung-in-der-fertigung.html

Erk, F. (2022). *SAP POWER*. Platform Industrie 4.0. https://www.plattform-i40.de/IP/Redaktion/DE/Anwendungsbeispiele/286-sap-power-man/sap-power.html

Etemadi, S. (2016). *Industrie 4.0 @RtP2: RFID WT-Tracking*. Plattform Industrie 4.0. https://web.archive.org/web/20161211100200/http:/www.plattform-i40.de/I40/Redaktion/DE/Anwendungsbeispiele/219-rfid-wt-tracking-bosch/beitrag-rfid-wt-tracking.html

Falkonry. (2022a). How Falkonry helped predict failures & avoid downtime. *Falkonry*. https://falkonry.com/use-cases/

Falkonry. (2022b). How Falkonry improves plant safety and compliance. *Falkonry*. https://falkonry.com/use-cases/

Fernandes, J., Reis, J., Melão, N., Teixeira, L., & Amorim, M. (2021). The role of industry 4.0 and BPMN in the arise of condition-based and predictive maintenance: A case study in the automotive industry. *Applied Sciences, 11*(8), 3438. https://doi.org/10.3390/app11083438

Figueira, G., Amorim, P., Guimarães, L., Amorim-Lopes, M., Neves-Moreira, F., & Almada-Lobo, B. (2015). A decision support system for the operational production planning and scheduling of an integrated pulp and paper mill. *Computers & Chemical Engineering, 77*, 85–104. https://doi.org/10.1016/j.compchemeng.2015.03.017

Flachmann, S. (2022). *Verfahrensgemischte Linienautomation im Werkzeugbau Phoenix Contact*. Platform Industrie 4.0. https://www.plattform-i40.de/IP/Redaktion/DE/Anwendungsbeispiele/149-digitaleverfahrensgemischte-automatisierte-prozessketten-phoenix-contact/beitrag-digitaleverfahrensgemischte-automatisierte-prozessketten-phoenix-contact.html

FleetPride keeps the wheels of commerce turning with seamless supply chain management. (2021, September 8). Ibm.Com. https://www.ibm.com/case-studies/fleetpride

Flynn, B. B., Schroeder, R. G., & Sakakibara, S. (1994). A framework for quality management research and an associated measurement instrument. *Journal of Operations Management, 11*(4), 339–366. https://doi.org/10.1016/S0272-6963(97)90004-8

Food Services Case Study | Sales Analytics Solution. (2022). https://www.saviantconsulting.com/data-analytics-case-studies/food-services-utilizes-cloud-based-analytics-platform.aspx

Frank, A. G., Dalenogare, L. S., & Ayala, N. F. (2019). Industry 4.0 technologies: Implementation patterns in manufacturing companies. *International Journal of Production Economics, 210*, 15–26. https://doi.org/10.1016/j.ijpe.2019.01.004

Fujitsu. (2018). *Yokohama Rubber has implemented FUJITSU Intelligent Dashboard to provide real-time, cross-sectional visibility of its Japanese and overseas factories' operations* (p. 2) [Customer Case Study]. https://www.fujitsu.com/global/Images/CS_2018Jun_Yokohama%20Rubber.pdf

Gapper, J. (2022). A remote village, a world-changing invention and the epic legal fight that followed. *Financial Times*. https://on.ft.com/3ABgXI0

Gartner. (n.d.). Definition of IT incident management—Gartner information technology glossary. *Gartner*. https://www.gartner.com/en/information-technology/glossary/it-incident-management

Gartner. (n.d.). Definition of demand forecasting—Gartner information technology glossary. *Gartner*. Retrieved April 19, 2021, from https://www.gartner.com/en/information-technology/glossary/demand-forecasting

Genc, E., Duffie, N., & Reinhart, G. (2014). Event-based supply chain early warning system for an adaptive production control. *Procedia CIRP, 19*, 39–44. https://doi.org/10.1016/j.procir.2014.04.076

Gerlach, S. (2022). *KapaflexCy*. https://www.plattform-i40.de/PI40/Redaktion/DE/Anwendungsbeispiele/096-kapaflexcy-selbstorganisierte-kapazitaetsflexibilitaet-fuer-die-industrie-4-0/beitrag-kapaflexcy-selbstorganisierte-kapazitaetsflexibilitaet-fuer-die-industrie-4-0.html

Gibson, I., Rosen, D., Stucker, B., Khorasani, M. (2021) *Additive manufacturing technologies* (3rd ed.). Springer International Publishing, https://doi.org/10.1007/978-3-030-56127-7

Gligor, M. (2014). The role of demand management in achieving supply chain agility. *Supply Chain Management: An International Journal, 19*(5/6), 577–591. https://doi.org/10.1108/SCM-10-2013-0363

Global eTrade Services (GeTS). (2020, August 19). *How does a supply chain platform work?* Global ETrade Services. https://globaletrade.services/blogs/how-does-supply-chain-platform-work

Goedert, J. D., & Sekpe, V. D. (2013). Decision support system-enhanced scheduling in matrix organizations using the analytic hierarchy process. *Journal of Construction Engineering and Management, 139*(11), 05013003. https://doi.org/10.1061/(ASCE)CO.1943-7862.0000734

Göttlich, R. (2020). *Fachkräftemangel. CHEMKON, 27*(1), 5–5. https://doi.org/10.1002/ckon.201900079

Graves, S. C. (1981). A review of production scheduling. *Operations Research, 29*(4), 646–675. https://doi.org/10.1287/opre.29.4.646

Grieves, M. (2015). *Digital twin: Manufacturing excellence through virtual factory replication.*

Gröger, C., Niedermann, F., & Mitschang, B. (2012). *Data mining-driven manufacturing process optimization*, 7.

Gruber, V. (2020, August 31). What is shop floor management & why is it important? *FORCAM.*

Hammerstein, R. (2019, May 23). The economic implications of agile manufacturing. *Fastems.* https://www.fastems.com/blog/the-economic-implications-of-agile-manufacturing/

Hänggi, R., Fimpel, A., & Siegenthaler, R. (2021). *LEAN production – einfach und umfassend: Ein praxisorientierter Leitfaden zu schlanken Prozessen mit Bildern erklärt.* Springer Berlin Heidelberg. https://doi.org/10.1007/978-3-662-62702-0

Harapko, S. (2020, November 11). *How COVID-19 impacted supply chains and what comes next.* EY. https://www.ey.com/en_gl/supply-chain/how-covid-19-impacted-supply-chains-and-what-comes-next

Harapko, S. (2021, February 18). *How COVID-19 impacted supply chains and what comes next.* EY. https://www.ey.com/en_gl/supply-chain/how-covid-19-impacted-supply-chains-and-what-comes-next

Herzberg, N. (2022). *Kaeser Kompressoren SE: Predictive maintenance.* platform industrie 4.0. https://www.plattform-i40.de/IP/Redaktion/DE/Anwendungsbeispiele/040-predictive-maintenance-kaeser-kompressoren/beitrag-predictive-maintenance-kaeser-kompressoren.html

Holtewert, P., & Bauernhansl, T. (2016). Increase of capacity flexibility in manufacturing systems by substitution of product functions. *Procedia CIRP, 57*, 92–97. https://doi.org/10.1016/j.procir.2016.11.017

HOMAG. (2020). *Demystifying the path towards batch size 1 production | HOMAG.* https://www.homag.com/en/news-events/news/article/demystifying-the-path-towards-batch-size-1-production.

Homebase. (2022). *Easy shift planning.* Homebase. https://home.kpmg/se/sv/home/tjanster/radgivning/management-consulting/digital-supply-chain/predictive-supply-chain-risk-management.html

Hops, B. (2022). *Mikroelektronik-Fertigung.* Platform Industrie 4.0. https://www.plattform-i40.de/IP/Redaktion/DE/Anwendungsbeispiele/001-mikroelektronik-fertigung-infineon-technologies/beitrag-mikroelektronik-fertigung-infineon-technologies.html

IBM. (n.d.). *What is data storage?* IBM. https://www.ibm.com/topics/data-storage

Infosys. (2021). *Infosys supply chain interventions and early warning solution* (p. 4). Infosys. https://www.infosys.com/services/data-analytics/documents/supply-chain-early-warning-solution.pdf

International Labour Organization. (2011, July 13). *World statistic* [Document]. http://www.ilo.org/moscow/areas-of-work/occupational-safety-and-health/WCMS_249278/lang--en/index.htm

IONOS. (2020, September 28). *RAM (Arbeitsspeicher): Was steckt dahinter?* IONOS Digitalguide. https://www.ionos.de/digitalguide/server/knowhow/was-ist-ein-arbeitsspeicher/

IQMS. (2019). *Best technology practices for traceability reporting* [Whitepaper]. https://www.iqms.com/products/whitepapers/best-technology-for-traceability-reporting.pdf

Jain, A. K., & Lad, B. K. (2019). A novel integrated tool condition monitoring system. *Journal of Intelligent Manufacturing, 30*(3), 1423–1436. Scopus. https://doi.org/10.1007/s10845-017-1334-2

Kagermann, H., Lukas, W.-D., & Wahlster, W. (2011, April 1). *Industrie 4.0: Mit dem Internet der Dinge auf dem Weg zur 4. industriellen Revolution - ingenieur.de*. ingenieur.de - Jobbörse und Nachrichtenportal für Ingenieure. https://www.ingenieur.de/technik/fachbereiche/produktion/industrie-40-mit-internet-dinge-weg-4-industriellen-revolution/

Katana. (n.d.). *Walk with me replaces excel with Katana and gets 65% boost*. Katana MRP. https://katanamrp.com/customers/walk-with-me/

Kehoe, D., & Boughton, N. (2001). Internet based supply chain management: A classification of approaches to manufacturing planning and control. *International Journal of Operations & Production Management, 21*(4), 516–525. https://doi.org/10.1108/01443570110381417

Keith Mobley, R. (1999). *Root cause failure analysis*. Butterworth-Heinemann.

Keith Mobley, R. (2002). *An introduction to predictive maintenance* (2nd ed.). Butterworth-Heinemann. https://1lib.ch/book/463597/c59f0b

Keren, D. N. (2010). Incident database-based framework for establishing industrial safety performance assessments. *Journal of Industrial Technology, 26*(2), 12.

Kharpal, A. (2018, October 30). *Firm linked to Alibaba opens China's biggest robot warehouse to help deal with Singles Day demand*. CNBC. https://www.cnbc.com/2018/10/30/alibaba-cainiao-chinas-biggest-robot-warehouse-for-singles-day.html

Khoshnevis, B. (2004, January 1) Automated construction by contour crafting—Related robotics and information technologies. *Automation in Construction, The best of ISARC 2002, 13*(1), 5–19. https://doi.org/10.1016/j.autcon.2003.08.012

Kiangala, K. S., & Wang, Z. (2018). Initiating predictive maintenance for a conveyor motor in a bottling plant using industry 4.0 concepts. *The International Journal of Advanced Manufacturing Technology, 97*(9), 3251–3271. https://doi.org/10.1007/s00170-018-2093-8

Kirmse, P. (2022). *Flexible Montage in der Fahrzeugproduktion*. Platform Industrie 4.0. https://www.plattform-i40.de/IP/Redaktion/DE/Anwendungsbeispiele/137-wandelbare-r8-manufaktur/beitrag-wandelbare-r8-manufaktur.html

Klein, K. (2022). *MotionEAP*. Platform Industrie 4.0. https://www.plattform-i40.de/IP/Redaktion/DE/Anwendungsbeispiele/064-motion-eap-ein-kontext-sensitives-assistenzsystem/beitrag-motion-eap-ein-kontext-sensitives-assistenzsystem.html

Kloosterman, J. (2019). *How to predict customer demand with machine learning on Oracle autonomous data warehouse*. Retrieved November 13, 2021, from https://blogs.oracle.com/ai-and-datascience/post/how-to-predict-customer-demand-with-machine-learning-on-oracle-autonomous-data-warehouse

Knappe, P. (2015). *Alles im Fluss in der Robotik und Automation*. https://www.ihk-krefeld.de/de/media/pdf/innovation/industrie/digitale-transformation-und-industrie-4.0-best-practice.pdf

Koch, M., Vickers, P., & Ritzmann, S. (2020). *Building supply chain resilience beyond COVID-19*. Deloitte. https://www2.deloitte.com/ch/en/pages/consumer-industrial-products/articles/building-supply-chain-resilience-beyond-covid-19.html

Kohlert, M. (2022). *Vernetzte Resourcen optimierte Produktion von Polymerfilmen*. Platform Industrie 4.0. https://www.plattform-i40.de/IP/Redaktion/DE/Anwendungsbeispiele/294-mondi-gronau-gmbh-vernetzte-ressourcen-optimierte-produktion/vernetzte-resourcen-optimierte-produktion-von-polymerfilmen.html

Köksal, G., Batmaz, İ, & Testik, M. C. (2011). A review of data mining applications for quality improvement in manufacturing industry. *Expert Systems with Applications, 38*(10), 13448–13467. https://doi.org/10.1016/j.eswa.2011.04.063

Kolassa, S., & Siemsen, E. (2016). *Demand forecasting for managers*.

KPMG. (2020). *Predictive solution for supply chain risk management—KPMG Sverige*. https://home.kpmg/se/sv/home/tjanster/radgivning/management-consulting/digital-supply-chain/predictive-supply-chain-risk-management.html

Krueger, W. (2020). *How one family machine shop used customer insights and reinvestment to grow*. https://www.xometry.com/resources/case-studies/how-one-family-machine-shop-used-customer-insights-reinvestment-to-grow/

Kumke, M. (2018) *Methodisches Konstruieren von additiv gefertigten Bauteilen*. Springer Fachmedien Wiesbaden. https://doi.org/10.1007/978-3-658-22209-3

Lam Research Organisation. (2019, April 24). *Record productivity achieved with Lam research self-maintaining equipment*. Lam Research Organisation. https://investor.lamresearch.com/news-releases/news-release-details/record-productivity-achieved-lam-research-self-maintaining

Lambert, D. M., & Cooper, M. C. (2000). Issues in supply chain management. *Industrial Marketing Management, 29*(1), 65–83. https://doi.org/10.1016/S0019-8501(99)00113-3

Last, M., & Kandel, A. (2001). Data mining for process and quality control in the semiconductor industry. In D. Braha (Ed.), *Data mining for design and manufacturing: Methods and applications* (pp. 207–234). Springer. https://doi.org/10.1007/978-1-4757-4911-3_9

Lee, C. K. M., Lv, Y., Ng, K. K. H., Ho, W., & Choy, K. l. (2018). Design and application of Internet of Things-based warehouse management system for smart logistics. *International Journal of Production Research, 56*(8), 2753–2768. https://doi.org/10.1080/00207543.2017.1394592

Lee, J. (2015). Smart factory systems. *Informatik-Spektrum, 38*(3), 230–235. https://doi.org/10.1007/s00287-015-0891-z

Lee, J., Ghaffari, M., & Elmeligy, S. (2011). Self-maintenance and engineering immune systems: Towards smarter machines and manufacturing systems. *Annual Reviews in Control, 35*(1), 111–122. https://doi.org/10.1016/j.arcontrol.2011.03.007

Lee, K.-H. (2011). Integrating carbon footprint into supply chain management: The case of Hyundai Motor Company (HMC) in the automobile industry. *Journal of Cleaner Production, 19*(11), 1216–1223. https://doi.org/10.1016/j.jclepro.2011.03.010

Lemmon, C. (2017). *OTTO uses AI to improve delivery times*. Retail Systems. https://www.retail-systems.com/rs/OTTO_Blue_Yonder_AI_Delivery_Times.php

Lepenioti, K., Bousdekis, A., Apostolou, D., & Mentzas, G. (2020). Prescriptive analytics: Literature review and research challenges. *International Journal of Information Management, 50*, 57–70. https://doi.org/10.1016/j.ijinfomgt.2019.04.003

Liao, S., Rüegg, A., & Hänggi, R. (2021). Deriving a global production network type in times of uncertainty—A simulation based approach. *Die Unternehmung, 75*(4), 552–575. https://doi.org/10.5771/0042-059X-2021-4-552

Lim, M. K., Bahr, W., & Leung, S. (2013). RFID in the warehouse: A literature analysis (1995–2010) of its applications, benefits, challenges and future trends. *International Journal of Production Economics, 145*, 409–430. https://doi.org/10.1016/j.ijpe.2013.05.006

Liu, P., Huang, S. H., Mokasdar, A., Zhou, H., & Hou, L. (2014, October 26). The impact of additive manufacturing in the aircraft spare parts supply chain: Supply chain operation reference (Scor) model based analysis. *Production Planning & Control, 25*(13–14), 1169–1181. https://doi.org/10.1080/09537287.2013.808835

Liu, R., Kumar, A., & van der Aalst, W. (2007). A formal modeling approach for supply chain event management. *Decision Support Systems, 43*(3), 761–778. https://doi.org/10.1016/j.dss.2006.12.009

Lorenz, R., Powell, D., & Netland, T. (2019, June 17–19). *Exploring the effect of digitalizing shop floor management*. 26th Annual EurOMA Conference: Operations adding value to society (EurOMA 2019), Helsinki, Finland. https://doi.org/10.3929/ethz-b-000352295

Lou, H. H., & Huang, Y. L. (2003). Hierarchical decision making for proactive quality control: System development for defect reduction in automotive coating operations. *Engineering Applications of Artificial Intelligence, 16*(3), 237–250. https://doi.org/10.1016/S0952-1976(03)00060-5

Luckert, M. (2022). *Werkzeugmaschinenlabor WZL der RWTH Aachen: Wertstromorientierte Produktionssteuerung*. Platform Industrie 4.0. https://www.plattform-i40.de/IP/Redaktion/DE/Anwendungsbeispiele/346-wlz-wertstromorientierte-produktionssteuerung/beitrag-wzl.html

Lustig, I., Dietrich, B., Johnson, C., & Dziekan, C. (2019). *The Analytics Journey* [Data set]. https://doi.org/10.1287/LYTX.2010.06.01

Maier, A. (2022). *Fraunhofer IOSB-INA*. https://www.plattform-i40.de/IP/Redaktion/DE/Anwendungsbeispiele/510-Fraunhofer_IOSB-INA/beitrag.html

Manz AG. (n.d.). *100-Prozent-Qualitätskontrolle mit Deep-Learning-Funktionen von HALCON*. MVTec. https://www.mvtec.com/de/unternehmen/success-stories/100-prozent-qualitaetskontrolle-mit-deep-learning-funktionen-von-halcon

Maravelias, C. T., & Sung, C. (2009). Integration of production planning and scheduling: Overview, challenges and opportunities. *Computers & Chemical Engineering, 33*(12), 1919–1930. https://doi.org/10.1016/j.compchemeng.2009.06.007

Marquez-Chamorro, A., Resinas, M., & Ruiz-Cortés, A. (2017). Predictive monitoring of business processes: A survey. *IEEE Transactions on Services Computing*, 1. https://doi.org/10.1109/TSC.2017.2772256

Martins, L., Varela, M. L. R., Fernandes, N. O., Carmo-Silva, S., & Machado, J. (2020). Literature review on autonomous production control methods. *Enterprise Information Systems, 14*(8), 1219–1231. https://doi.org/10.1080/17517575.2020.1731611

Meisenzahl, M. (2020, November 12). Alibaba's singles day is over. Now comes the task of delivering $74 billion of merchandise ordered on world's largest shopping holiday. *Business Insider*. https://www.businessinsider.com/alibaba-singles-day-shipping-logistics-robot-photos-2020-11

Mentzer, J. T., DeWitt, W., Keebler, J. S., Min, S., Nix, N. W., Smith, C. D., & Zacharia, Z. G. (2001). Defining supply chain management. *Journal of Business Logistics, 22*(2), 1–25. https://doi.org/10.1002/j.2158-1592.2001.tb00001.x

METTLER TOLEDO Gruppe. (2018). *Personal library* [Success Story]. https://www.mt.com/pl/pl/home/registered_user_services/profile-dashboard/personallibrary.html

Miller, C. G. (2017, November 6). The volume of data are increasing [E-Magazine]. *Quality Magazine*. https://www.qualitymag.com/articles/94336-the-volume-of-data-is-increasing?v=preview

Mitra, A. (2008). *Fundamentals of quality control and improvement* (3rd ed.). Wiley. https://doi.org/10.1002/9781118491645

Mobley, R. (2004). Maintenance fundamentals. *Elsevier*. https://doi.org/10.1016/B978-0-7506-7798-1.X5021-3

Mortenson, M. J., Doherty, N. F., & Robinson, S. (2015). Operational research from Taylorism to Terabytes: A research agenda for the analytics age. *European Journal of Operational Research, 241*(3), 583–595. https://doi.org/10.1016/j.ejor.2014.08.029

National Safety Council. (2022). *Work safety introduction*. National Safety Council Injury Facts. https://injuryfacts.nsc.org/work/work-overview/work-safety-introduction/

Ngo, T. D., Kashani, A., Imbalzano, G., Nguyen, K. T. Q., & Hui, D. (2018, June 15). Additive manufacturing (3D Printing): A review of materials, methods, applications and challenges. *Composites Part B: Engineering, 143*, 172–196. https://doi.org/10.1016/j.compositesb.2018.02.012

Ngoh, S. (2020). *Case study: Medical robot built by Galen robotics prototyping with Xometry*. https://www.xometry.com/resources/case-studies/case-study-medical-robot-built-by-galen-robotics-prototyping-with-xometry/

Nguyen, S., Mei, Y., & Zhang, M. (2017). Genetic programming for production scheduling: A survey with a unified framework. *Complex & Intelligent Systems, 3*(1), 41–66. https://doi.org/10.1007/s40747-017-0036-x

Nosenko, V., Silaev, A., Efremkin, S., & Grednikov, S. (2019). Study of the assembly manufacturing automated traceability system identification tools. *MATEC Web of Conferences, 297*, 01005. https://doi.org/10.1051/matecconf/201929701005

O'Rahilly, P. (2021, June 21). Council post: Redefining the future of manufacturing to plug the skills gap. *Forbes*. https://www.forbes.com/sites/forbesbusinesscouncil/2021/06/21/redefining-the-future-of-manufacturing-to-plug-the-skills-gap/

Ogura, T., Atsuki, K., Motonobu, S., & Chetan, G. (2021). Value chain coordination service for manufacturing and distribution in the new normal. *Hitachi Review, 70*(4), 432–433.

Oliva, R., & Watson, N. (2009). Managing functional biases in organizational forecasts: A case study of consensus forecasting in supply chain planning. *Production and Operations Management, 18*(2), 138–151. https://doi.org/10.1111/j.1937-5956.2009.01003.x

Opitz, T. (2013). Supply chain platform. Gaining visibility and control over global supply networks. https://studylib.net/doc/8821483/case-study

OrderFox. (n.d.). Orderfox—The digital marketplace for the manufacturing procurement. *Orderfox*. Retrieved June 27, 2022, from https://www.orderfox.com/

Pahwa, D., Starly, B., & Cohen, P. (2018). Reverse auction mechanism design for the acquisition of prototyping services in a manufacturing-as-a-service marketplace. *Journal of Manufacturing Systems, 48*, 134–143. https://doi.org/10.1016/j.jmsy.2018.05.005

Pfohl, H., Yahsi, B., & Kurnaz, T. (2016). Concept and diffusion-factors of industry 4.0 in the supply chain. *LDIC*. https://doi.org/10.1007/978-3-319-45117-6_33

Pham, H., & Wang, H. (1996). Imperfect maintenance. *European Journal of Operational Research, 94*(3), 425–438. https://doi.org/10.1016/S0377-2217(96)00099-9

Piegert, S. (2022) SIEMENS AG: 3D-Druck von Gasturbinenkomponenten. https://www.plattform-i40.de/PI40/Redaktion/DE/Anwendungsbeispiele/012-3d-druck-von-gasturbinenkomponenten-siemens/beitrag-3d-druck-von-gasturbinenkomponenten-siemens.html.

Pietri, A. (2020). *The self-optimizing plant: A new era of autonomy, powered by industrial AI*. Aspen Technology Inc. https://www.aspentech.com/en/resources/executive-brief/the-self-optimizing-plant-a-new-era-of-autonomy-powered-by-industrial-ai

Pilat, Z., Klimasara, W., Pachuta, M., & Słowikowski, M. (2019). Some new robotization problems related to the introduction of collaborative robots into industrial practice. *Journal of Automation, Mobile Robotics & Intelligent Systems, 13*(4), 91–97. https://doi.org/10.14313/JAMRIS/4-2019/42

Pinedo, M. L. (2012). Introduction. In M. L. Pinedo (Ed.), *Scheduling: Theory, algorithms, and systems* (pp. 1–10). Springer International Publishing. https://doi.org/10.1007/978-1-4614-2361-4_1

Pinedo, M. L. (2016). *Scheduling: Theory, Algorithms, and Systems* (5th ed.). Springer International Publishing. https://doi.org/10.1007/978-3-319-26580-3

PLANETTOGETHER. (2020). *5 components of production scheduling in manufacturing*. PLANETTOGETHER. https://www.planettogether.com/blog/4-components-of-production-scheduling-in-manufacturing

PLANETTOGETHER. (n.d.). *Innovative flexible packaging manufacturer goes from spreadsheets to smart factory without adding ERP*. PLANETTOGETHER. https://www.planettogether.com/hubfs/PlanetTogether%20Flexible%20Packaging%20Case%20Study%20-%20Bema%20Print.pdf

Platform Industrie 4.0. (2022). *Air supply*. https://www.plattform-i40.de/PI40/Redaktion/DE/Anwendungsbeispiele/269-airbus/air-supply.html

proALPHA Software GmbH. (2015). *proALPHA Software: Qualitätssicherung durch ERP-System*. VDMA. https://www.arburg.com/fileadmin/redaktion/sonstiges/vdma_brochure_industry_4.0_2015_04.pdf

Qi, Q., & Tao, F. (2018). Digital twin and big data towards smart manufacturing and industry 4.0: 360 degree comparison. *IEEE Access, 6*, 3585–3593. https://doi.org/10.1109/ACCESS.2018.2793265

Quality Management System. (n.d.). *The history of quality management—Quality management system*. Quality Management System. https://qualitymanagementsystem.com/total-quality-management/the-history-of-quality-management/

Queiroz, M. M., Ivanov, D., Dolgui, A., & Fosso Wamba, S. (2020). Impacts of epidemic outbreaks on supply chains: Mapping a research agenda amid the COVID-19 pandemic through a structured literature review. *Annals of Operations Research*. https://doi.org/10.1007/s10479-020-03685-7

Oliveira, R., Cubo, C., Estrada, R., Fernandes, A. C., Afonso, P., Carvalho, M. S., Sampaio, P., Roque, J., & Rebelo, M. (2019). A composite indicator for supply chain performance measurement: A case study in a manufacturing company. *IEEE International Conference on Industrial*

Engineering and Engineering Management (IEEM), 2019, 1611–1615. https://doi.org/10.1109/IEEM44572.2019.8978598

Radkowski, R. (2015). Investigation of visual features for augmented reality assembly assistance. In R. Shumaker & S. Lackey (Eds.), *Virtual, augmented and mixed reality* (pp. 488–498). Springer International Publishing. https://doi.org/10.1007/978-3-319-21067-4_50

RDI Technologies. (n.d.). *Case study: Power generation.* https://rditechnologies.com/wp-content/uploads/2020/07/Power-Case-Study.pdf

Renishaw. (2016). Metal additive manufacturing technology used for customised mountain bikes. Renishaw PLC. https://www.renishaw.com/en/metal-additive-manufacturing-technology-used-for-customised-mountain-bikes--39741

Rexhausen, D., Pibernik, R., & Kaiser, G. (2012). Customer-facing supply chain practices—The impact of demand and distribution management on supply chain success. *Journal of Operations Management, 30*(4), 269–281. https://doi.org/10.1016/j.jom.2012.02.001

Riemann, V. (2022). *Traceability von Elektronikgeräten.* Plattform Industrie 4.0. https://www.plattform-i40.de/PI40/Redaktion/DE/Anwendungsbeispiele/322-lenze-traceability-von-elektronikgeraeten/lenze.html

Romagnoli, J. A., & Palazoglu, A. (2005). *Introduction to process control.* CRC Press. https://doi.org/10.1201/9781420056259

Rüedy, A., Hänggi, R., Budde, L., Ehrig, F., Wick, C., & Schwendemann, D. (2021). Implementation of industry 4.0 in injection molding and compounding. Part 1 of the series: Use cases based on strategic priorities in plastics processing. *Kunststoffe*, 7. https://www.ost.ch/fileadmin/dateiliste/3_forschung_dienstleistung/institute/iwk/spritzgiessen/fachbeitraege/kunststoffe_07-2021_implementation-industry-4.0.pdf

Sai, B. (2022). *Identifying optimal setting parameters—Fraunhofer IPA.* Fraunhofer Institute for Manufacturing Engineering and Automation IPA. https://www.ipa.fraunhofer.de/en/expertise/factory-planning-and-production-management/autonomous-production-optimization/identifying-optimal-setting-parameters.html

Saeidi, W. (2022). *Einsatz von Datenbrillen-gestützter Werkerführung beim Automobilzulieferer WS System GmbH*. Platform Industrie 4.0. https://www.plattform-i40.de/IP/Redaktion/DE/Anwendungsbeispiele/261b-werkerfuehrung-in-der-produktion-ws-system-ubimax-gmbh/einsatz-von-datenbrillen-gestuetzter-werkerfuehrung.html

SAFETY4SEA. (2019, May 26). ILO: 2.78 million workers die from occupational accidents annually. *SAFETY4SEA.* https://safety4sea.com/ilo-2-78-million-workers-die-from-occupational-accidents-annually/

Schaeffler. (n.d.). *Optimierung durch Instandhaltung 4.0 für Intralogistik 4.0.* Schaeffler. Retrieved July 21, 2021, from https://www.schaeffler.de/content.schaeffler.de/de/produkte-und-loesungen/industrie/industrie-4.0/kundenprojekte-geschaeftsmodelle/instandhaltung-4.0-distributionszentrum-mitte/index.jsp

Schilling, M. (2022) Additive Fertigung von Großteilen. https://www.plattform-i40.de/PI40/Redaktion/DE/Anwendungsbeispiele/248-additive-fertigung-von-grossteilen/beitrag-additive-fertigung-von-grossteilen.html.

Schleich, B., Anwer, N., Mathieu, L., & Wartzack, S. (2017). Shaping the digital twin for design and production engineering. *CIRP Annals—Manufacturing Technology, 66*, 141–144. https://doi.org/10.1016/j.cirp.2017.04.040

Schmeltzpfenning, K., & Krämer, B. (2019). Innovative warehouse-management-Systeme im Kontext von Industrie 4.0. In M. ten Hompel, B. Vogel-Heuser, & T. Bauernhansl (Eds.), *Handbuch Industrie 4.0: Produktion, Automatisierung und Logistik* (pp. 1–22). Springer. https://doi.org/10.1007/978-3-662-45537-1_105-1

Schmidt, C., Weber, P., Hocke, T., & Denkena, B. (2018). Influence of prepreg material quality on carbon fiber reinforced plastic laminates processed by automated fiber placement. *Procedia CIRP, 67*, 422–427. https://doi.org/10.1016/j.procir.2017.12.236

Schnellecke Logistics. (n.d.). *Prozessverbesserungen mit X-Pick*. Schnellecke. https://www.schnellecke.com/de/schnellecke-logistics/smart-logistics/digitalization/wearable-computing/artikel-xband

Schuh, G., Anderl, R., Gausemeier, J., & Wahlster, W. (2017). *Industrie 4.0 Maturity Index*. 62.

Schwerdtfeger, B., Reif, R., Gunthner, W. A., Klinker, G., Hamacher, D., Schega, L., Bockelmann, I., Doil, F., & Tumler, J. (2009). *Pick-by-vision: A first stress test*. 2009 8th IEEE International Symposium on Mixed and Augmented Reality, 115–124. https://doi.org/10.1109/ISMAR.2009.5336484

Seebo. (2018, September 2). Case study—Food manufacturing: How an international manufacturer of breads, crackers and biscuits reduced quality & waste losses by up to 70%. Seebo. https://iot.seebo.com/process-optimization-case-study-ty-0

Seeliger, A., Netland, T., & Feuerriegel, S. (2022). Augmented reality for machine setups: Task performance and usability evaluation in a field test. *Procedia CIRP, 107*, 570–575. https://doi.org/10.1016/j.procir.2022.05.027

Sharma, A., & Singh, D. (2020). Evolution of industrial revolutions: A review. *International Journal of Innovative Technology and Exploring Engineering, 9*, 66–73. https://doi.org/10.35940/ijitee.I7144.0991120

Sharma, S., & Patil, S. V. (2015). Key indicators of rice production and consumption, correlation between them and supply-demand prediction. *International Journal of Productivity and Performance Management, 64*(8), 1113–1137. https://doi.org/10.1108/IJPPM-06-2014-0088

Sieber, V. (2022). *Montage-Assistenzsystem auf Basis von Bewegungserkennung*. Platform Industrie 4.0. https://www.plattform-i40.de/IP/Redaktion/DE/Anwendungsbeispiele/128-montage-assistenzsystem-auf-basis-von-bewegungserkennung/beitrag-montage-assistenzsystem-auf-basis-von-bewegungserkennung.html

Siegwart, R., Nourbakhsh, I. R., & Scaramuzza, D. (2011). Introduction to autonomous mobile robots. *Robotica, 23*(2). https://doi.org/10.1017/S0263574705221628

Silvestri, L., Forcina, A., Introna, V., Santolamazza, A., & Cesarotti, V. (2020). Maintenance transformation through Industry 4.0 technologies: A systematic literature review. *Computers in Industry, 123*, 103335. https://doi.org/10.1016/j.compind.2020.103335

Singh, A., Mishra, N., Ali, S. I., Shukla, N., & Shankar, R. (2015). Cloud Computing technology: Reducing carbon footprint in beef supply chain. *International Journal of Production Economics, 164*, 462–471. https://doi.org/10.1016/j.ijpe.2014.09.019

Snap Schedule. (2020). *Cloud-based Manufacturing Shift Scheduling Software*. Snap Schedule. http://www.snapschedule.com/industry/manufacturing/

SnapFulfil, S. (n.d.). *Manufacturing warehouse management case study: Comfy quilts*. Snapfulfil. http://www.snapfulfil.com/case-studies/comfy-quilts-improves-inventory-management-and-warehouse-operations/

Soltanpoor, R., & Sellis, T. (2016). *Prescriptive analytics for big data*. Australasian Database Conference, 245–256.

Sorostinean, R., Gellert, A., & Pirvu, B.-C. (2021). Assembly assistance system with decision trees and ensemble learning. *Sensors, 21*(11), 3580. https://doi.org/10.3390/s21113580

Spinnarke, S. (2015, February 10). *Qualitätssicherung in der Fertigung*. Produktion Online. https://www.produktion.de/technik/qualitaetssicherung-in-der-fertigung-304.html

Spotsee. (2020, June). *RFID tracking labels are the future of packaging*. Spotsee. https://spotsee.io/blog/rfid-tracking-stickers-are-the-future-of-packaging

Startups Against Corona. (2020). *Startups against corona—Startup solutions for Corona problems*. Startups Against Coronavirus. https://startupsagainstcorona.com/

State of Flux. (n.d.). *Data tool offers early warning of supply chain risk*. State of Flux. Retrieved November 10, 2021, from http://www.stateofflux.co.uk/ideas-insights/articles/data-tool-offers-early-warning-of-supply-chain-ris

Staying on top of your supply chain. (2022). GEP. https://www.gep.com/case-studies/supply-chain-dashboard-case-study

Stefanovic, N., & Stefanovic, D. (2011). Supply chain performance measurement system based on scorecards and web portals. *Computer Science and Information Systems, 8*(1), 167–192. https://doi.org/10.2298/CSIS090608018S

Steinbauer, M. (2022). *Lifetime analyse*. Platform Industrie 4.0. https://www.plattform-i40.de/IP/Redaktion/DE/Anwendungsbeispiele/031-neue-serviceangebote-durch-die-digitalisierung-verfuegbarkeitsgarantie-und-vermeidung-von-ausfallzeiten/beitrag-neue-serviceangebote-durch-die-digitalisierung-verfuegbarkeitsgarantie-und-vermeidung-von-ausfallzeiten.html

Steinberg, G. (2019, November 14). *How reinventing the supply chain can lead to an autonomous future*. EY. https://www.ey.com/en_gl/consulting/how-reinventing-the-supply-chain-can-lead-to-an-autonomous-future

Straßer, S., Tripathi, S., & Kerschbaumer, R. (2018). *An approach for adaptive parameter setting in manufacturing processes*. Proceedings of the 7th International Conference on Data Science, Technology and Applications (DATA 2018), 24–32. https://doi.org/10.5220/0006894600240032

SupplHi. (2021, April 18). With SupplHi's Carbon Tracker, set reduction targets and track the carbon footprint of your industrial B2B Suppliers (Scope 3 emissions)—Blog SupplHi. *SupplHi*. https://www.supplhi.com/with-supplhis-carbon-tracker-set-reduction-targets-and-track-the-carbon-footprint-of-your-industrial-b2b-suppliers-scope-3-emissions/

Supply Chain Analytics. (2022). *Tableau*. https://www.tableau.com/solutions/supply-chain-analytics

SupplyOn—The Supply Chain Business Network. (2013, April 25). *AirSupply: The joint supply chain management platform for the European aerospace industry*. https://www.youtube.com/watch?v=gvqd-Psrtq8&ab_channel=SupplyOn-TheSupplyChainBusinessNetwork

Tao, F., & Zhang, M. (2017). Digital twin shop-floor: A new shop-floor paradigm towards smart manufacturing. *IEEE Access, 5*, 20418–20427. https://doi.org/10.1109/ACCESS.2017.2756069

Tao, F., Cheng, J., Qi, Q., Zhang, M., Zhang, H., & Sui, F. (2018). Digital twin-driven product design, manufacturing and service with big data. *The International Journal of Advanced Manufacturing Technology, 94*(9), 3563–3576. https://doi.org/10.1007/s00170-017-0233-1

Tao, F., Cheng, Y., Cheng, J., Zhang, M., Xu, W., & Qi, Q. (2017). Theory and technologies for cyber-physical fusion in digital twin shop-floor 数字孪生车间信息物理融合理论与技术. *Computer Integrated Manufacturing Systems, 23*, 1603–1611. https://doi.org/10.13196/j.cims.2017.08.001

Taqi, H. M. M., Ahmed, H. N., Paul, S., Garshasbi, M., Ali, S. M., Kabir, G., & Paul, S. K. (2020). Strategies to manage the impacts of the COVID-19 pandemic in the supply chain: Implications for improving economic and social sustainability. *Sustainability, 12*(22), 9483. https://doi.org/10.3390/su12229483

Tauber, M., Gallmetzer, A., Rauch, E., Brown, C. A., & Matt, D. T. (2019). Concept design of a digital shop floor information system for assembly operators in machine industry. *MATEC Web of Conferences, 301*, 00017. https://doi.org/10.1051/matecconf/201930100017

Techopedia. (2013, December 1). *What is data transmission?* Techopedia. http://www.techopedia.com/definition/9756/data-transmission

Techopedia. (2018, February 9). *What is data acquisition?* Techopedia.Com. http://www.techopedia.com/definition/30000/data-acquisition

The Numina Group. (2016, July 12). *Warehouse voice picking success story*. Numina Group. http://numinagroup.com/2016/07/12/warehouse-voice-picking-success-story/

Thomas Warburton. (n.d.). *What are C-parts and what is inventory management?* https://www.warburtons.com.au/en/warburton/c_parts_management/what_are_c_parts.php

Tönshoff, H. K., & Inasaki, I. (Eds.). (2001). *Sensors in manufacturing* (1st ed.). Wiley. https://doi.org/10.1002/3527600027

Treichel, K. (2022). *ABB AG: Intelligentes alarm management*. https://www.plattform-i40.de/PI40/Redaktion/DE/Anwendungsbeispiele/180-intelligentes-alarm-management-abb/beitrag-intelligentes-alarm-management-abb.html

Trump, M. (2022). *Rota Yokogawa GmbH & Co. KG: Termingetreue Versandsteuerung mittels digitalem Pull-System*. https://www.plattform-i40.de/PI40/Redaktion/DE/Anwendungsbeispiele/338-rota-yokogawa-versandsteuerung/beitrag-rota-yokogawa-versandsteuerung.html

Tung, T. V., & Yang, B.-S. (2009). Machine fault diagnosis and prognosis: The state of the art. *International Journal of Fluid Machinery and Systems, 2*(1), 61–71. https://doi.org/10.5293/IJFMS.2009.2.1.061

Ullrich, G. (2015). Automated guided vehicle systems. *Springer*. https://doi.org/10.1007/978-3-662-44814-4

van der Aalst, W. (2016). Process mining. *Springer*. https://doi.org/10.1007/978-3-662-49851-4

van der Velden, H. (2022). *Bossard Deutschland GmbH: Smart Factory Logistics*. https://www.plattform-i40.de/IP/Redaktion/DE/Anwendungsbeispiele/425-bossard-smart-logistics/beitrag-bossard.html

VDI. (2005). *VDI 2510—Fahrerlose Transportsysteme (FTS)*. https://www.vdi.de/richtlinien/details/vdi-2510-fahrerlose-transportsysteme-fts

Verenich, I., Dumas, M., La Rosa, M., & Nguyen, H. (2019). Predicting process performance: A white-box approach based on process models. *Journal of Software: Evolution and Process, 31*, e2170. https://doi.org/10.1002/smr.2170

Veza, I., Mladineo, M., & Gjeldum, N. (2015). Managing innovative production network of smart factories. *15th IFAC Symposium On Information Control Problems In Manufacturing, 48*(3), 555–560. https://doi.org/10.1016/j.ifacol.2015.06.139

Vieira, G. E. (2006). A practical view of the complexity in developing master production schedules: Fundamentals, examples, and implementation. In J. W. Herrmann (Ed.), *Handbook of production scheduling* (Vol. 89, pp. 149–176). Kluwer Academic Publishers. https://doi.org/10.1007/0-387-33117-4_7

Vorraber, W., Gasser, J., Webb, H., Neubacher, D., & Url, P. (2020). Assessing augmented reality in production: Remote-assisted maintenance with HoloLens. *Procedia CIRP, 88*, 139–144. https://doi.org/10.1016/j.procir.2020.05.025

Walters, N., Anderson, N., & Kozyrenko, O. (2020). COVID-19, meet Industry 4.0. *Supply Chain Management Review, 24*(6), 50–53.

Wang, J. (2014). *RFID as a key enabler of the Internet of Things: Localization and communication.*

Wang, Y., & Zhang, F. (Eds.). (2017). *Trends in control and decision-making for human–robot collaboration systems*. Springer International Publishing. https://doi.org/10.1007/978-3-319-40533-9

Weckenborg, C., Kieckhäfer, K., Müller, C., Grunewald, M., & Spengler, T. S. (2020). Balancing of assembly lines with collaborative robots. *Business Research, 13*(1), 93–132. https://doi.org/10.1007/s40685-019-0101-y

Weckenmann, A., Akkasoglu, G., & Werner, T. (2015). Quality management—History and trends. *The TQM Journal, 27*(3), 281–293. https://doi.org/10.1108/TQM-11-2013-0125

Wenking, M., Flaeschner, O., & Torbjørn, N. H. (2021). *Unlocking value with production network optimization: A strategic perspective*. Springer International Publishing.

Werner, W. (2022). *Plattform Industrie 4.0—BlueRain Software GmbH & Co. KG - ID4i*. Plattform Industrie 4.0. https://www.plattform-i40.de/IP/Redaktion/DE/Anwendungsbeispiele/489-bluerain/beitrag-bluerain.html

Werning, S., Konusch, D., & Ickerott, I. (2020). Pick-by-Vision: Potenziale in der Unterstützung der Kommissionierung durch Smart Glasses. In O. Thomas & I. Ickerott (Eds.), *Smart Glasses: Augmented Reality zur Unterstützung von Logistikdienstleistungen* (pp. 168–189). Springer. https://doi.org/10.1007/978-3-662-62153-0_10

Why prescriptive analytics and decision optimization are crucial. (2020, April 14). Journey to AI Blog. https://www.ibm.com/blogs/journey-to-ai/2020/04/why-prescriptive-analytics-and-decision-optimization-are-crucial/

Wick, C., Ehrig, F., & Schuster, G. (2020). Data driven injection moulding. In C. Hopmann & R. Dahlmann (Eds.), *Advances in polymer processing 2020* (pp. 128–136). Springer. https://doi.org/10.1007/978-3-662-60809-8_11

Wick, C., Ehrig, F., Schwendemann, D., Hänggi, R., & Rüedy, A. (2021). On the way to the smart injection molding factory. Part 2 of the series: Integrated data acquisition for the injection molding production. *Kunststoffe, 9*. https://www.ost.ch/fileadmin/dateiliste/3_forschung_dienstleistung/institute/iwk/spritzgiessen/fachbeitraege/kunststoffe_09-2021_smart-injection-molding-factory.pdf

Wickel-Bajak, D. (2022). *Plattform Industrie 4.0—Floor Care Zukunft*. Platform Industrie 4.0. https://www.plattform-i40.de/IP/Redaktion/DE/Anwendungsbeispiele/084-floor-care-zukunft/beitrag-floor-care-zukunft.html

Wiegand, M., Stolpe, M., Deuse, J., & Morik, K. (2016). Prädiktive Prozessüberwachung auf Basis verteilt erfasster Sensordaten. *at - Automatisierungstechnik, 64*(7), 521–533. https://doi.org/10.1515/auto-2016-0013

Wirth, S. (2022). *Energietransparenzsystem*. https://www.plattform-i40.de/PI40/Redaktion/DE/Anwendungsbeispiele/037-energietransparenzsysteme-festo/beitrag-energietransparenzsysteme-festo.html

World Business Council for Sustainable Development. (2021). *Value chain carbon transparency pathfinder: Enabling decarbonization through Scope 3 emissions transparency* [Publication]. World Business Council for Sustainable Development. https://www.wbcsd.org/Programs/Climate-and-Energy/Climate/SOS-1.5/Resources/Value-Chain-Carbon-Transparency-Pathfinder-Enabling-decarbonization-through-Scope-3-emissions-transparency

World Economic Forum (WEF). (2020). *How to rebound stronger from COVID-19. Resilience in manufacturing and supply systems* [Whitepaper]. World Economic Forum. https://www.weforum.org/whitepapers/how-to-rebound-stronger-from-covid-19-resilience-in-manufacturing-and-supply-systems

Wuest, T., Kusiak, A., Dai, T., & Tayur, S. R. (2020). Impact of COVID-19 on manufacturing and supply networks—The case for AI-Inspired digital transformation (SSRN Scholarly Paper ID 3593540). *Social Science Research Network*. https://doi.org/10.2139/ssrn.3593540

Wurms, G. (2017). *Use cases | Tracking of floor conveyors and goods in logistics*. https://www.infsoft.com/use-cases/tracking-of-floor-conveyors-and-goods-in-logistics

Würth Industrie Service GmbH & Co. KG. (2022). *Intelligentes, vernetztes C-Teile-Management*. https://www.plattform-i40.de/IP/Redaktion/DE/Anwendungsbeispiele/035-intelligentes-vernetztes-c-teile-management-wuerth-industrie/beitrag-intelligentes-vernetztes-c-teile-management-wuerth-industrie.html

Wuthrich, R., & Hof, L. A. (2019). Low batch size production of glass products requiring micrometer precision. *IFAC-PapersOnLine, 52*(10), 319–322. https://doi.org/10.1016/j.ifacol.2019.10.050

Young, R. I. M., Popplewell, K., Jaekel, F.-W., Otto, B., & Bhullar, G. (2015). Intelligent systems configuration services for flexible dynamic global production networks. In M. Lauras, M. Zelm, B. Archimède, F. Bénaben, & G. Doumeingts (Eds.), *Enterprise interoperability* (pp. 85–90). Wiley. https://doi.org/10.1002/9781119081418.ch11

Yusuf, B. (2021, June 15). *Digital factory use case: 5S and Gemba Walks with NavVis IVION Go*. https://www.navvis.com/blog/digital-factory-use-case-5s-and-gemba-walks-with-navvis-ivion-go

Zemann, P. (2022). *Nutzung von KI zur Optimierung von Qualitätskontrollen in Elektro-Fertigung*. Industrie 2025. https://www.industrie2025.ch/wissen-industrie-40/use-cases/use-cases/nutzung-von-ki-zur-optimierung-von-qualitaetskontrollen-in-elektro-fertigung

Zheng, L., Liu, X., An, Z., Li, S., & Zhang, R. (2020). A smart assistance system for cable assembly by combining wearable Augmented Reality with portable visual inspection. *Virtual Reality & Intelligent Hardware, 2*(1), 12–27. https://doi.org/10.1016/j.vrih.2019.12.002

Žižek, S. Š., Nedelko, Z., Mulej, M., & Čič, Ž. V. (2020). Key performance indicators and industry 4.0—A socially responsible perspective. *Naše Gospodarstvo/Our Economy, 66*(3), 22–35. https://doi.org/10.2478/ngoe-2020-0015

Zonta, T., da Costa, C. A., da Rosa Righi, R., de Lima, M. J., da Trindade, E. S., & Li, G. P. (2020). Predictive maintenance in the Industry 4.0: A systematic literature review. *Computers & Industrial Engineering, 150*, 106889. https://doi.org/10.1016/j.cie.2020.106889

Chapter 4
Smart Factory Implementation Process

As the use case patterns show, the smart factory provides a bundle of new opportunities to drive performance and increase competitive advantages for a company. New technologies require new skills, in particular digital skills. Additionally, and even more importantly for implementing smart manufacturing, use cases become the holistic understanding of any kind of process from end-to-end. The combination of new technologies with the existing knowledge about processes and pain points is a challenging task. Many interdependencies need to be considered, and knowledge from different functional departments has to be intelligently exploited to create value. Implementing the required changes is a major challenge, with many hurdles. This section shows how to overcome them successfully.

Difficulty to Prioritize and Identify Suitable Use Cases

Often, the problem is not a lack of ideas. In most cases, there are many ideas existing in any company. The varying levels of maturity, the missing comparability, the uncertainty about potential benefits and the risks of financial investments all make prioritization of potential use cases difficult. The key issue in defining and implementing use cases for the smart factory is selecting the right use case. Objective orientation and a focused approach are fundamental here.

Missing Organizational Anchoring

Use cases do not stop at functional departments. An end-to-end perspective is needed to exploit the full potential. We observed that many companies fail to establish an organizational anchoring that allows more cross-functional activities. Relevant functional departments need to be involved in the use case selection and design process. The project team driving digitalization should have the ability and backing of top management to enable this. All organizational units affected by the use case must be included in the process.

L. Budde et al., *Smart Factory Navigator*,
https://doi.org/10.1007/978-3-031-17254-0_4

Silo Thinking

Smart factory use cases do not know system boundaries in the sense of functional company departments. Rather, they process the knowledge, data or product from different areas into a whole. Therefore, a common understanding of the definition of a use case provides the basis for a discussion. With the help of a use case pattern, a basis will be created that will enable the company to identify potential departmental boundaries. Different departments have different needs and prioritize investments differently. In addition, the diversity, complexity and great potential of the various Industry 4.0 technologies in relation to a use case must be understood and used efficiently. Only use cases that fit the corporate strategy will be beneficial in the long term.

Waste in the Processes

Digital processes often function in different ways, and existing processes need to adapt to this. But before digitalizing processes, potential waste has to be identified. These organizational and process deficiencies need to be addressed first. Companies must learn to simplify and optimize the processes greatly before digitalizing them. Otherwise, they will end up with a digitally enhanced mess and not a better process that saves money.

Competence Gap

With the advance of digitalization, existing IT and production technologies must be adapted to the new world. This change is costly. Therefore, emerging IT and technology investments must be deployed in a focused approach. The challenge is to keep track of potentially applicable technologies and to derive their benefits for individual use cases for one's own company. This applies not only to the IT infrastructure but also to the required human resources. The empowerment of employees or the development of central competencies in the company is a central point to ensure sustainable success in the implementation of use cases.

Underestimated Relevance of the Underlying Database

In most circumstances, the use cases form the basis for the data acquisition requirements and the development of the necessary IT infrastructure. This view of the use cases essentially reveals what data needs to be processed and what the quality of this data must be. The foundation thus created shows companies how to learn from data in specific use cases. Before the company starts working on use cases with a high level of digital maturity, it should be ensured that the underlying basis for data capturing, transmission and storage is guaranteed. If this is not the case, there is a risk that the project will not deliver the expected results.

Complexity of Projects

The process of identifying suitable smart factory use cases associated with Industry 4.0 can be challenging. Usually, different technological approaches can be used to implement a use case. The interaction of these technologies with the use cases is another difficulty when selecting appropriate use cases for a company. The variety of possible technologies can overwhelm small and medium-sized enterprises, particularly when it comes to identifying the use cases that are relevant for them. Topics such as the scope in which an initial implementation takes place should be defined as early as possible.

To overcome these hurdles, we introduce you to a process that guides you through this transformation towards the smart factory. The following process outlines a structured procedure for companies to implement suitable use cases.

The Implementation Process

In order to develop towards becoming a smart factory, we provide an established process (see Fig. 4.1) and a checklist to guide this transformation and make it a success. The detailed use case patterns presented in Chap. 3 serve as a starting point for identifying suitable use cases in your company. A common understanding and also a common language foundation are important. The discussion and debate that takes place will result in a focused approach. The framework leads to the identification of potential use cases for your company. With the help of the provided generic use cases, a common understanding can be built and the question of the appropriate implementation of digitization measures can be discussed.

The procedure for the successful implementation of smart factory use cases involves seven steps. Figure 4.1 provides an initial overview of the procedure. As presented in Fig. 4.2, the seven steps can be divided into two sequences. The first sequence with its six steps deals with the framework and its generic use cases. It leads from the selection of a generic use case, its alignment with strategic priorities and the company-specific scenario to its implementation. Step six refers to the first deployment of the use case in a factory, a production line, a machine, or a process.

After implementing the above, based on the selected use case, the second sequence leads to a possible scaling and global introduction of the chosen use case. In this sequence you will have to deal, in addition to implementing the enabling technologies, with the relevant IT architecture. They will need to be addressed strategically and often on a global scale and across organizational boarders. To enable this, a stepwise approach is also recommended. The first project will give you some guidance on what it means to implement the smart factory in regard to the long-term needs of your IT architecture. Consequently, you will learn a lot, and you will probably implement certain changes to your actual IT set up. Following the first successful smart factory steps in the first sequence, you need to discuss the potential future IT-landscape. This will give you the necessary foundation for a future rollout.

This chapter highlights the individual steps as well as the specific procedure and results that ought to be developed after each step.

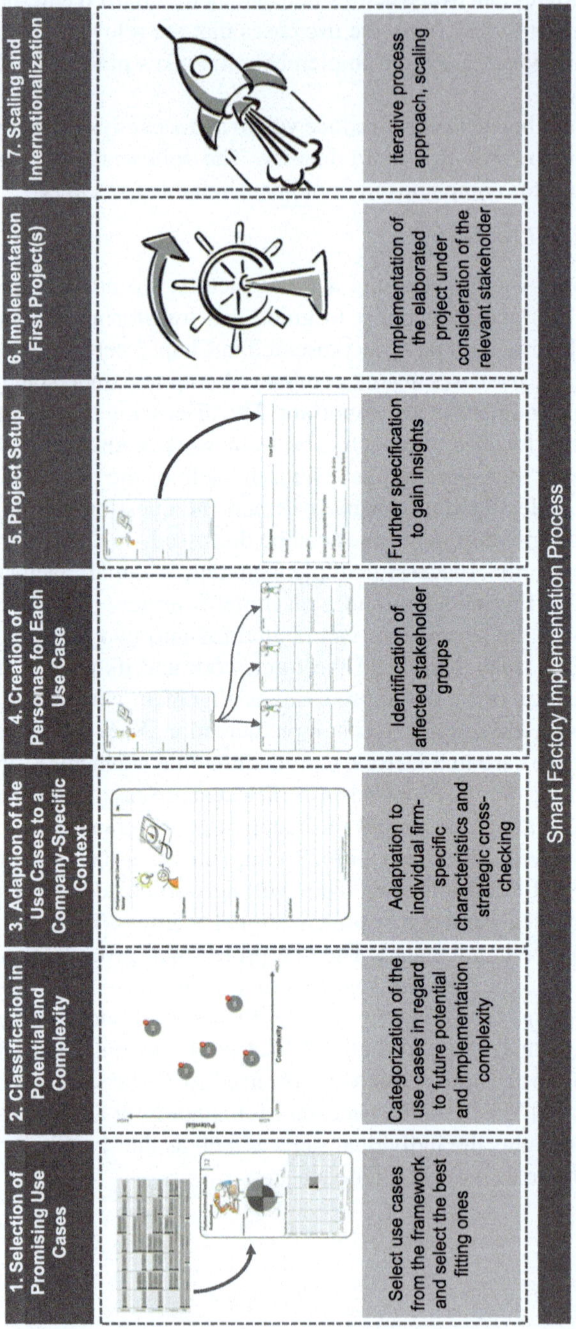

Fig. 4.1 Smart factory implementation process

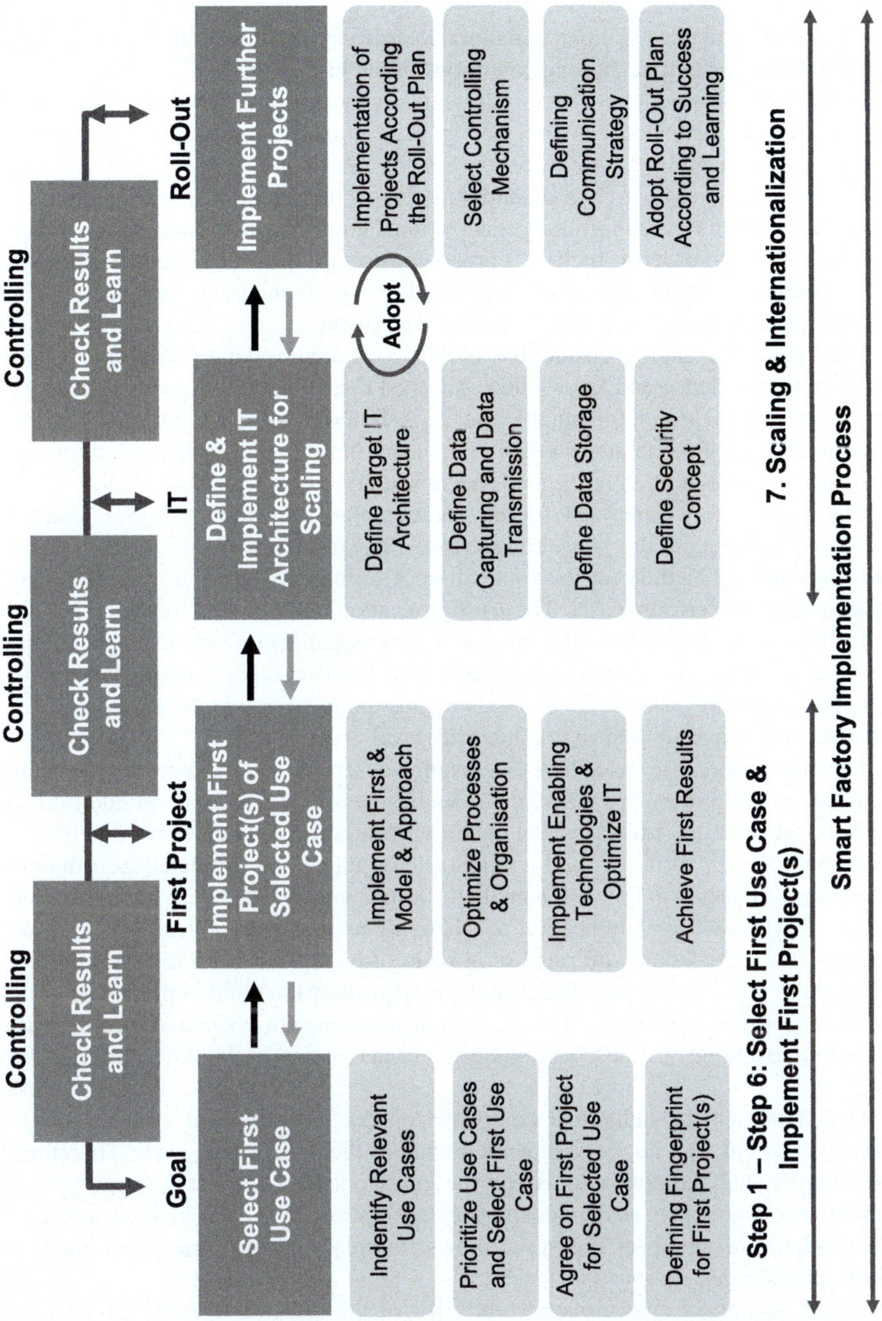

Fig. 4.2 Model for roll-out

4.1 Step 1: Selection of Promising Use Cases

The search for suitable use cases is an iterative process that ensures that the company adapts to the constantly changing conditions of the market. Not only the market can trigger this reassessment but also the strategic orientation of the company. This is an iterative process that should be repeated by the company periodically to respond to changing influences.

For the subsequent search of potential use cases, it is useful to assemble an interdisciplinary team. Accordingly, all organizational units that are of importance should be represented. In order to include a broad spectrum of knowledge and experience in the evaluation, both C-level and subject matter experts should be represented. The size of the team should be between six and ten people.

Care should be taken to ensure that all the people involved have their say and can share their experience and knowledge. An open discussion at all levels is decisive for the subsequent search for suitable use cases. It makes sense for one participant to take on the role of moderator and guide the team through the subsequent steps and intervene if the discussion in the steps becomes unfocused.

The smart factory framework for the selection of a promising use case is depicted in Fig. 4.3 and contains the 44 generic use cases. The framework provides a structured and fast selection method for the application of suitable use cases in your company based on these generic use cases. These use cases should be assessed for their potential and relevance, including how they align with your organization's strategic goals. The use cases are categorized into two dimensions: "process stages" and the "degree of analytics foresight". The horizontal axis relates the use cases to the core processes and the vertical axis to their digital maturity level.

For each generic use case, there is a detailed description that allows the company to decide whether they are relevant for its business (see Chap. 3). In addition to the description of the use cases, the generic potential benefits that can be achieved through implementation are also highlighted. The potential is evaluated according to four factors, depicted in Fig. 4.4, namely the impact on cost, delivery, flexibility, and quality. This provides an indication of where the use case will potentially generate benefits. The four factors are rated on a scale of one to ten, with ten representing the highest possible impact. The benefit is also described and summarized. The technologies to be applied may differ according to the specific context of the company. Therefore, examples of these technologies are also given in the description of the use cases.

Depending on the individual company context, an individual use case could already exist but may not be assignable to any of the generic use cases. Therefore, the company has the option to extend the generic use cases. The description will contain the same topics as the generic use cases, allowing a clear understanding of the distinction in retrospect. Figure 4.5 shows a template of a blank card for individual cases and their description.

At the end of the first process step, a list of the selected and, therefore, most relevant use cases for the company will be available. Care should be taken to ensure

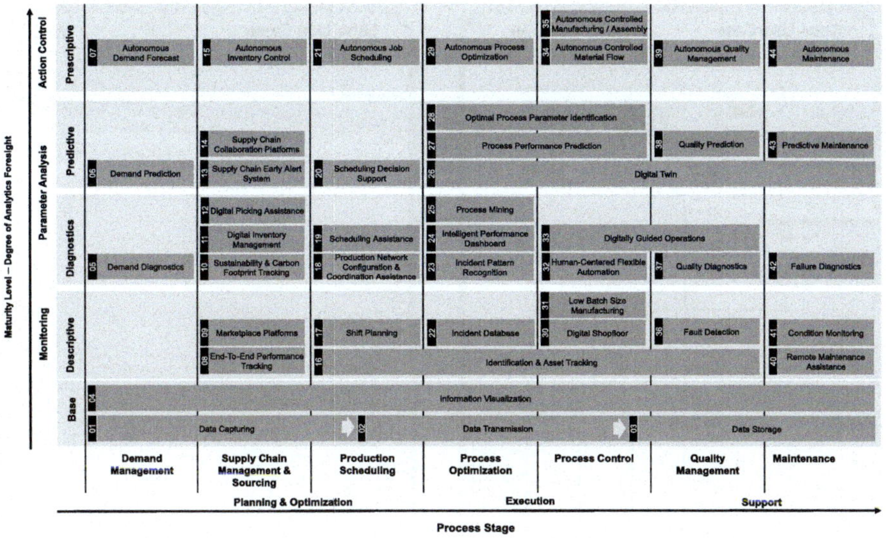

Fig. 4.3 Smart factory framework with 44 generic use cases

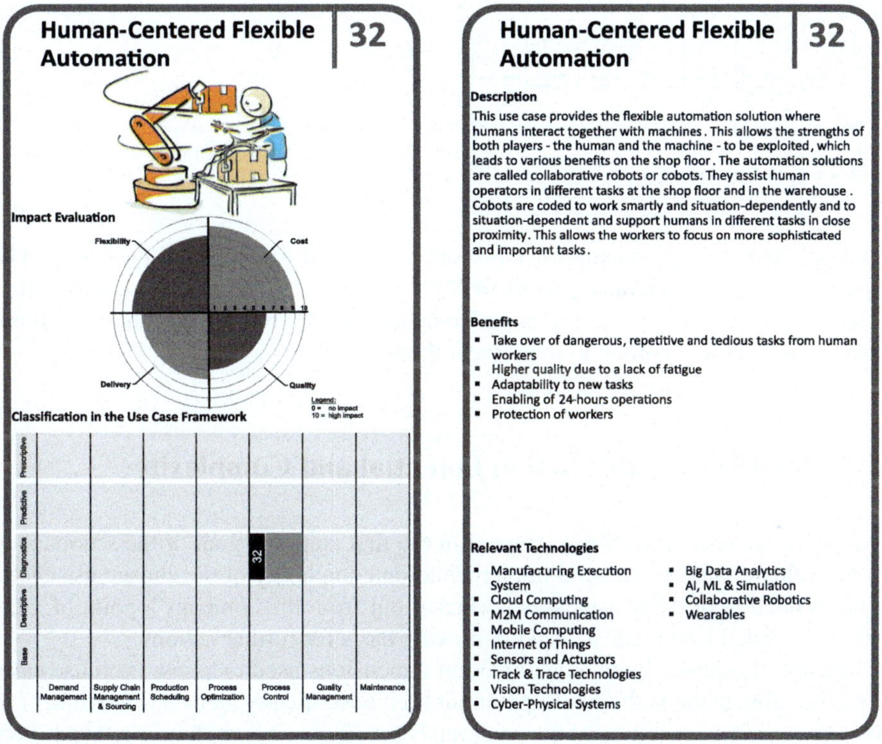

Fig. 4.4 Generic use case card

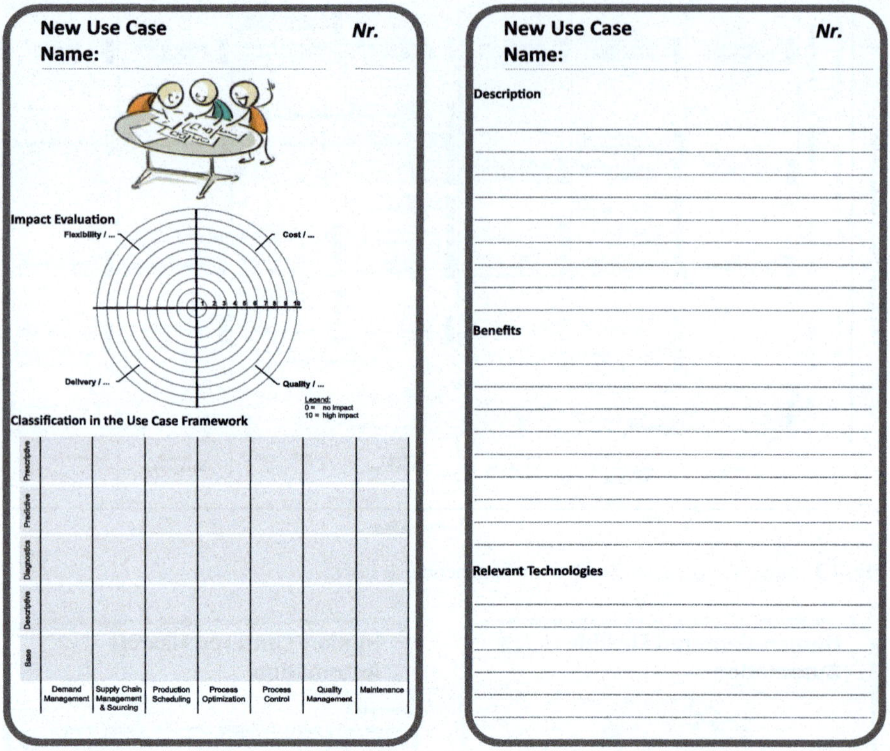

Fig. 4.5 Specific use case card

that only the most promising use cases are considered for the subsequent step. The selection of use cases may consist of both generic and individual use cases. It is important that there is a common understanding of the reasons for choosing these use cases and the intention to implement them.

4.2 Step 2: Classification in Potential and Complexity

After the use cases have been selected in the first step, the goal of the second step is to evaluate the potential and implementation complexity of the chosen use cases. The evaluation of these use cases is carried out from the company's point of view and is intended to provide a decision-making basis for further actions.

Figure 4.6 shows the two qualification dimensions used to assess each use case. The first dimension is the future potential (i.e., cost, quality, delivery, flexibility) of the use case. An estimation of the potential is key. If possible, it should be quantified. For this purpose, clear metrics of the expected potentials can be defined based on KPIs. For each competitive priority, possible influencing factors that can be used

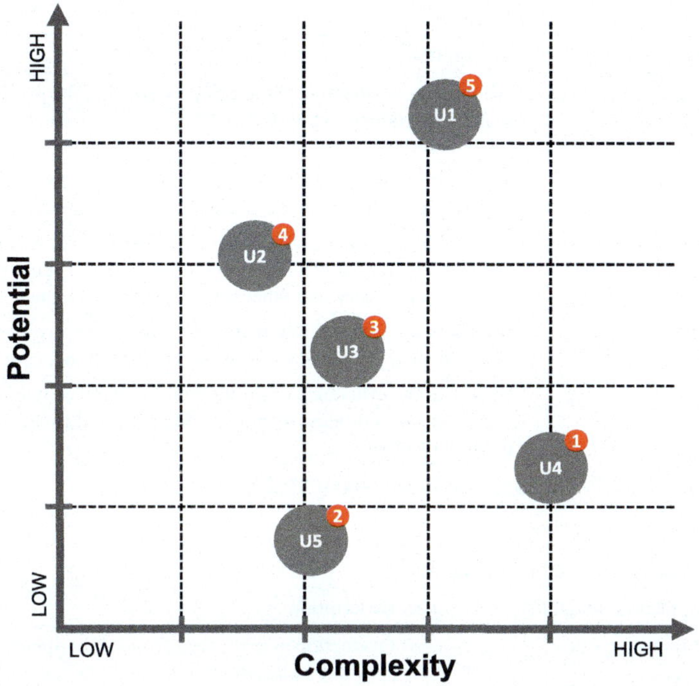

Fig. 4.6 Potential-complexity map (PCM)

to assess the impact are listed below (see Tables 4.1, 4.2, 4.3, and 4.4). This list of influencing factors is not to be considered exhaustive and is intended only as a guide. When evaluating the competitive priorities, the quantification should be carried out in such a way that a comparison of the use cases is ensured. All use cases should be evaluated with the same KPIs and their expected impact. This procedure ensures that the use cases can be classified according to their potential.

The second dimension refers to the complexity of the implementation of the use case. Topics such as the necessary personnel competencies, required personal resources, uncertainties about investments and potential benefits or the existing IT infrastructure serve as categories for discussions. This dimension should also be quantified.

If financial quantification is not possible at this stage, you can also give qualitative ratings along a Likert-Scale (1–5) such as "low", "medium" or "high". In such situations, this can also take you a step further at this stage. It is important that you can, based on facts, discuss the differences between use cases. The more detailed and larger the number of facts, the better. But the key is the joint discussion and not the exact financial figure.

Both dimensions should be assessed taking into account the relevant technologies. With the help of the positioning on the potential-complexity map (PCM), the use cases can be visually distinguished and compared. In order to properly position each use

Table 4.1 Influencing factors on costs

Influencing factors on costs	Description
Direct manufacturing costs	All directly attributable resources consumed in the process of making a product, i.e., direct material costs and direct labor costs but no overhead costs
Labor productivity	The number of goods and services that a group of workers produce in a given amount of time, i.e., output per labor input
Capacity utilization	The amount of a factory's production capacity that is currently utilized, i.e., how much of the potential output is being met in terms of machinery and other resources like personnel
Low inventory	Inventory includes all raw materials, work-in-progress or semi-finished products and finished goods within a company
Productivity	Measures the efficiency of production as output of two input used in a production process over a specific period of time, i.e., output per unit of input

Sources Ward et al. (1998), Boyer and Pagell (2000), Hänggi et al. (2021)

Table 4.2 Influencing factors on quality

Influencing factors on quality	Description
High right first time rate	Production as error-free as possible right from the start due to low scrap rates and low rework rates, i.e., high share of good parts in total production
Zero defects to customer	No defect part leaves the factory, e.g., through well-performing production processes or quality control measures
Traceability of product history	In case of a customer claim, the complete production and product history is available
Effective and efficient testing procedures	Providing testing resources for quality purposes at low cost and high accuracy preferably by inline testing during production
Automated process control	Providing efficient process control with minimal human effort, e.g., by statistical process control (SPC)

Sources: Ward et al. (1998), Boyer and Pagell (2000), Hänggi et al. (2021), Flynn et al. (1994), Flynn et al. (1995), Noble, (1995), Saraph et al. (1989), Zhang et al. (2014)

case, the project team should discuss and reach a consensus. For the final assessment, the results of the discussions in the project team should be recorded.

As soon as all use cases have been added to the PCM, a further, more refined selection will be made. The evaluation necessary for further refinement of the selection is carried out by means of a vote. The voting is based on a scoring system. During the scoring process, participants can assign points to the individual use cases to express their opinion on the respective cases. The following rules apply for voting in the project team:

Table 4.3 Influencing factors on delivery description

Influencing factors on delivery	Description
Order fulfilment time	The time between the customer order and receipt of the product or service by the customer
Delivery on due date	Share of finished goods or services delivered to customers on time (ship on time), i.e., the on-time share of all delivered products
Production cycle time	The time needed to convert raw materials into finished goods including process, move, inspection and queue times
Delivery as promised	Share of finished goods or services delivered to customers in the complete ordered quantity and with the promised reference/quality
Order handling time	The time between the customer order and the shipment of the finished product to the customer

Sources Ward et al. (1998), Boyer and Pagell (2000), Hänggi et al. (2021)

Table 4.4 Influencing factors on flexibility description

Influencing factors on flexibility	Description
Design changes in production	Ability to make design changes in or on the product after production has started, i.e., the production processes have to be adaptable
Flexibility to change product mix	Ability to change over quickly to other products produced in a facility, as required by demand shifts in product mix, e.g., of a product line
Flexibility to change output volume	Ability of the production processes to quickly accommodate large variations in production levels, e.g., in- or decreasing the production volume through a flexible workforce
Delivery flexibility	Ability to act flexibly to customer demands from a logistics perspective, e.g., by flexible delivery frequencies in logistics
Large number of product options	Ability to handle a large number of product options in the product portfolio, e.g., through complex management of manufacturing processes

Sources Ward et al. (1998), Boyer and Pagell (2000), Hänggi et al. (2021), Kraemer & Kraemer (2010)

- Each participant receives a limited number of points [p] to be assigned to the respective use cases.
- The number of points per participant [p] is calculated according to the following formula and the result is rounded up to the next higher integer number:

$$p = \frac{\#\ of\ Use - Cases}{3} * 2\ where\ it\ holds\ that\ p > 2$$

- Each point assigned has the same weight.
- To weigh your selection, a maximum of two points per use case should be applied.

Once all votes have been cast in the form of points, a prioritization of the use cases can be made. Subsequently, only the use case with the highest score will be considered in detail in the following steps three to five. If several use cases receive the same score, the project team must decide on one of these use cases.

Based on our experience in applying the concept of classifying use cases, four generic strategies are common (see Fig. 4.7).

Quadrant I: These projects need to be addressed immediately. You must act and allocate resources to implement this use case.

Quadrant II: These projects have significant potential but are also complex to implement. It is worth tackling these projects, so you need to plan it well and often break it down into smaller steps.

Quadrant III: There you have use cases that have some potential and minimal complexity to implement. Look for ongoing activities in your organization and try to incorporate these use cases into already defined and ongoing activities. Often this is a viable approach by shifting the scope of these initiatives with lessons learned from previous smart factory use cases.

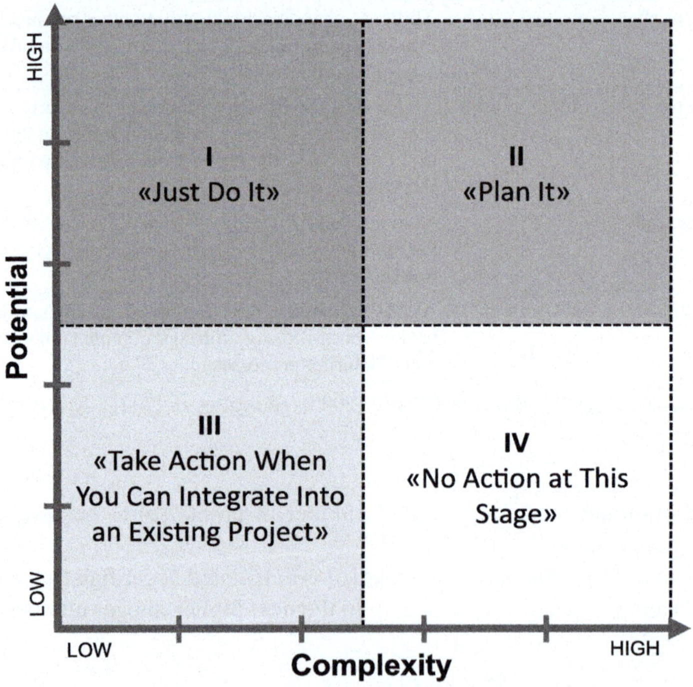

Fig. 4.7 Potential-complexity map (PCM) decision-making

Quadrant IV: Do not act. It is important that you decide to do nothing and communicate this. The decision to do nothing is very important, because projects without a goal often distract the focus in a company.

4.3 Step 3: Adaption of the Use Case to a Company-Specific Context

Step three focuses on the adaptation of the selected use cases to the company-specific context. More specifically, this means that the selected generic use case from the previous step is adapted to the company's circumstances. The following four points, shown in Fig. 4.8, are provided to help create a clear understanding of the intended application of the use case.

Situation

The future or existing challenges in the company's environment usually provide an initial starting point. Whether this is derived from requirements in the market

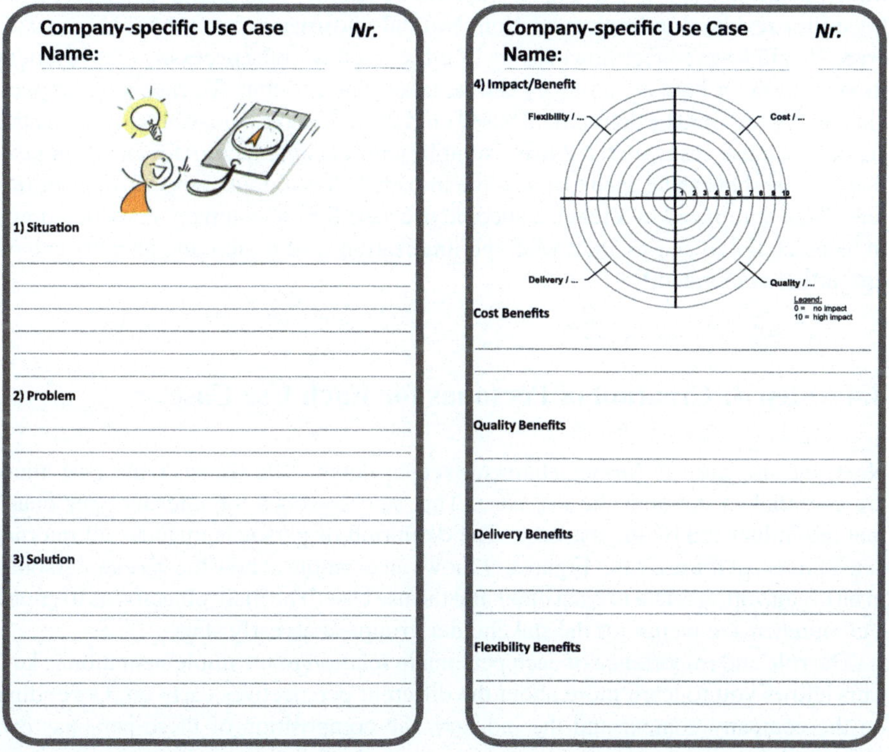

Fig. 4.8 Company-specific use case card

environment, or the corporate strategy is not decisive. It is important that this point should not describe a problem but rather the motivation to develop the company in a specific area of this use case.

Problem

What problems does the company intend to solve with the use case? For example, problems caused by a lack of traceability information of a product can have internal or external origins. It is important that the problems can be clearly outlined. A specific product, process, or department can make a delimitation, for example. The focus is on the hurdles that need to be overcome in this specific context in order to move the company forward.

Solution

The solution should describe how the company intends to deal with the initial situation and the associated problems. The focus should be on the use case and its delimitation. A clear scope helps to define the intended goals.

Impact

Possible impacts can be assessed based on the competitive priority (cost, quality, flexibility, and delivery). For example, improved customer satisfaction can also be the hoped-for result of an implementation. Not only positive aspects should be considered. The findings concerning missing IT infrastructure or the necessity of investment in know-how or human resources can be taken into account. To assess the impact, clear metrics should be determined based on KPIs that make it possible to evaluate the success of a use case. Sect. 4.2 lists exemplary influencing factors for each competitive priority that can be used to assess the impact. Once all four competitive priorities have been completed as a team, a specific use case for the company is written down in as much detail as possible. With the finalization of this step, the insights gained are further elaborated.

4.4 Step 4: Creation of Personas for Each Use Case

Next, the use case is further characterized in detail. The users or the customers are essential in defining the use case. This step searches for relevant "personas" that are influenced by the use case. The definition of personas includes all internal "customers" of the use case. Figure 4.9 shows an example of how the various personas in an organization relate to each other and the use case. The main objective is to create a comprehensive picture of the stakeholder groups at an early stage.

The role and importance of each persona in relation to the implementation is key. This allows you to learn more about the different perspectives early on. Depending on the company context and the industry, the composition of these personas can and possibly will vary. As in the previous steps, the following four points, shown in Fig. 4.10, should be used to support the definition of the personas:

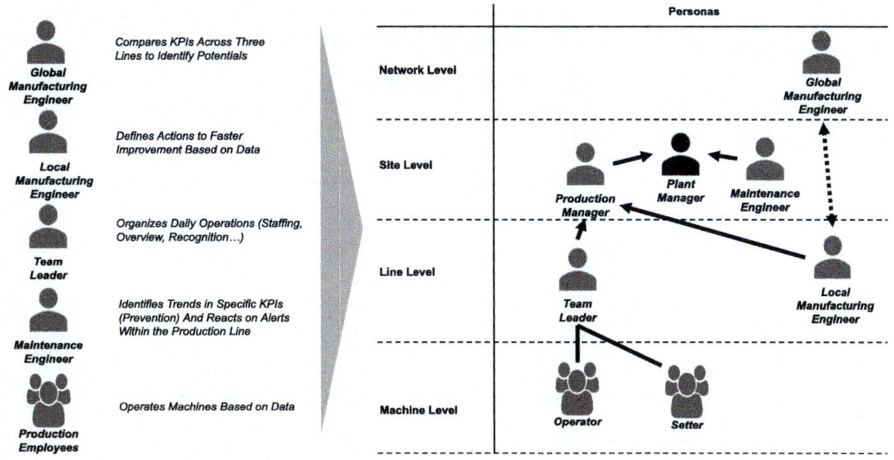

Fig. 4.9 Persona map

Name

Personas are fictional characters based on specific roles in the company to represent the different customers of the use case who use a service or product in a similar way. Creating personas helps to understand needs, experiences, behaviours, and goals. It can also help to recognize that different stakeholders have different needs and expectations. To identify the personas, an appropriate name is given. This name can also reflect the person's role in the company, for example. The requirements of the personas for the use case can differ greatly from one another. Aspects such as user-friendliness, data preparation, and availability of data or others can be included.

Goals (What Should Achieve or What Does the Persona Want to Achieve?)

The use cases have clear goals to be achieved. They need to be defined. Different organization levels might see things differently and wish to focus on other topics or measurements. These differences need to be transparent. To steer the different use cases strategically different functions and hierarchy levels will have different needs regarding the required information. Similarly, from a content point of view, the different personas will also have different needs. On the shop floor, the needs are more product and quality-oriented. They focus on achieving operational efficiency. The management level, however, focuses more on financial and strategy content. Understanding these differences is critical as they will have a large impact on the implementation of the use cases.

Activities (Which Activities Does the Persona Perform?)

The activities or jobs and work that the different personas perform are important to understand. This will also highlight potential shortcomings you need to address while implementing the use case. These activities are also closely linked to the goals the different personas perform.

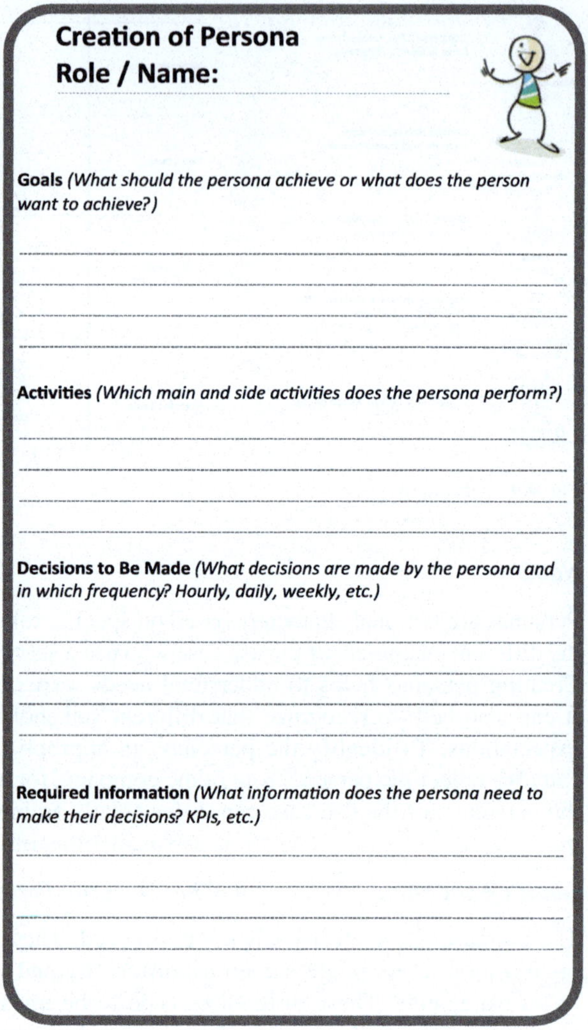

Fig. 4.10 Definition of the different personas

Decisions to be Made (What Decisions are Made by the Persona and in Which Frequency? Hourly, Daily, Weekly, etc .)

An important requirement for the implementation of the use cases is the information needed by the different personas to make decisions. Frequency is a critical point, but also information content is important to understand. Decisions like planning of order sequences in production or product allocation decisions might differ in the frequency of occurrence, which is important information for the design and implementation of the use cases.

Required Information (What Information does the Persona Need to Make their Decisions? KPIs, etc .)

Detailed data are critical. Be as specific as possible. What kind of information is needed exactly? How often is this information needed? In what form is the information needed? These are all relevant questions that you need to answer. This specification is a critical requirement of every use case and results in added value for the respective stakeholder group.

4.5 Step 5: Project Setup

In this step all relevant results of the previous steps are compiled and transferred to the project fingerprint. The goal of the project fingerprint is to develop the framework conditions for a project that implements the use case in a stepwise approach. In practice, different projects will implement the use case over time. Thus, the project fingerprint should address the first step or the first project of the use case implementation. It should not consist of a multi-year and multi-plant project or program. Every project in implementing the use case needs a new fingerprint document.

In defining how to implement the use case, you need to share the vision of your use case. Step one to step four helped you define and describe the relevant use case. This is the basis for you to plot the journey to the vision into a simple drawing with six fields. In these fields, you describe the journey of e.g., a customer, an employee or the department after you have implemented your use case. Draw a picture of how you will work or how a process will function after the use case has been implemented. Draw pictures e.g., of how your maintenance department will work after the artificial intelligence algorithm has predicted the breakdown of a production machine in the future. You describe how the future will look, what the different people will do, which tools will have to be implemented or how the communication flow will work. We propose you draw and describe the future in six simple pictures on a flip chart (see Fig. 4.11). This is just a guide. You can definitely change the structure. It is not important that you describe your use case in six pictures. The key is to envision the future in a team approach. Graphical elements are normally a great help in this process.

You now have a common vision of your work after you have fully implemented the selected use case. This helps you greatly in defining the first implementation. Now you need to get concrete. Therefore, you need to define the first project. We show you how to do this in the following example.

The company in our example produces plastic injection moulded parts as part of its product portfolio. The products that are manufactured in production differ in shape and purpose. In the run-up, the company decided to elaborate on the use case "optimal process parameter identification". Based on the findings of the preceding process steps, the use case was adapted to the company-specific context, taking all

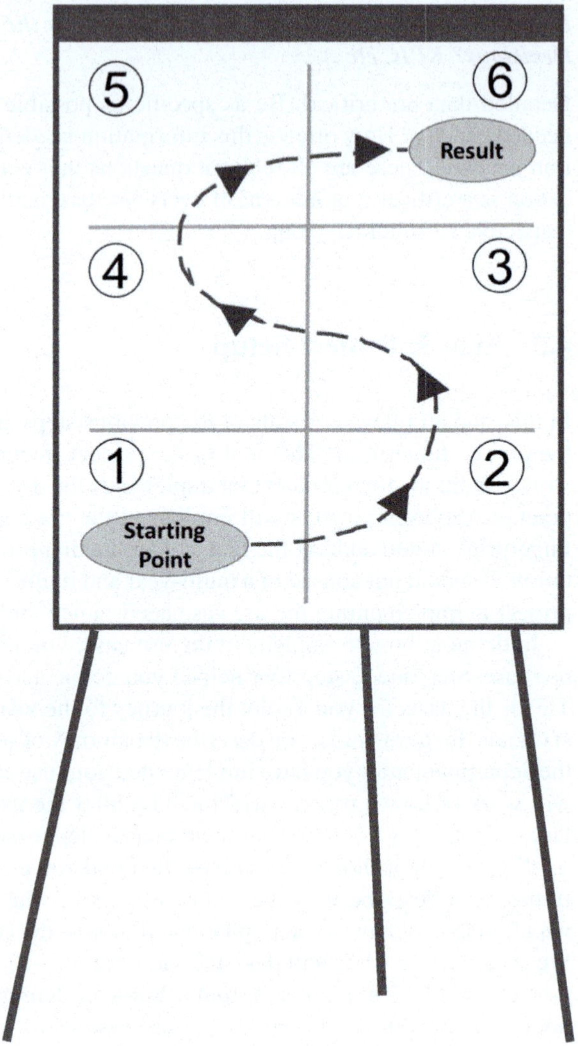

Fig. 4.11 Approach to describe the future

relevant personas into account. With the implementation of the use case, the company wanted to achieve the following goals:

- A reduction of scrap and a better allocation of resources through optimized parameters
- An improved process stability enabling a reduction of the safety stock
- A decreased cycle times
- Greater flexibility in switching between products through analysis and adjustment of the process parameters.

Accordingly, real-world operating data such as machine, production pieces and process data were collected, analyzed and used to simulate various alternative process configurations in the future.

The company described the use case "optimal process parameter identification" (see also generic use case 28) in the above six fields (see Fig. 4.12). This drawing was done in close corporation between the different departments and functions such as the production department, IT, quality, planning and service and sales. In addition, an external IT-consultant was involved. Although the discussion on depicting the future was heavy, all parties were involved. It was a great achievement that everybody shared the same vision.

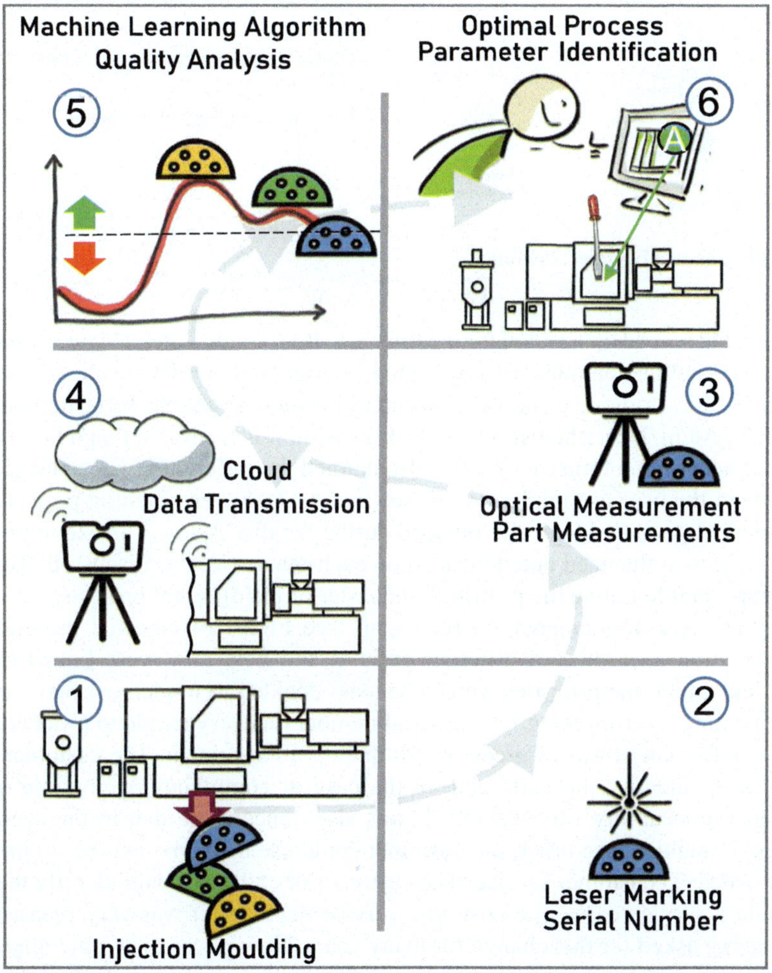

Fig. 4.12 Case study

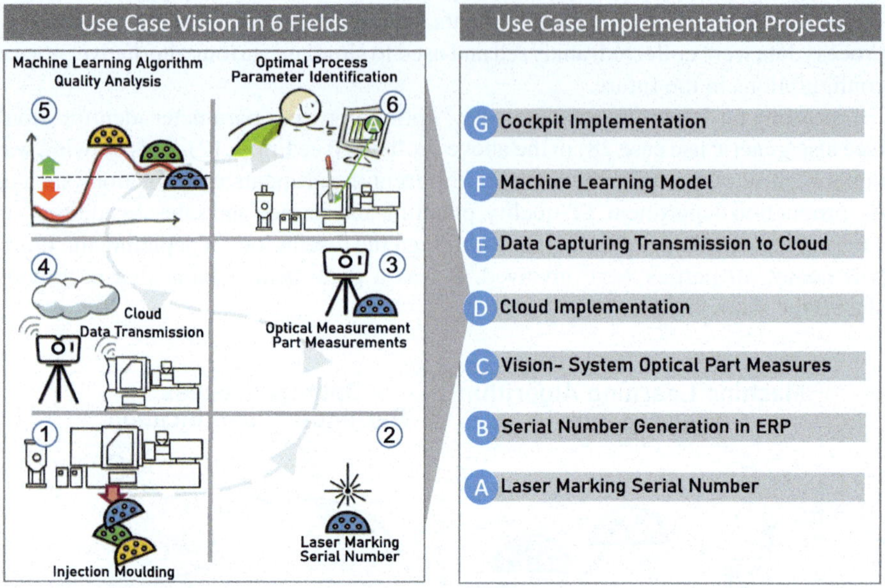

Fig. 4.13 Case study project definition

The team agreed on a vision for the future. In this process, detailed questions were addressed, and the team had to stay highly focused and not debate already detailed implementation topics e.g., how the laser marking process gets the serial number from the ERP system. Nevertheless, the wide discussion was fruitful in that it not only led to agreement on the future vision but also defined the different projects needed over time to get the use case implemented (see Fig. 4.13). Seven potential projects were identified, aligned and logically pursued during the discussion. The extent to which these projects influenced and depended on each other was also discussed. This was very important to define the priorities and order of the different projects.

The use case implementation road map (see Fig. 4.14) showed the graphical representation over time of different projects and how they were linked to each other, and where the priorities were. The team decided to implement first a project implementing laser marking for the serial number on every moulded product.

This information was of great use for quality management. The customer could more easily identify the parts and, in the case of complaints or possible quality problems, traceability was ensured. It was also conceivable that in the event of a service or maintenance order, the customer could use the serial number to initiate a repeat order via customer service. The customer or end user could identify the parts easier in a service case. The customer service department was very pleased with this, having asked for this change for many years. Now it was to be implemented as the first step of this smart factory use case. The team was then asked to fill in the fingerprint document for this first project.

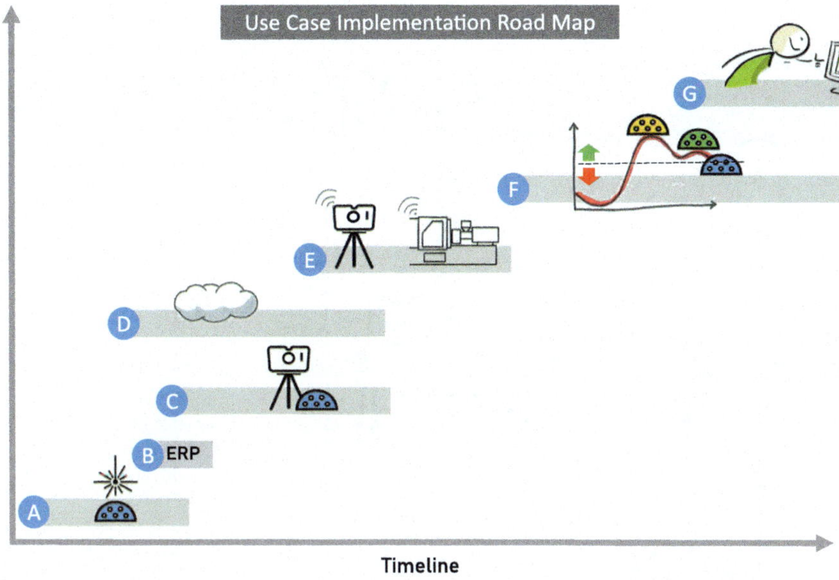

Fig. 4.14 Case study road map definition

The following fingerprint template was kept short (see Tables 4.5 and 4.6), as the result contained only the most important points that were relevant for an internal project application and plan of attack.

Use cases that interact with customers often require an additional process description to define the potential projects. It is vital for all parties involved to understand and agree on the interaction process of the respective use case (see Fig. 4.15).

In addition to all previous aspects such as use case description, vision, strategic goals or needs by the different personas, a project set up and a concrete schedule for implementation must be created. This schedule contains milestones with measurable results that represent the progress of the project. Even if the whole procedure seems relatively abstract and detailed, it is very important to agree on the resource allocation due to the interdisciplinary nature of such projects. Usually, unexpected hurdles with significant investment need to be overcome, and this is common in such projects. Alignment with top management is important and needs to be defined. The use cases are strongly linked to the benefit of using new technologies. Existing knowledge gaps must be closed. This is essential in order to pool limited resources at an early stage. In most cases, companies have existing legacy systems that are anchored in the IT infrastructure. Therefore, you need to connect new technologies and use cases to past systems or change the past systems and create new foundations. In this case, it is important to also measure project results along the way in order to keep focusing on results. In the following, the individual components of the project fingerprint will be discussed.

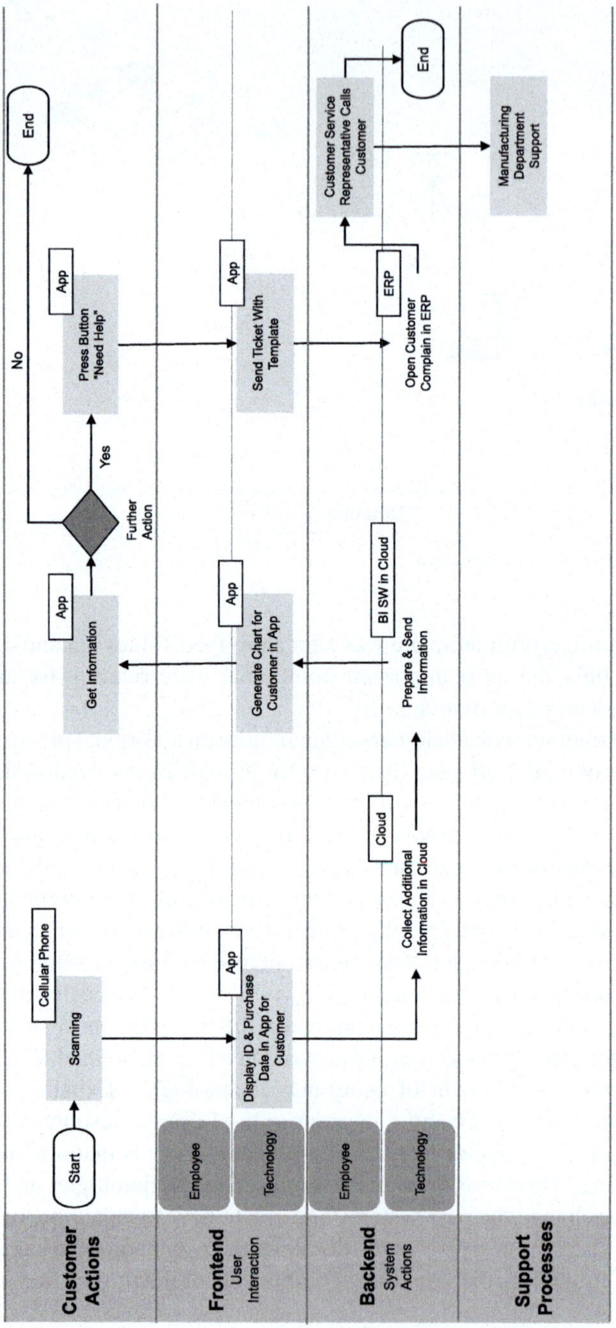

Fig. 4.15 Case study process map

Table 4.5 Project fingerprint—Page 1

Project fingerprint		Project name	
(1–2 sentences max per response)		Date/version	
Project team	Project sponser	The project sponsor, in the role of internal financier, approves the project and represents the vision of the project to the management. In this role, the sponsor ensures the acceptance of the project within the company	
	Project manager	The project manager takes over the project management and reports the results as well as the current interim status to the project sponsor	
	Project team	The project team includes all relevant competencies needed to implement the project	
	Steering board	The steering board is usually composed of members of the management and knowledge bearers in the company. The project sponsor and the project manager are also part of this steering board as well as any external partners	
	External partners	The external partners provide their contribution to topics that the company itself cannot work on or where it needs a neutral point of view	
Project scope	Our vision	The vision presents the implementation of the use case in the context of the strategic company goals	

(continued)

Table 4.5 (continued)

	Our initial scope	The scope describes the delimitation of the project. In this context, the scope of the implementation is clearly defined
	Our competitive advantage	What benefits will result from the implementation of the use case, and how will this affect the success of the company?
	Potential scalability	Based on the project scope, it should be shown in which form the scaling is made possible
Project target	Who/what we are targeting	What are the overall goals of the project?
	The problem we are solving	Which existing problems or challenges are addressed by the project?
Situational analysis (SWOT)	Internal factors	
	Strengths (+)	Weakness (−)
	Which strengths will come into play in the project that the company can already draw on today?	What weaknesses in terms of know-how, resources, IT infrastructure and others can already be identified today?
	External factors	
	Opportunities (+)	Threats (−)
	What opportunities arise from the use case?	What are the risks associated with the project?

Table 4.6 Project fingerprint—Page 2

Project impact	Cost benefits	What monetary benefits are achieved by the use case?
	Quality benefits	How is quality influenced by the use case?
	Delivery benefits	How is delivery time or process efficiency influenced?
	Flexibility benefits	What benefits in terms of flexibility can be achieved?
Project budget	Internal cost-resources	What costs are incurred during implementation? This aspect includes all internal costs such as personnel, resources or system/infrastructure costs
	External cost	What external costs are incurred, for example, through the support of external partners?
	Budgeted period/yearly	What annual budget is needed to realize the implementation?
Interdependencies/interferences with other projects		
Project name		Description
		What existing project is directly or indirectly related to and may affect the project?
Measures along the process/results		
Objectives we paln to achieve in a given time frame and how to measure them		
	M1	What measurable milestones are decisive for the success of the project?
	M2	
	M3	
	M5	

(continued)

Table 4.6 (continued)

Project plan	Year/month		2023												2024					
			Jan	Feb	Mar	Apr	May	Jun	Jul	Aug	Sep	Oct	Nov	Dec	Jan	Feb	Mar	Apr	May	Jun
	Topics and work packages																			
	Sub project	Work package																		
		Work package																		
		Work package																		
		Work package																		
		Work package																		

The fingerprint is the decision-making foundation for the implementation of the first project in the company that leads, hopefully, to a promising start of many projects towards building a smart factory. Therefore, it is very important that the scope of the project has a certain size so that the results are visible. The scope of the first project should also be realistic and have a clear duration (max. 1 year). The team for the implementation of the first project should consist of employees from different departments, and partners must be closely involved.

4.6 Step 6: Implementation First Project(s)

Implementing the smart factory has to be treated as a large change program. The introduction of the smart factory is also closely related to any lean transformation in your company. Waste, organization silos, wrong priorities, and also wrong project setups need to be addressed. Digitalization and lean transformation depend on each other and are closely linked. Stable and waste-free processes are the basis of a smart factory, while digitalization supports the lean program in defining the right initiatives.

Change and people are the critical topics in any transformation. Organizational changes, status shifts but also fear of job losses are constant concerns in the minds of employees. Will I lose my job? Will I have the right competencies for the future? Will I lose power in the organization? All such thoughts restrict change. Even when not explicitly expressed, they are always present. You will need to address them by involving people very early on. This is important even though it might slow down the speed of implementing the smart factory. This can be a problem for top management, boards and shareholders.

Besides this key topic of involvement and alignment, we have identified the following key factors necessary for a successful transformation (see also Hänggi et al., 2021).

Change Is Never Finished—Stepwise Implementation/Cut Complexity into Pieces

The use case describes the end situation, e.g., it predicts the regular maintenance of the tool for the product based on the wear. To get there, lots of information, processes, IT Infrastructure and mathematical models, are required. You will only be successful if this complexity is sliced into pieces over time. This allows you to learn, digest the investment and achieve fast results.

Iterative Steps Towards the Vision

The high uncertainties and needs of the smart factory journey can only be managed by an iterative approach. You need to measure results and learn from every project, from every investment in your IT architecture and challenge yourself constantly to ensure that you are on the right track.

Involvement of All Relevant Stakeholders

We have seen that different personas not only have different needs but also bring different competencies to the table. Implicit knowledge, process-wise and machine-wise by the operators, needs to be coupled with specific IT knowledge. The following example illustrates the importance of this. In a construction company, the job site people explained to the data analytics people how concrete is made. They worked hand in hand to implement a new model-based digital process. It took some time for them to get acquainted with each other and performed well as a team. This involvement is fundamental to understanding and digesting change. Stakeholders can also be external partners, such as suppliers or technology providers or even customers.

Management Commitment and Constant Support—Also When Problems Occur

The implementation of these use cases is complex, and learning occurs constantly. This requires flexibility and adaptability. Changes need to be discussed closely with management. Expectations, results, and investments always need to be transparent to ensure the full long-term support of management.

Stable Processes are the Base

"Do not digitize your mess" is an often-heard statement. It sounds logical but is often overseen or not observed. Time pressure and pressure to achieve the potential by the implementation of the use cases are both good but should not negate the necessity of putting your operative bases in order first. Stable processes are the key.

Fact-Based Implementation—Measure Results Along the Way

The implementation of the different use cases requires often significant investments and personnel resources. The technology and IT risks will challenge your implementation path constantly. To measure and derive fact-based decisions in such an environment is key to keeping commitment to the transformation throughout the whole organization.

Workshop Based Implementation Plan—No Big Project Plan in All Details—It Will Change Anyway

Stepwise implementation works best with a constant flow of workshops. Team orientation, adaptability, and wide involvement of different experts is an integral part of this approach. The workshop agenda and approach should follow the SCRUM methodology of the IT industry.

Availability of Internal and External Resources Is Key

Without the right resources, any implementation process does not work. Digitalization technologies change at a speed that is difficult to absorb for industrial companies, particularly for small and medium-size companies (SMEs) this is a significant challenge. Sourcing of know-how and resources is a potential way. This works only if internal resources are also available to manage them. In addition, there needs to be an internal investment in IT training. You need to be able to assess different technology possibilities and paths for implementation.

Keep Focusing on the Smart Factory Project, Despite Heavy Resource Requirements for the Operational Business

The full benefit of these use cases normally takes some time. This is a risk of implementation. Short-term priorities tend to be more important than the long-term benefit of the different use cases. Often the large benefits of the smart factory implementation are some years down the road. You need to understand this and also communicate it. This will help you keep focusing and allocate the necessary resources to the smart factory initiatives.

Control and Steer the Project— Involve Management in Short Updates

Complex projects tend to be "monsters" and the costs often overrun the budget. Close steering with the involvement of management leads to good process. This requires significant management capacities. This is required, details matter.

The Rollout Plan Needs to be Defined, but Can Still Change

The first use case will bring you forward. After the successful implementation, however, you need a plan to move forward. This rollout plan needs to address topics such as the implementation of additional use cases, or implementation of the same use case with other machines or in other manufacturing sites. The rollout plan is closely linked to your overall IT infrastructure.

IT Architecture Is Relevant and Needs to Be Defined, but Can Change over Time

The definition of the future IT architecture is relevant. It needs to be defined after a (some) first successful project(s). It influences also the rollout plan. In addition, the IT architecture is a guide and needs to be reviewed often. Changes can happen due to the continuous development of your rollout plan.

Detailed Understanding of All Data and Definitions Avoids Unnecessary Implementation Costs

Data acquisition, storing, and using them is complex, and investments and operating costs can be significant. Therefore, a thorough detailed understanding of the details limits costs. Detail definition is mandatory.

Target IT Architecture

Before you debate which IT System will cover which function, you will need to define the data sovereignty of each data field. It is critical that you do not have any duplication of data, but also make your processes and later your IT infrastructure clear. Questions like, "where do you report when you have produced goods and parts for an order?" are fundamental. Depending on your business and organization, you can report in an MES system or directly into your ERP system. Both are possible, but only at one place, and it needs to be defined. We would really urge you to reduce speed when creating this IT architecture plan and wait until you have implemented the first use case. You need to learn as an organization what a smart factory really means for you. Every company is different, and some generic IT tool specifications do not bring you much further. You need to specifically understand this for your business and your company. It is important to mention that the target IT architecture and the further rollout plan are interconnected. Iteration and adaption are both normal (see also Fig. 4.2).

Enabling Technologies e.g., IoT, Augmented Reality and Others

Every machine is different. Even the same machine from the same manufacturer changes over time. New controls, different sensors or software make every machine with an internet connection a project of its own in your factory. You can benefit from other projects, but you will never be able to simply copy them. Keep that in mind. To build this kind of data collection for machine learning, some challenges have to be overcome. For example, it can become apparent that device-specific solutions still have to be implemented so that the data can be exported and recorded in the desired quality. This is a hurdle for smaller manufacturing companies in particular since such interdisciplinary know-how is often not available. Another difficulty is the synchronization of data from different machines and devices.

The respective challenges are closely related to any use case and all the different associated technologies. The successful implementation of a use case requires a solid understanding of the required enabling technologies. This detailed understanding is critical to the successful implementation of the use case. You do not need to build this knowledge up yourself. You can collaborate with technology companies or other technology providers.

Data-Lake

Relevant data generated in your different IT systems, but also data from the machines you use, needs to be stored. The question concerning central vs. a decentral storage strategy is important. If a centralized approach applies, connecting different data are easier, but requires a more structured and coordinated approach. Storage in the cloud via the Internet is included in this discussion.

Security Concept

IT security for a smart factory is extremely important. The threats are wide. Technical topics, organizational topics and people topics need to be addressed. Additionally, potential legacy systems play an important role. You will need to understand the needs of your use case based on your existing IT infrastructure. Understanding the potential risks need to be added. This guides you in defining what needs to be done.

4.7 Step 7: Scaling and Internationalization

Often the discussion focuses on how quickly the smart factory will be adopted on a global scale across all product lines, departments, or even the global production network. It is clear that this ultimate vision will bring many benefits. Therefore, the roll-out plan after the first successful implementation of a use case is critical for you (see Fig. 4.16). Transparency through data and efficient digital processes allow you the resource allocation and resource shifts across the production network. To achieve this goal, a step-by-step roll-out process has proven to be the way to go. A roll-out plan that takes the different possible scaling ways into account is critical. Transparency of the steps helps in planning the required resources, keeps management focused, and helps reduce costs by transferring knowledge.

Generally speaking, you have three major options to roll-out the smart factory (see Fig. 4.16). These generic ways are in practice never sharply differentiated.

Option I: Take One Use Case, Then Expand to Other Machines and Then Go to the Other Plant

You scale the first use case in one plant first. You extend the use case to the auxiliary equipment of this machine (integrated or semi-integrated line), then to other machines or lines, and then to the full factory. After this has been accomplished, you start the implementation with a further manufacturing site.

Option II: Implement the First Use Case on the Same or Similar Machines Plant by Plant

Another scaling process would be to implement the same use case on the same or very similar machines in other plants. Scaling goes to different plants and potentially also to other countries and regions.

Option III: Implement the First Use Case and Then Move to the Next Selected Use Case

You choose another new use case. You implement the new use case after the first successful implementation. You can choose the same machine or another machine. The focus is on the new use case and its new benefits and not on the machine. You can also switch plants for this new use case.

Which option to choose depends on technology, strategy, and the availability of resources. The three options to stagger allow you to learn and incorporate the different needs of your operations organization into the process. Therefore, learning and adapting are the key words. You not only need to integrate all the knowledge and avoid costly multiple mistakes but also take into account the different needs and priorities of different organizations and locations.

These scaling and internationalization steps should always follow the above-described use case framework process. It provides strategic guidance and is a rigid approach so that you never lose sight of the financial benefit of your investment—no shortcuts.

The proposed scaling and internationalization approach is a bottom-up digital path and is focused on results step-by-step. This approach is often highly debated due to its preference for a bottom-up global and standardized roll-out of a specific enterprise—IT-solution. We are not against global standards, but when it comes to implementing lots of different use cases for many plants globally, the specific requirements and

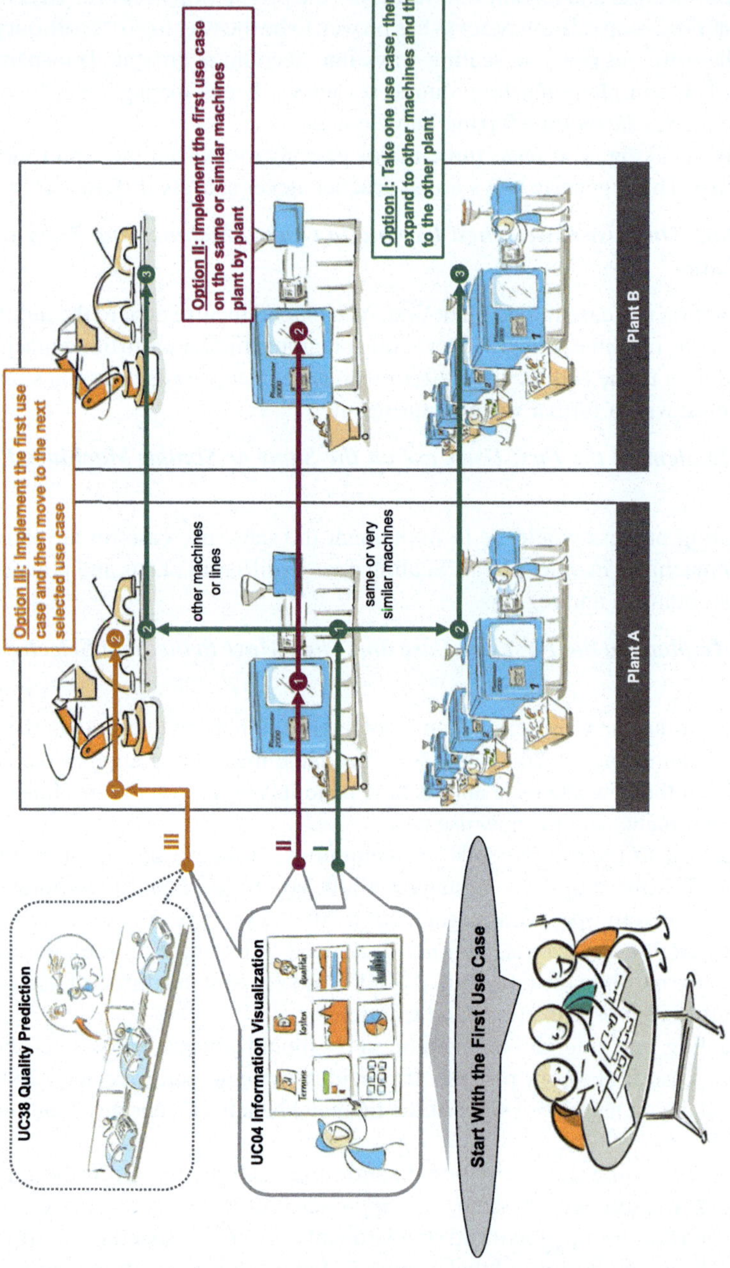

Fig. 4.16 Rollout-plan options

manufacturing technologies of the plant should be the driving force. You should standardize and should learn from and between the plants. However, you need to do this in a step by step, bottom-up and result oriented manner. In this way, you fulfil the different needs of the plants. Then the question often arises do I need one global IT-solution (e.g., an MES-System) for this global roll-out? There is no clear answer to this. As long as you drive it from a "use case" and result oriented point of view, your best future IT architecture is the result of this approach. Today's digital possibilities and software approaches can offer a more decentralized and flexible IT approach than one global IT system. This does not mean that standardization cannot be reached. A standard tool set is possible and often required to optimize IT platforms and solutions. The discussions on IT architecture for scaling is complex and cannot be copied from one company to another. Gaining speed and traction in implementing all the different use cases in many different plants is a critical requirement in this discussion.

References

Boyer, K. K., & Pagell, M. (2000). Measurement issues in empirical research: Improving measures of operations strategy and advanced manufacturing technology. *Journal of Operations Management, 18*(3), 361–374. https://doi.org/10.1016/S0272-6963(99)00029-7

Flynn, B. B., Schroeder, R. G., & Sakakibara, S. (1994). A framework for quality management research and an associated measurement instrument. *Journal of Operations Management, 11*(4), 339–366. https://doi.org/10.1016/S0272-6963(97)90004-8

Flynn, B. B., Schroeder, R. G., & Sakakibara, S. (1995). The impact of quality management practices on performance and competitive advantage. *Decision Sciences, 26*(5), 659–691. https://doi.org/10.1111/j.1540-5915.1995.tb01445.x

Hänggi, R., Fimpel, A., & Siegenthaler, R. (2021). LEAN Production – einfach und umfassend: Ein praxisorientierter Leitfaden zu schlanken Prozessen mit Bildern erklärt. *Springer, Berlin Heidelberg*. https://doi.org/10.1007/978-3-662-62702-0

Kraemer, A. K., & Kraemer, J. (2010). Flexibility of delivery frequency in logistics competition. *SSRN Electronic Journal*. https://doi.org/10.2139/ssrn.1583015

Noble, M. A. (1995). Manufacturing strategy: Testing the cumulative model in a multiple country context. *Decision Sciences, 26*(5), 693–721. https://doi.org/10.1111/j.1540-5915.1995.tb01446.x

Saraph, J. V., Benson, P. G., & Schroeder, R. G. (1989). An instrument for measuring the critical factors of quality management. *Decision Sciences, 20*(4), 810–829. https://doi.org/10.1111/j.1540-5915.1989.tb01421.x

Ward, P. T., McCreery, J. K., Ritzman, L. P., & Sharma, D. (1998). Competitive priorities in operations management. *Decision Sciences, 29*(4), 1035–1046. https://doi.org/10.1111/j.1540-5915.1998.tb00886.x

Zhang, L., Luo, Y., Tao, F., Li, B. H., Ren, L., Zhang, X., Guo, H., Cheng, Y., Hu, A., & Liu, Y. (2014). Cloud manufacturing: A new manufacturing paradigm. *Enterprise Information Systems, 8*(2), 167–187. https://doi.org/10.1080/17517575.2012.683812

Chapter 5
Use Case Framework to Identify Digital Services for Commercialization

Section 2.1 described two ways towards digital services (see Fig. 5.1). The first way drives the development of the smart factory based on the operational strategy. So, you drive digitalization in your factory by collecting data or implementing base technologies and first use the cases to optimize operational excellence on your shop floor. These implemented use cases (e.g., track and trace solution for a product) are then the base and starting point for a digital service towards the customer. You then reap the first benefits in your plant (e.g., by identifying the products uniquely and linking your production data to every specific item produced).

The second way is to derive digital services directly from the corporate strategy. Here you design a solution for your market, primarily to generate growth. These new product lines, often smart services, add to your portfolio of products. The use case framework can also be applied here. The framework then represents the customer perspective, for example, a machine supplying company. If you are a company delivering machines and other equipment (e.g., transportation logistics) the generic smart manufacturing use cases can be applied to develop digital services supporting customers in doing their activities in their factories. A digital service of, for example, a machine manufacturer then supports the customer in gaining efficiency on its shop floor. This step is the logical closure of our approach, designing new digital services on top of smart factory use cases to gain operational excellence for an industrial company.

In this chapter, we show how to use the framework and how to commercialize these services. The commercialization process is relevant and is an additional step. This step is not needed when implementing the smart factory use cases. The identification process of potential digital services follows the smart factory implementation process as introduced in the previous Chap. 4. Additionally, however, we are adding three specific steps (DS I–DS III): the identification of the customer value, the quantification of the customer value and the derivation of the pricing options. These steps should be added after stage 4 of our factory implementation process, where the personas for the use cases are created (see Fig. 5.2).

L. Budde et al., *Smart Factory Navigator*,
https://doi.org/10.1007/978-3-031-17254-0_5

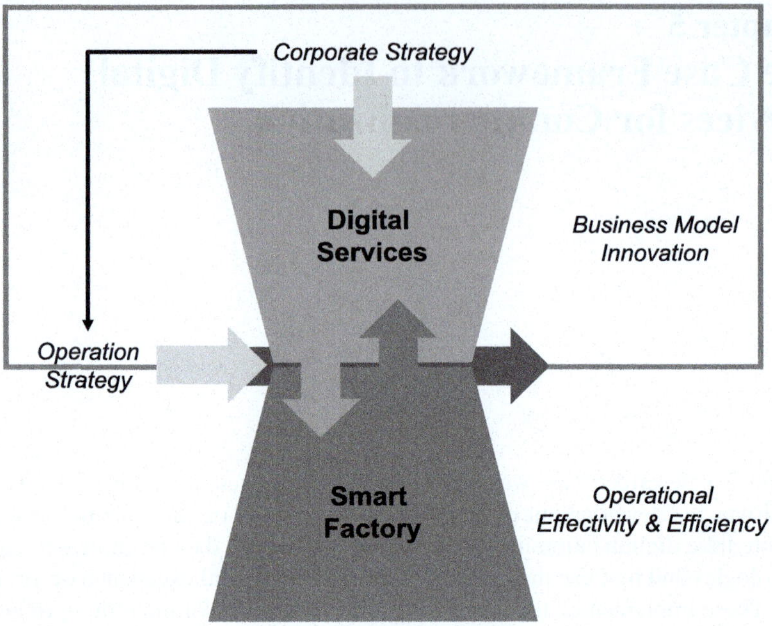

Fig. 5.1 Digital services driven from the corporate strategy

When defining digital services, we sometimes change the perspective towards our smart factory framework. We are looking now at the framework from the customer's viewpoint that is potentially using these digital services in its factory. Thus, the framework represents the potential of the use cases for the customer. The digital services support the customer along the path towards the implementation of the smart factory. So, for example, a machine tool manufacturer discusses the services provided to its customers in its factory using the production machine of the tool manufacturer. The use case model then represents the customer factory of, for example, the machine tool manufacturer. If the digital service of the machine tool manufacturer is successful, it adds value to one of the use cases towards the smart factory of its customer.

When discussing potential digital services from the customer's perspective, often internal services surface, affecting internal organizations and processes such as, for example, the installation support by augmented reality. This digital service does not add value for the customer but leads to benefits for internal operations of, for example, the machine tool manufacturer. To capture both types of digital services, the framework for digital services now consists of two viewpoints: the customer and the internal view. This perspective is added to the smart factory framework (see Fig. 5.3).

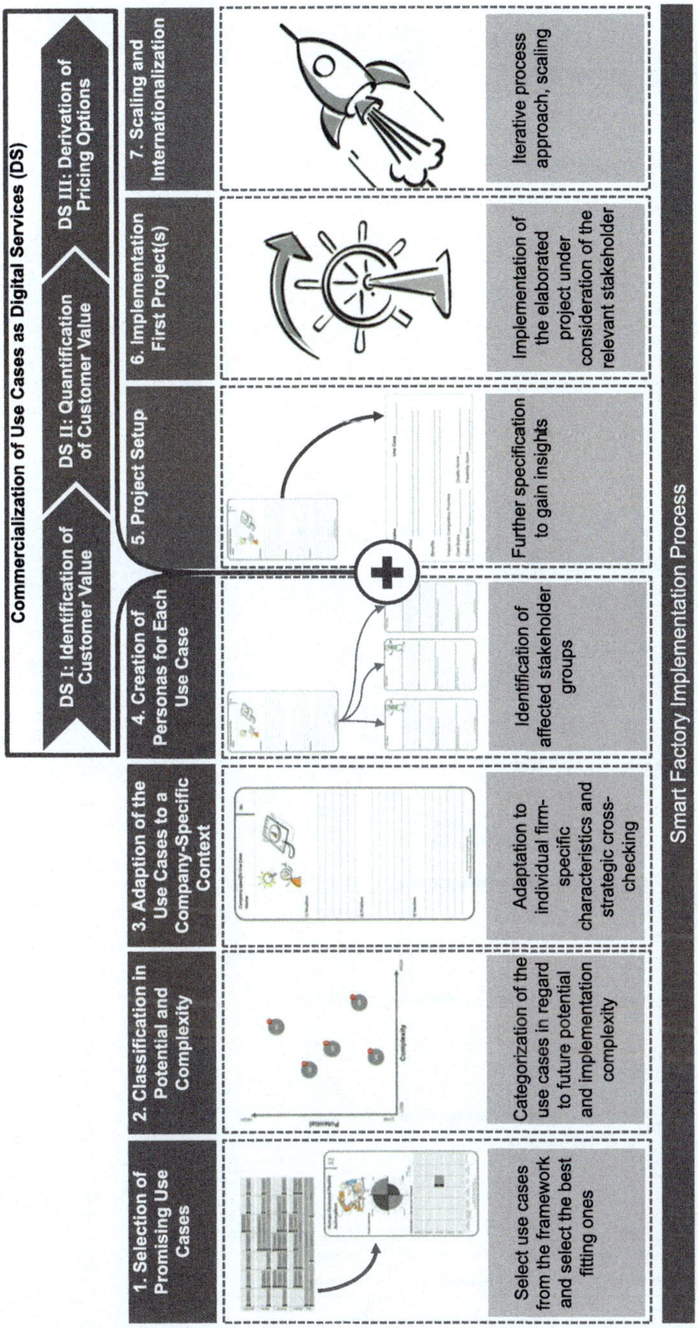

Fig. 5.2 Smart factory implementation process for commercialization of use cases as digital services

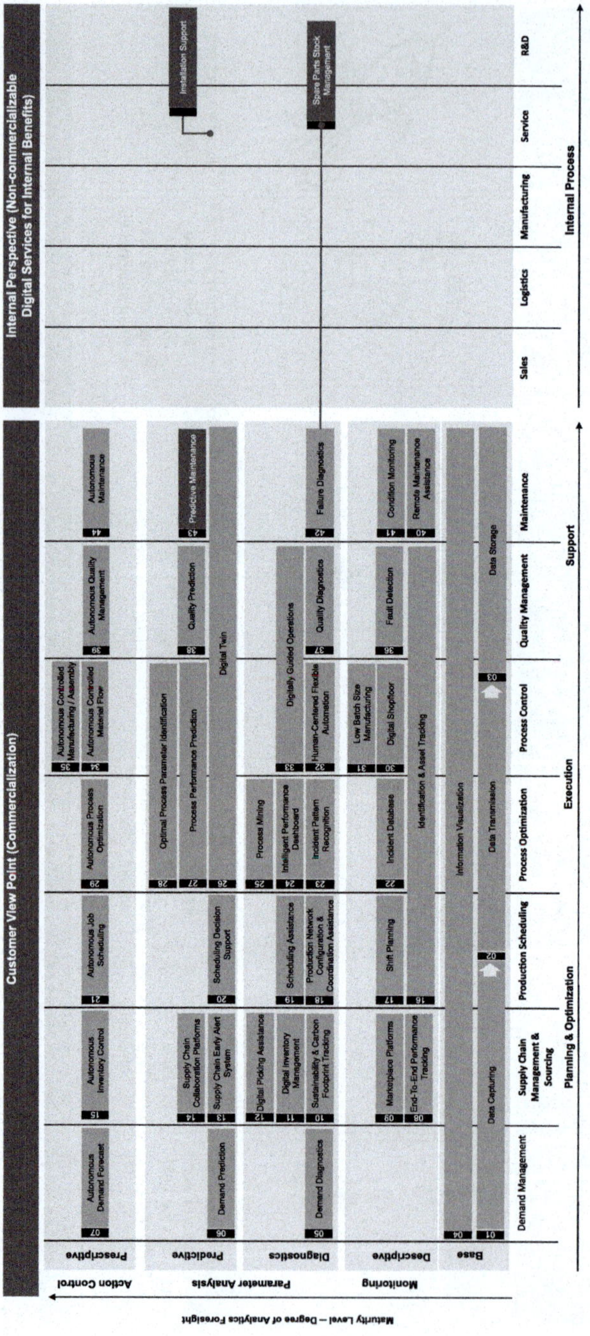

Fig. 5.3 Smart factory framework expanded with internal perspective

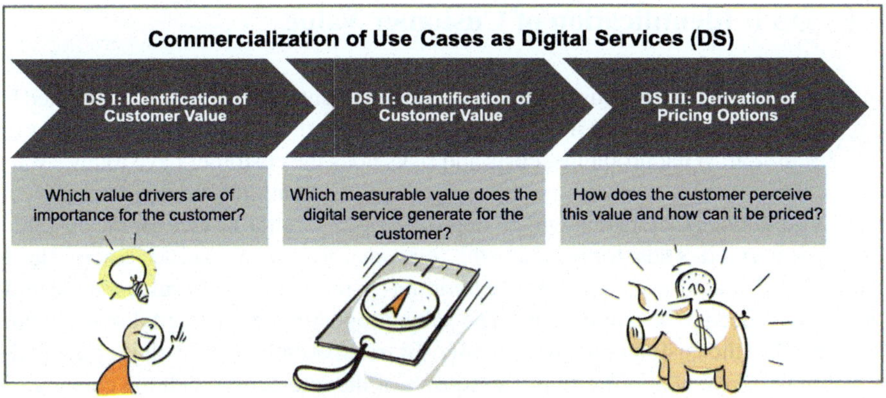

Fig. 5.4 Commercialization of use cases as digital services

The payable digital service for "the customer" could be a "predictive maintenance service package for a machine". The customer no longer has to worry about machine breakdowns and the unavailability of spare parts. The machine provider ensures the machine is always ready for use. Based on this added value, the customer is willing to pay.

A digital service for "internal purposes" could be, for example, "the installation support by augmented reality for internal staff" to improve the efficiency of the installation process in the customer's factory. This reduces time and thus costs for the machine manufacturer. However, the customer, as the user of the machine, is not willing to pay for this service because it benefits the manufacturer. The framework for digital services is adapted as follows (see Fig. 5.3). And both digital service types will surface.

Often there is a link between added use cases from an internal perspective and use cases from the customer's perspective. For example, installation support can be extended over time to become a predictive maintenance service, which is then billable. Any customer use case should be discussed from the internal perspective of your company and the potential benefits marked in the framework (e.g., optimizing the stock management of spare parts).

Having identified the use cases as digital services and selecting the promising use cases, a fundamental additional step is to commercialize the identified digital services. As mentioned previously, this involves three main steps (DS I–DS III): the identification of customer value, the quantification of the customer value and the derivation of pricing options for the digital service (see Fig. 5.4).

5.1 DS I: Identification of Customer Value

In the first step (DS I), the customer values must be identified that might be affected by the selected digital services. The customer's "issues" should be the starting point for this discussion. What problems, costs and risks does the customer have today? What can be improved for the customer? Hereby the consideration of both quantitative as well as qualitative factors that will be improved by the digital service for the customer must be observed. Thus, for instance, the "elimination of manual process activities to increase efficiency" and "competence building"—to be able to optimize the machine processes—need to be emphasized. The value proposition of the digital service must be aligned to the customer's requirements. To transparently discuss these potentials, a mapping of actual and future process flows of the interaction with the customer is of great value. We discussed the method process map in Sect. 4.5 (see Fig. 4.11),

Many companies often make mistakes in this early phase of the process, not knowing the customer and market needs sufficiently and the details of the customer's processes. Therefore, the selected customer should be involved in the development of the new digital services. This enables an understanding of the customer's processes, the concrete problems and the customer's needs. A solid understanding of the requirements through orientation on concrete problems of the customer is the key. Knowing the problems of a specific customer enables you to generalize later and work for a larger customer group with similar problems and needs.

5.2 DS II: Quantification of Customer Value

The actual generated value is often over- or underestimated. Digital services are often priced according to the value that is generated for the customer. A critical point in selling digital services is the quantification of this value. Therefore, the realized customer value needs to be monitored and a mechanism for the value improvement must be created. A clear, measurable, and concrete assessment of the value is mandatory. If the quantification of the value is not demonstrable to the customer, it will be difficult to sell the digital service. Therefore, it is crucial to show the differences in the customer process between the current and the possible future state using the digital service (see Fig. 5.5). You can only do this when you understand in detail the customer's process.

Figure 5.6 highlights the questions necessary to quantify the customer value (step DS II). Based on these findings, and with the defined assumptions, the quantification model can be built to estimate the impact on customer KPIs. This is all done before the customer has a digital service in use. Even though this is only an estimation, the model helps to split the digital services into features and enables a prioritization related to the customer benefits and consequently his willingness to pay. This transparency and debate is vital to proceed. If you still miss information, you need to obtain additional

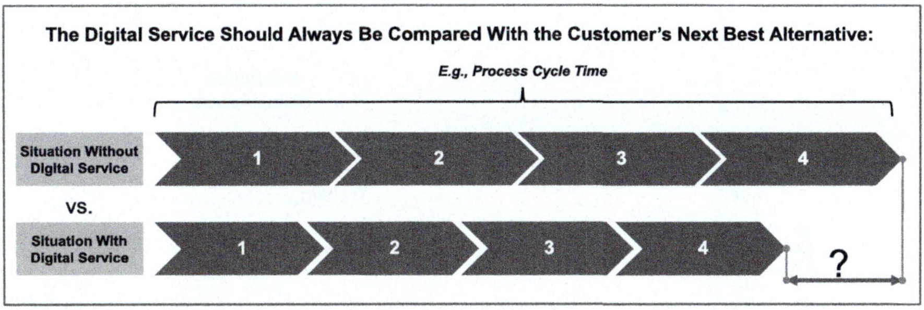

Fig. 5.5 Digital service compared with the next best alternative

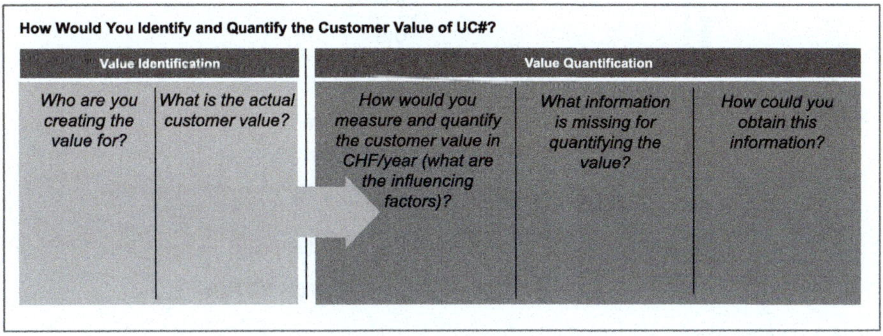

Fig. 5.6 Customer value quantification model of a use case

information. This might slow down the process but is critical. The time you "lose" now is regained many times at later stages.

The generated quantified values now enable the classification of the value of the different potential digital services to the customer. Which services might the customer be willing or not be willing to pay for? Building up a classification scheme (see Fig. 5.7), divided into two dimensions such as "perceived customer value" and "feature type" for any digital service use case, supports this discussion and is the basis for the pricing and packaging strategy of your digital services.

Once the digital service has been selected and positioned, it should be checked with selected customers to generate data in use and allow the quantification model to be confirmed and optimized. This is a fundamental step in the validation of the quantified benefits and serves as a basis for the pricing strategy. Having completed this step DS II, the pricing strategy can be derived in step DS III. How to do this is explained in the next chapter.

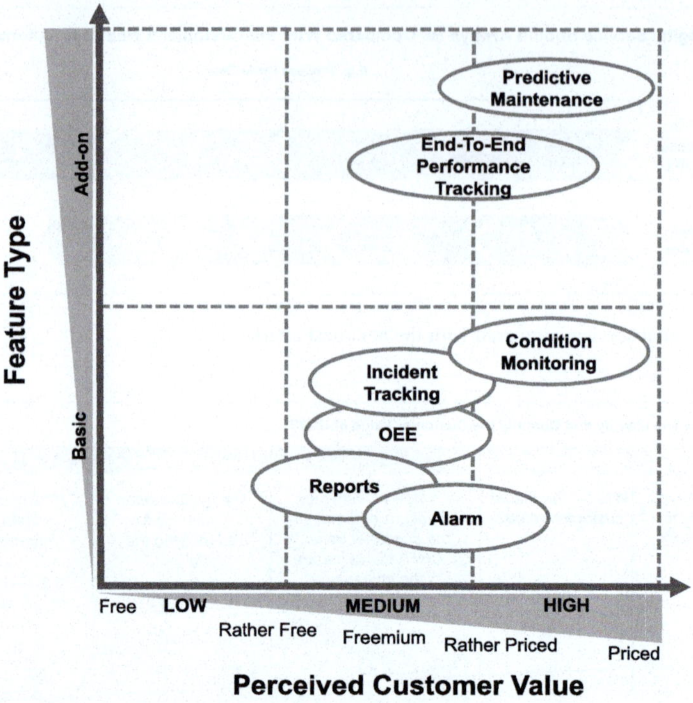

Fig. 5.7 Digital service classification scheme

5.3 DS III: Derivation of Pricing Options

Business models of companies are changing, especially in terms of the revenue model. More and more companies are not selling their machines to the customer; they are renting them. That means, for instance, that the customer only pays for the output of the machine. Well-known examples of this are the turbines for airlines. Airlines only pay for the hour run (pay-per-hour). But this is not limited to the airline business; it can be found within industrial machines as well. These revenue models, known as "pay-per-use" or "pay-per-output" models, are becoming increasingly attractive because these models allow companies to better integrate and access the user lifecycles of their customers. In doing so, the machine supplying companies are suddenly in the position to get to know how customers are really using their machines. This availability of "data in-use" is paving the way for a change in the pricing model.

How then should these digital services be priced? Three elements need to be considered: offer structure, revenue model and contract terms. Figure 5.8 provides you with an overview of the different possible combinations of digital services and supports you in identifying the optimal one for the digital service selected. The model

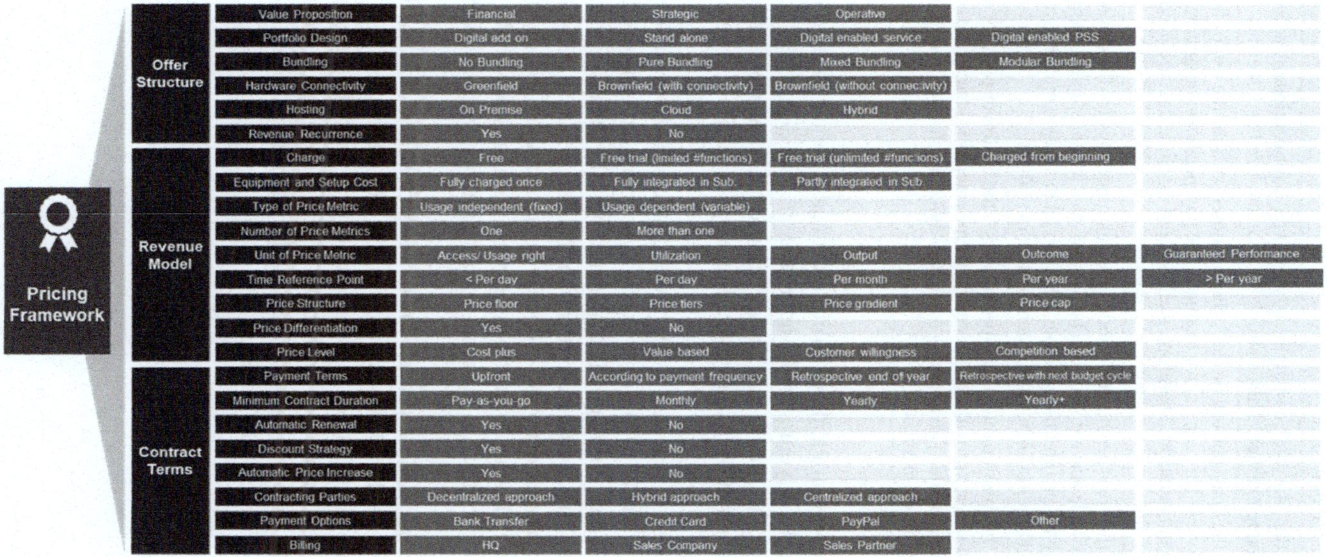

Fig. 5.8 Pricing framework (ITEM-HSG PDP project 2021)

Table 5.1 Four types of pricing strategies

Pricing Strategies	Description
Flat rate	In the "flat rate" model, the customer pays for continuous access to the product/service. This is simple and standardized and does not require data
Pay-per-use	In the "pay-per-use model", the customer pays for the use of the available service. This can be, for example, the pay-per-hour for spindle use or for the lamp operation. Therefore, standardized machine data are needed but not in real-time
Pay-per-output	In the "pay-per-output" model, the customer pays for the result of the available service. This can be pay-per-part for sheet metal stamped parts or a pay-by-picture, for example. This model has a dynamic change over time and requires individual machine data in real-time
Pay-per-success	In the fourth model, "pay-per-success", the customer pays for the success of the service provided. This can be the payment for guaranteed energy saving, for example. The model is subject to dynamic changes, requires KPIs and a definition of success

is understood as a morphological box. You select and combine the characteristics of the pricing framework for your specific approach.

The choice of pricing model you chose represents a trade-off between pricing potential and risk assumption. Four types of pricing strategies can be distinguished (see Table 5.1):

When pricing digital services, you have to additionally consider the initial "onboarding" of customers to your services. Your main target is to get customers willing to share their data with you. The ownership of the data is thereby one of the most crucial factors. This is often a critical hurdle that needs to be overcome. This topic needs to be addressed early on, reflected in your pricing model and ultimately in your sales approach. For any industrial company getting access to the data of your machine or product is critical to develop further digital services and improve your internal processes.

This closed-loop life cycle approach sown in Fig. 5.9 closes the link back to your factory (Nyffenegger et al., 2018). In other words, digital services can be a critical link in getting additional data from all your products in the field (product instances) to improve your internal processes further. To get this information we return to one critical use case of the smart factory—"Use Case 16: Identification & Asset Tracking" of our smart factory framework (see Fig. 4.3).

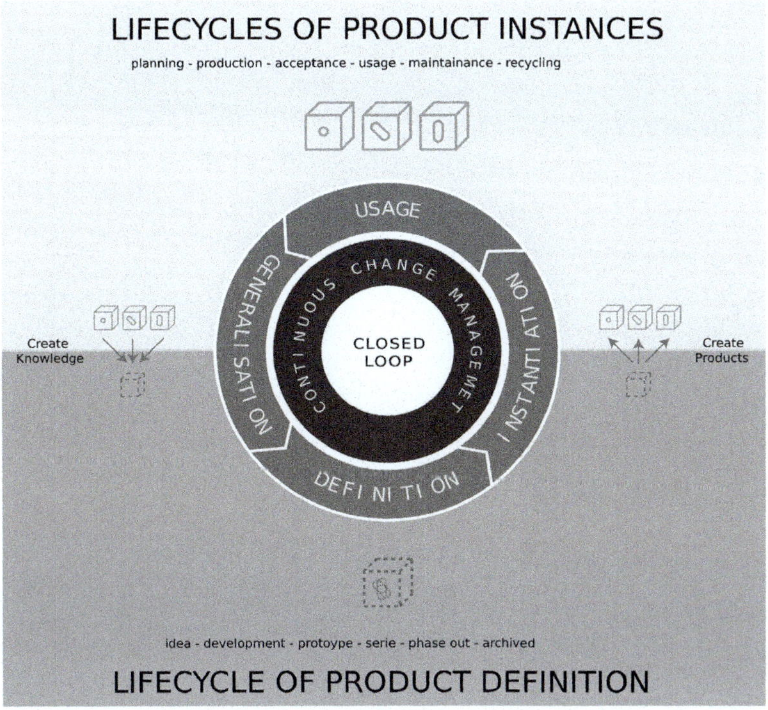

Fig. 5.9 Lifecycle of product instances in the field (Nyffenegger et al., 2018)

Reference

Nyffenegger, F., Hänggi, R., & Reisch, A. (2018). A reference model for PLM in the area of digitization. In P. Chiabert, A. Bouras, F. Noël, & J. Ríos (Hrsg.), *Product lifecycle management to support industry 4.0* (Bd. 540, S. 358–366). Springer International Publishing. https://doi.org/10.1007/978-3-030-01614-2_33

Chapter 6
Learning from Case Studies and Outlook

We have been using the smart factory navigator and the implementation process in different projects with industrial companies since 2017. The "Smart Factory Framework" as well as the "Implementation Process" has already proven itself in more than 20 research and consulting projects with industrial companies, from large multinational to small and midsize companies. The projects have included the manufacturing of machining parts, the railway industry, the plastics processing industry, construction suppliers, logistics companies, machinery and plant engineering, and automotive suppliers. In some of the projects, we also used the framework to design new digital services that can be further commercialized. Some of these digital services have been successfully implemented and launched. The experience gained from all these projects has been directly incorporated into the validation of the presented procedure.

Three major factors for a successful digital transformation can be highlighted[1]. First, use cases should be aligned to the strategy (strategy fit) of the manufacturing company. Use cases supporting the strategic targets should be prioritized. Second, lean should be the basis before any digital use case is implemented. Inefficient processes need to be improved and reinvented, before digitalizing. Third, all these digitalization activities can only be successful when humans are involved and are an integral part of the entire transformation. By having humans on board in this journey, companies ensure the buy-in for the change. A smart factory is one in which humans play a crucial role and where higher levels of performance are only be achieved when humans team up with new technologies.

---

[1] Budde, L., Benninghaus, C., Hänggi, R. & Friedli, T. (2022). Managerial Practices for the Digital Transformation of Manufacturers. Digital 2 (4), 463–483. https://www.mdpi.com/2673-6470/2/4/25

L. Budde et al., *Smart Factory Navigator*,
https://doi.org/10.1007/978-3-031-17254-0_6

To sum up, the assimilated knowledge learnt is as follows:

1. It is critical that you describe your use case specifically based on the 44 generic use cases. Every company is different. The specifics of your company need to be incorporated. This alignment within the company makes sure you are tackling the right focuses, and that everybody understands the use case in the same way.
2. The selection of use cases can only be done as a team involving different functions and hierarchical levels. You need to bring the consolidated views and content together. Apply a strategic combination of top-down and bottom-up approaches to facilitate digital transformation.
3. Often the different use cases are linked to or even built on each other. Discussing and highlighting them in the use case navigator, makes them transparent. Often the logic is seen immediately. This helps you greatly in defining your roadmap for implementation.
4. Prioritizing the different use cases is critical. Often there are many interesting possibilities in a company. Too many opportunities spread resources too thinly, initiatives take too long, and results are not reached in a reasonable time.
5. Do not be shy of defining additional use cases—even though the 44 use cases provide a good summary and have been shown to be comprehensive, they can never be the definitive solution for all companies.
6. Be critical after you have selected your use cases. Have you thought enough about the "future"—or have you only focused on the current situation and the base use cases? By highlighting the selected use cases in the navigator, the answer will be apparent.
7. Discussing your specific use case in relation to its impact and benefits early on is important. This gives you a clear indication of where the use case can achieve a measurable benefit. It will also highlight the strategic fit. It is critical to talk about the financial benefits, even with a first business case. However, you have to know when the financial modelling should be stopped. By constantly developing and detailing the business cases, the use cases don't become more successful. The input data or assumptions for your business cases are often too vague anyway.
8. Different hierarchical levels and different functions (e.g., different personas) have different needs regarding information. This needs to be spelt out and to be made transparent. It has significant implications for the technical implementation of the use cases.
9. The implementation of a use case needs to start with a clear vision of what the final result will be. For example, how your manufacturing department would work if machines predicted the upcoming maintenance schedule. If you have a clear view of the final result, the steps to get there will fall into place immediately. Try to illustrate the vision pictorially, not only in words. By doing so, everybody can envision the future easier.

10. Be very specific in detailing the project fingerprint. Financial, personnel resources, and the timing of the project must be committed and realistic. These complex projects need focus—this can only be achieved by transparency and commitment based on a detailed plan.
11. Data-driven use cases should be implemented and tested in iterative small steps to facilitate digital transformation. Often industrial companies are organized and led by "hardware thinking" and data do not get the required focus. Data transparency can revolutionize the way of working and decision-making in manufacturing.
12. Foster collaboration with customers, suppliers and other external partners to facilitate digital transformations.
13. Measure results along the way. Test early and try to get results early on. Feedback on first prototypes or models is also a result.
14. Scaling and the rollout must be planned carefully. But only after you know if the project has been successfully completed. Involve other departments or plants early on. This will help the transfer of knowledge.
15. Focus on the use cases and their stepwise implementation. This should be the priority and not discussions like the "right" IT and IT supplier. We have quite often seen that such never-ending discussions about IT prevent progress and hinder the benefits of the digitalization potential.

In each factory there exists three main IT systems: PLM, ERP and MES. We all know that every company does not need all three systems simultaneously. A subject often debated is the logic between smart factory use cases, IT systems and lean. Based on our research, we identified a feasible logic that integrates these different dimensions into one methodological flow (see Fig. 6.1). This does not mean that you have to strictly follow the defined six steps. However, it provides you with a guide to create common ground for the further development of the interplay between smart factory use cases, IT and lean processes. At the end of the day, all elements must be focused on the operational excellence of your work.

The "Smart Factory Framework" and the "Implementation Process" are integral parts of the teaching and executive education program at the OST Eastern Switzerland University of Applied Sciences and HSG—University of St. Gallen. Students in different faculties will bring this knowledge to you—back to the factories—and will make sure your plants stay competitive.

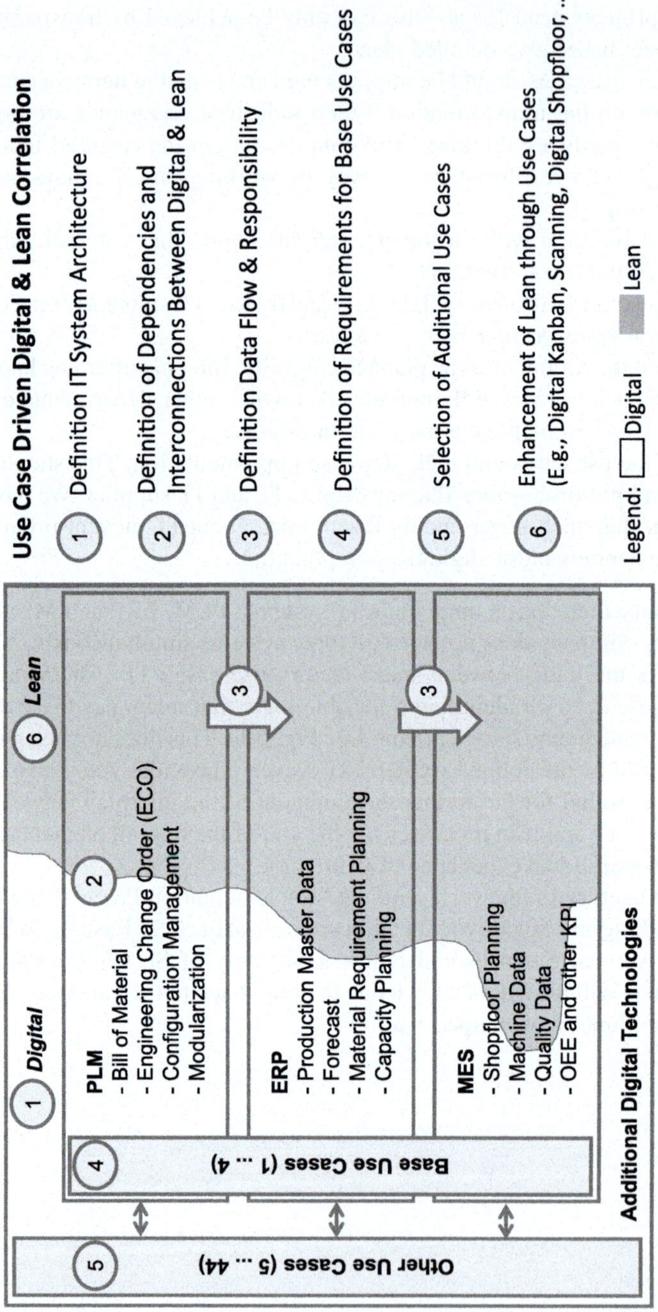

Fig. 6.1 Use-case driven digital & lean correlation

The manufacturer's authorised representative in the EU is Springer Nature Customer Service Centre GmbH, Europaplatz 3, 69115 Heidelberg, Germany. If you have any concerns regarding our products, please contact ProductSafety@springernature.com

Printed and bound by CPI Group (UK) Ltd, Croydon, CR0 4YY
01/10/2026
02225872-0006